Mission Statement

"As one of the Army's two on-call heavy contingency force divisions, the First Team has an on-order mission to deploy to a designated contingency area of operations by sea, air or land, conduct reception, staging, onward movement and integration, and on order, conduct combat operations and redeployment."

The roots of the 1st Cavalry Division go deep and wide, protecting the basic rights established by the founders of this great nation. They span the United States and extend to theaters around the world where the Division fought in the Second World War, the Korean War, the Vietnam War and the Persian Gulf War. In these conflicts the Division protected the interests of the United States and defended other countries from aggression. The recorded exploits of the 1st Cavalry Division are rich with individual and unit heroics of attack, defense and daring rescue. The Division, with superb leadership, employed devastating firepower to overcome the aggressors. In newer roles, the First Team has served as a model of an international force in maintaining peace between waring factions in Southwest Asian countries and Bosnia. The theaters of operations for the numerous worldwide deployments of the 1st Cavalry Division are highlighted below:

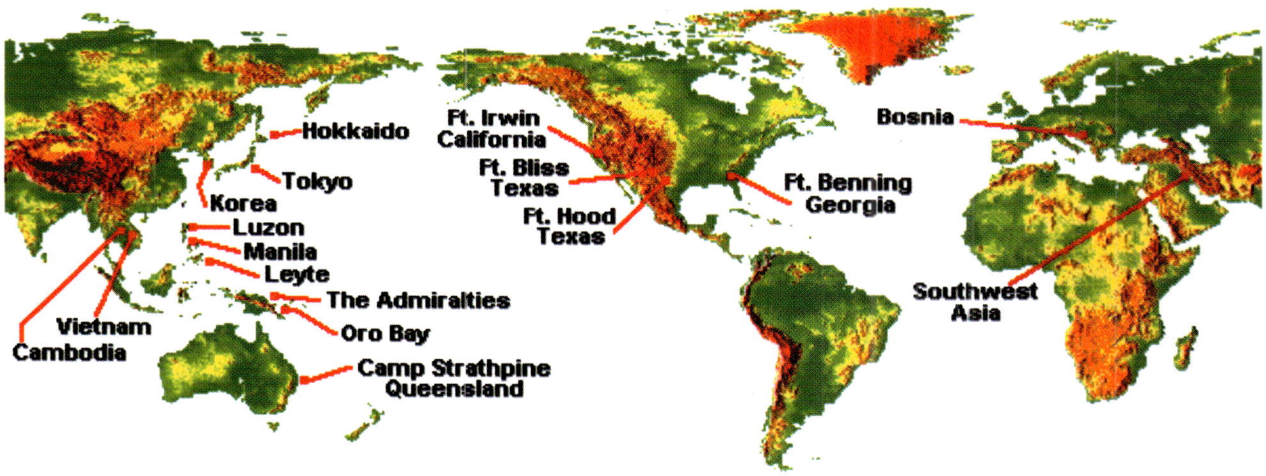

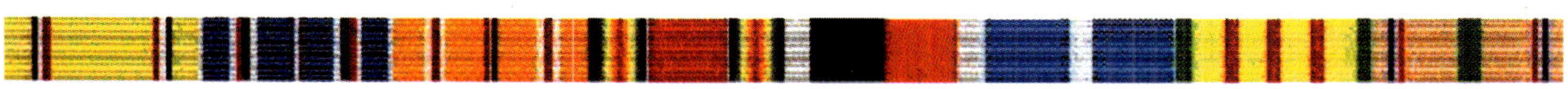

The First Team is the only continuous in-service US Army combat force to extend its exploits and reach from the Pacific Islands, Japan, Korea and the Southeast Asian countries of Vietnam, Cambodia and across the wide expanse of the oceans and continents of the earth to the Southwest Asian countries of Kuwait, Saudi Arabia, and Iraq. The documentation of this history, covering the peacetime and war experiences of the 1st Cavalry Division, beginning with its constitution and organization in 1921 to today's environment, is dedicated to all of those who have served, are presently serving, and who will serve in the future.

First Team

1st Cavalry Division

1st Cavalry Division

A Spur Ride Through the 20th Century
"From Horses to the Digital Battlefield"

Turner Publishing Company

Turner Publishing Company
Publishers of America's History
412 Broadway • P.O. Box 3101
Paducah, KY 42002-3101
(270) 443-0121

Turner Publishing Company Staff
Editor: Herbert C. Banks II
Designer: Tyranny J. Bean

Library of Congress Control No. 2002100538
ISBN: 1-56311-785-1

Printed in the United States of America.

Additional copies may be purchased from
Turner Publishing Company.

Limited Edition.

Table Of Contents

Introduction

General Edward Charles Meyer

Each of us who served in the 1st Cavalry Division since its organization and activation in 1921 can be proud of our division, our unit and our service. This book chronicles that service.

In peace and in war, troopers of the First Team have been in the forefront of war fighting and of peacekeeping. Past books have highlighted specific eras in our history. This book is all inclusive. It reminds us that every trooper has been important in support of our nation's military efforts throughout the 20th Century, and will be equally important in the new millenium as described in the last Chapter - "The Next Generation, 2000 - 2020".

Edward C. Meyer

Biography

Edward Charles Meyer graduated from West Point in 1951. Following the Associate Infantry Officers course at the Infantry School, he went directly into combat in the opening months of the Korean War. From April 1952 to November 1953 Meyer served with the 224th Infantry Regiment as platoon leader, executive officer, company commander and battalion intelligence officer. After the war he was assigned to the 1st Officer Candidate Regiment, Fort Benning, Georgia, followed by appointment as Aide de Camp to the Assistant Commandant of the Infantry School. After completing the Infantry Officers Basic and Advanced Airborne courses at Fort Benning, Georgia, he served as commanding officer of two airborne companies from November 1957 to August 1959.

Meyer entered the Command and General Staff College at Fort Leavenworth, Kansas in August 1959, then went to Europe for duty in the office of the Assistant Chief of Staff for Personnel at Headquarters US Army Europe. In June 1961 he was appointed Assistant Executive Officer and Aide-De-Camp to General James E. Moore, Chief of Staff of SHAPE. Returning to the US in 1963, Meyer graduated from the Armed Forces Staff College and then served in the office of the Chief of Staff, US Army until July 1965. A Lieutenant Colonel since May 1964, Meyer next served as Deputy Commander of the 3rd Brigade, 1st Cavalry Division prior to its deployment to Vietnam and during the Battle of Ia Drang. From December 1965 to July 1966 he was Commander of the 2nd Battalion, 5th Cavalry in Vietnam.

Meyer attended the National War College 1966 to 1967, earning a masters degree at George Washington University, concurrently. From August 1967 to September 1969 he was a staff officer with the Objectives, Plans and Programs Division of the Joint Chiefs of Staff, earning promotion to Colonel in November 1968. Meyer returned to Vietnam in September 1969 as Commander of the 2nd Brigade, 1st Air Cavalry Division in Phuoc Long Province, followed by duty as Chief of Staff of the 1st Air Cavalry Division. While in Vietnam he was notified of early selection to Brigadier General.

Returning to the US in September 1970, he was assigned briefly as a Federal Executive Fellow at the Brookings Institution before taking over as Assistant Division Commander of the 82nd Airborne Division at Fort Bragg, North Carolina in March 1971. In August he pinned on a first star. General Meyer was Deputy Commandant of the Army War College at Carlisle Barracks, Pennsylvania from August 1972 to August 1973, when he received a second star and returned to Europe, first as Deputy Chief of Staff for Operations at US Army Europe, then Commanding General of 3rd Infantry Division until August 1975.

Assigned once again to the Pentagon in 1975, he was Assistant Deputy Chief of Staff for Operations 1975 to 1976, then Deputy Chief of Staff for Operations, with promotion to Lieutenant General, from 1976 to 1979. On June 22, 1979, General Meyer was promoted to full General and assumed responsibility for the US Army as Chief of Staff, the youngest four-star chief in history.

Chapter 1: The Early Years

1921-1941

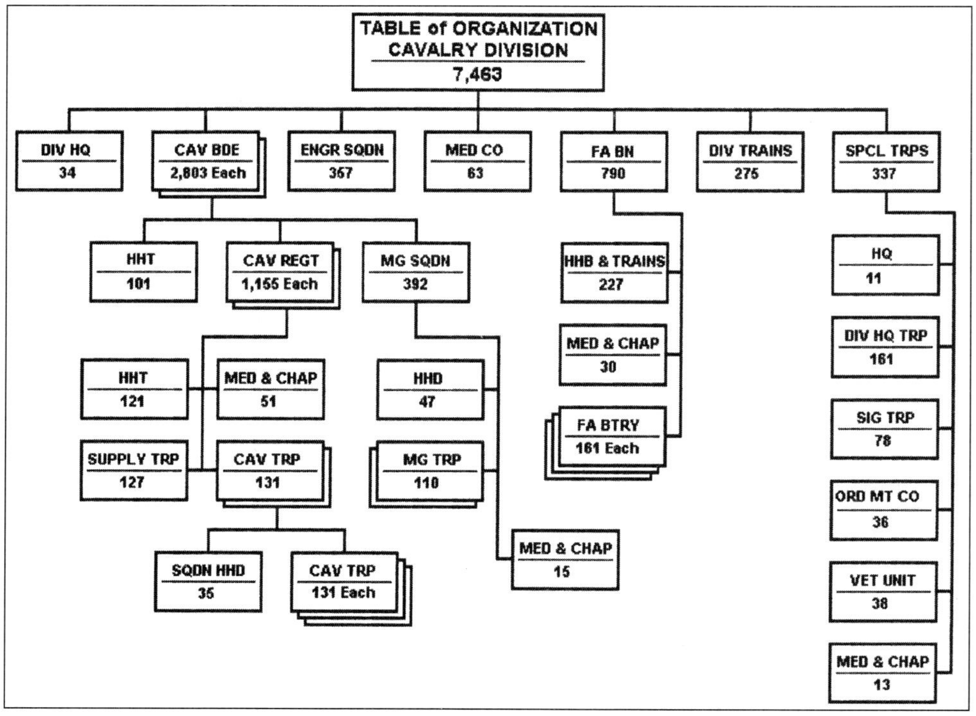

Cavalry Division Organization. (1921)

On 13 September 1921, with the initiation of the National Defense Act, the 1st Cavalry Division was formally activated at Ft. Bliss, Texas. With almost a century of service behind the oldest of its regiments and sixty five years of service for its youngest, the units that had already ridden and fought their way into the pages of history were to be organized into a newly formed divisional structure. The four regiments were now to fight side by side. The first unit of the 1st Cavalry Division, the 1st Cavalry Regiment, had been preassigned to the 1st Cavalry Division on 20 August 1921, nearly a month before the formal divisional activation date. Upon formal activation, the 7th, 8th and 10th Cavalry Regiments were assigned to the new division. Other units initially assigned to the 1st Cavalry Division in 1921 included the 1st and 2nd Machine Gun Squadrons, Weapons Troops, 10th Light Tank Company, 13th Signal Troop, 15th Veterinary Company, 27th Ordnance Company, 43rd Ambulance Company, 82nd Field Artillery Battalion (Horse) and the 1st Cavalry Quartermaster Train which later was redesignated as the 15th Replacement Company. Major General Robert L. Howze was assigned as the first division commander. The 5th Cavalry Regiment was assigned on 18 December 1922, relieving the 10th Cavalry Regiment. It would not be until 03 January 1933 that the 12th Cavalry Regiment, organized in 1901, would join the 1st Cavalry Division, relieving the 1st Cavalry Regiment.

In the fall of 1923 the 1st Cavalry Division assembled at Camp Marfa, Texas to stage its first divisional-level maneuvers since its organization. The maneuvers were held in the Marfa-Shafter-Alamito area of the Big Bend District, Texas. The line of march was Fabens, Ft. Hancock, Sierra Blanca, Hot Wells, Lobo Flats, and Valentine. The wagon trains, all drawn by four mules (no motorized vehicles yet), seemed endless. A terrain covering an area of 900 square miles was obtained through the generosity and public spirit of ranch owners. The enormous tract was mapped and marked by a detachment from the 8th Engineer Battalion.

The actual maneuvers consisted of both one-sided and two-sided problems with brigade against brigade and included the division as a whole. Aircraft were employed

1st Cavalry Division barracks and stables, Camp Marfa, TX, (08 August 1923)

The 1st Cavalry Division crosses the railroad tracks and moves out over the harsh desert terrain to the next staging area in the Big Bend Area. (1923)

extensively and every detail was made to conform with war conditions. Since this was the first major United States (US) Army training exercise since WWI, the maneuvers were attended by representatives of several foreign governments.

Over the next four years, elements of the Division were stationed at Camp Marfa, Ft. Bliss, and Ft. Clark, all located in Texas. The early missions of the Division were comprised of rough riding, patrolling the Mexican border, and constant training. Operating from horseback, the cavalry was the only viable force capable of piercing the harsh terrain of the desert to halt the band of smugglers that operated along the desolate Mexican border. In spite of the lack of ample funding and the limited availability of new equipment, priorities were placed on readiness evaluation by extensive field maneuvers.

On 16 June 1926, a Provisional Squadron (Headquarters Detachment, "C", "E" and "F" Troops of the 7th Cavalry Regiment) departed Ft. Bliss by rail for the Crow Indian Agency, Montana to participate in the Semi-Centennial celebration of the Battle of Little Big Horn. The Provisional Squadron returned to Ft. Bliss on 01 July, having traveled 1,800 miles.

In 1927 the 1st Cavalry Division carried out the second divisional field maneuvers and readiness testing in the Marfa area. Following the maneuvers in October, the division added the capability of aerial observation by the assignment of the 1st Observation Squadron, US Army Air Corps. The unit remained with the division until the end of its subsequent organizational changes in February 1929. Today the 1st Squadron operates the high altitude SR-71 (YS-12) Reconnaissance Aircraft.

Early in 1928, Major General Herbert C. Cosby, Chief of Cavalry, faced with budget restraints, reduced the size of the cavalry regiments. Lettered troops of the regiments were reduced from six to four along with the integration of the separate Machine Gun Squadrons of each brigade under the regimental commands. Troops "A" and "B" of each regiment formed the 1st Squadron and "E" and "F" formed the 2nd Squadron. This new organizational change was designed to reduce command overhead and increase force mobility by the addition of three 1 1/2 ton trucks and three modified automobiles, called "light cross country cars". In addition, the change allowed easy expansion to war strength while retaining the capability to take to the field and deliver powerful and flexible firepower by the machine gun troops whose Table of Organization and Equipment (TO&E) had been increased from four machine guns to eight.

On 01 February 1928, the Army Staff, seeking to increase the usefulness of the wartime cavalry division, published new TO&E for an even larger unit (on paper). The new structure involved increasing the size of the signal troop and expanding the medical unit to a squadron. A divisional aviation section, an armored car squadron, and a tank company were added. The field artillery battalion was expanded to a regiment. Divisional wartime strength rose to 9,595. Although the new tables had little impact on the peacetime cavalry structure, the 1st Cavalry Division did eventually receive one troop of an experimental armored car squadron, and a field artillery regiment replaced its field artillery battalion.

Having learned from the experiences of World War I and to prepare for the mechanization of the future, the 1st Armored Car Squadron became part of the Division in November 1928. This new and experimental unit had been organized and trained at the Aberdeen Proving Grounds in

Officer review of 1st Cavalry Division troops, Marfa, TX. (1923)

1st Cavalry Division Maneuvers - Camp Marfa, TX. (1923)

Aberdeen, Maryland. The 1st Armored Car Squadron remained with the Division through its various reorganizations until it was redesignated the 91st Reconnaissance Squadron and was relieved from assignment to the 1st Cavalry Division in 1943 to join in the African campaigns against the Germans and Italians in the Kasserine Pass and later, battles in Tunisia and Sicily.

On 03 March 1929, the Escobar Revolution broke out in Mexico when the colonels and generals of ten Mexican states rebelled. The 1st Cavalry Division was alerted and positioned along the Mexican border from Douglas, Arizona to El Paso, Texas. Their mission was to seal off the border and halt the passage of weapons and refugees into and out of Mexico. Jose Gonzalo Escobar, a Brigade General in the Mexican Federal Army, led the unrest against former president Plutarco Elias Calles, who politically and militarily dominated the country. Control of Juarez, directly across the Rio Grande River from El Paso, became a top priority of the insurgents because they could collect all the import and export duties to support the continuing rebellion effort. A Chihuahua insurgent force of 2,000 rebels was sent by Escobar to capture the town.

On 08 March, the battle for the town commenced as thousands of El Pasoans clustered on roof tops and river levees to view the struggle. As reported by a local newspaper, "El Paso is the only section of the United States trained to appreciate warfare as a neighborhood spectacle." By mid morning, the rebels had pushed the Mexican Federal Troops to the banks of the Rio Grande. General George Van Horn Mosely, Commander of Ft. Bliss, arranged for the Mexican Federal Troops to take asylum at Ft. Bliss. After checking in with the United States Immigration Authorities, the Federal Troops surrendered all their ammunition, but kept their own guns. After a month at Ft. Bliss, the Federal Troops shipped out by the Southern Pacific Railroad to Naco, Arizona, where they crossed the Mexican border to help put down the rebellion.

In 1929 the 1st Cavalry Division carried out its third divisional field maneuvers. The depression of the 1930s forced thousands of unemployed workers into the streets. As a means to minimize the economic effects of the large unemployed work force, the government established the Civilian Conservation Corps (CCC) to serve as a worker pool for use in the development of local, state, and government projects. From 1933 to 1936, the 3,300 troopers of the 1st Cavalry Division provided training and leadership for 62,500 members of the CCC in the Arizona-New Mexico District. When World War II broke out, many of those who had been trained in the CCC were well prepared for the rigors of military training.

In the early thirties, the strength and composition of the Division was principally governed by the total strength of the Army, numbers of active regiments, and the desire to maintain a troop level large enough to be an effective fighting force. The Division had two authorized strengths: a wartime strength of 11,485 officers and enlisted men and a peacetime strength of 7,970. During peacetime all elements of the Division were to be filled except for the division headquarters and military police companies, which were combined. Other divisional elements required only enlisted personnel to bring them up to wartime manning levels. For

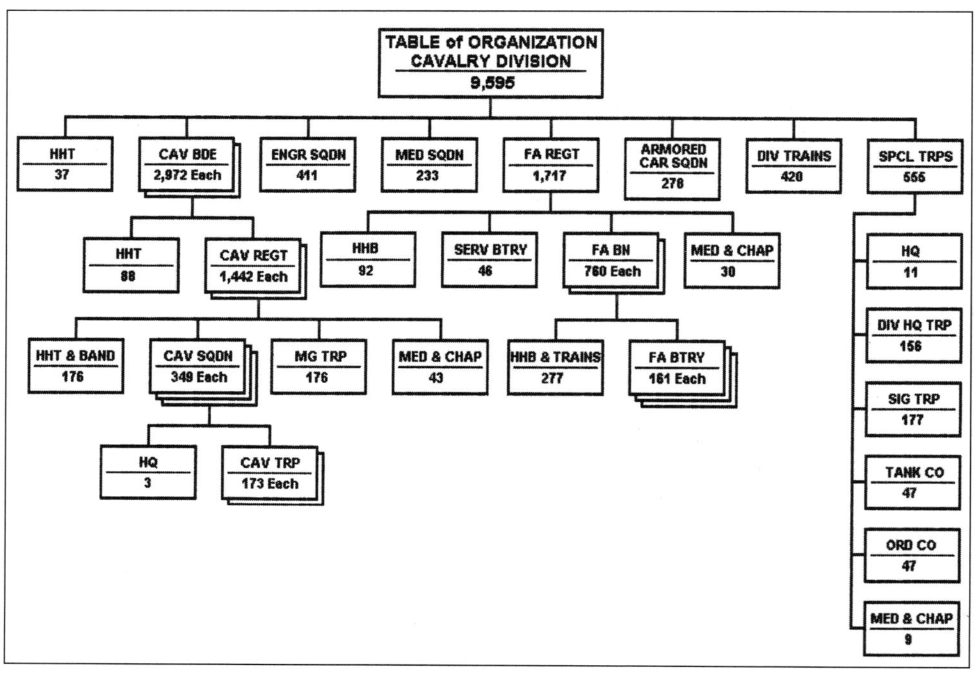

Cavalry Division Organization (1928)

The real "Cavalry" (1930s)

the first World War, one of fast maneuver. One of the experimental development vehicles produced for evaluation was the four wheel drive Model T4, built by the James Cunningham Company of Rochester, New York in 1932.

Within the Division, armored cars and scout cars continued to blend into the formations of tanks and horses in the post-war cavalry. Typical of these vehicles was the six-wheeled vehicle of "A" Troop, 1st Cavalry Regiment. The designation of the unit was indicated by the traditional crossed sabers on the door and a stylized version of the distinctive regimental insignia, a black hawk on a gold eight-pointed star, to recall the heritage of the First Regiment of Dragoons.

On 03 January 1933 the 1st Cavalry Regiment was relieved from the Division and transferred to Ft. Knox, Kentucky where it was reorganized and redesignated as a mechanized unit using the assets of the Independent Mechanized Force. The new mission of the 1st Cavalry Regiment was to expand on the evaluations, develop and test mechanized combat vehicles, continuing the mechanization era. The process of replacing horses with machines accelerated and the first mechanized cavalry organization came into being. Cavalrymen from the First Team became the nucleus of new armored forces for America. Concurrent with the relief of the 1st Cavalry Regiment, the 12th Cavalry Regiment was assigned to the 1st Cavalry Division.

In 1936, the Modernization Board, which was performing an evaluation of overall Army operations, began an evaluation of the 1st Cavalry Division. Most officers still envisioned a role for the horse, because it could go places inaccessible to motorized and mechanized equipment. Taking into account recommendations from the Eighth Corps Area, the Army War College and the Command, and General Staff School of the Army, the board recommended a new, smaller "triangular" cavalry division.

that period, a typical regiment was composed of 690 troops: a headquarters troop with 78 men; a band with 28 men; four rifle troops with 119 men each, and a machine gun troop with 108 men. Each rifle troop was organized into a troop headquarters, three rifle platoons of three squads each, and a machine gun platoon of three squads.

Recovery from the world wide depressions had planted the seeds of isolation and nationalism in many countries. In searching for solutions which would appeal to their own national interests and detract from the issues of discontent, governments began to concentrate on utilizing their resources in building armies and considering territorial expansion. In Asia, Japan had already demonstrated its desire to expand the Nipponese empire in the Pacific, extending into China and along the Pacific Rim to Australia. In September 1931, Japan began the first stage of its expansion plans by the invasion and occupation of the neighboring country of Manchuria. The US, still attempting to concentrate on recovery from its own depression, only condemned the invasion and refused to recognize the Japanese continued occupation.

In light of these events, Washington expressed a new urgency for modernization and recognized that the entire Army must expand and acquire new equipment. Faster and lighter medium tanks were assigned to both cavalry and infantry units. The mobile 105mm howitzer became the chief artillery piece of the Army divisions.

By 1930, the Army had formed a Mechanized Force at Ft. Eustis, Virginia to conduct a sufficient number of trial exercises using armored cars and tanks to prove the benefits of mechanization. When General Douglas MacArthur became Chief of Staff in 1931, in order to accelerate the utilization of armored vehicles into the Army, he moved the Mechanized Force from Ft. Eustis to Ft. Knox Kentucky to form the basis for a mechanized cavalry regiment.

In the twenties a small group of officers in the Cavalry foresaw that the next war would be mechanized and, unlike

Cavalry Armored Scout Car

1st Cavalry Division Maneuvers - Toyahvale, TX. (1938)

In July 1937, initiating the second stage of expansion, Japan launched a major invasion of northern and central China. After a costly resistance, the ill-prepared Chinese armies were forced back from eastern China and in December 1937 the Nationalist capital, Nanking, was subjected to an orgy of rape and destruction. At this time the rest of the world remained neutral, and some western countries, including the US, were still selling scrap materials to Japan, which were converted into armaments for use in additional expansion plans. Further, Nazi Germany had annexed Austria and was now threatening to seize Czechoslovakia.

During the fall of 1937, the 1st Cavalry Division, along with other troops stationed at Ft. Bliss engaged in the Provisional Infantry Division (PID) Tests at Ft. Sam Houston and Camp Bullis, Texas. The PID Tests developed the partially motorized triangular division that was employed in World War II. Following the completion of the tests, the Division returned to Ft. Bliss for a brief period to rehabilitate the horses. In the early spring of 1938, the Division moved out again and, against the background of growing international tensions, began its fourth divisional maneuvers staged in the mountainous and desert areas near Balmorhea and Toyahvale, Texas. In these maneuvers, the 1st Cavalry Division conducted tests and evaluated the many organizational recommendations for a Provisional Cavalry Division made by the Modernization Board.

Following the tests, a board of 1st Cavalry Division officers, headed by Brigadier General Kenyon A. Joyce, rejected the three-regiment division and recommended retention of the two-brigade (four-regiment) organization. The latter configuration allowed the Division to deploy easily in two columns, which was an accepted standard cavalry tactic. However, the board advocated reorganizing the cavalry regiment along triangular lines, which would give it a headquarters and headquarters troop, a machine gun squadron with special weapons and machine gun troops, and three rifle squadrons, each with one machine gun and three rifle troops. No significant change was made in the field artillery, but the test showed that the engineer element should remain a squadron to provide the divisional elements greater mobility on the battlefield. It further demonstrated that the special troops concept should be extended to include the division headquarters, signal, ordnance troops, quartermaster, medical, engineer, reconnaissance and observation squadrons, and a chemical warfare detachment. One headquarters would assume responsibility for all the administration and disciplinary control for these forces.

The results of the study did not lead to a general reorganization of the 1st Cavalry Division. However, on 01 December 1938 the wartime cavalry regiment was restructured to consist of a headquarters and headquarters troop, machine gun and special weapons troops, and three squadrons of three rifle troops each. The special troops remained as structured in 1928, and no observation squadron or chemical detachment was added to the Division. With the paper changes in the cavalry divisions and other minor adjustments, the strength of a wartime divisional force was set at 10,680.

The winds of war and mutual agreements continued to swirl. On 23 August 1939, in the presence of Stalin, the Foreign Minister of Germany, Joachim von Ribbentrop, and the Commissar of Foreign Affairs of Union of Soviet Socialist Republics (USSR), Vyacheslav Molotov, signed the Russo-German Pact. The pact, a complete surprise to France and England, was not only a non-aggression pact, but also contained a secret protocol providing for the division of Eastern Europe between the spheres of influence German and Soviet. This mutual agreement was a marked contrast in the ideology of Hitler, who from the beginning of his political life, had declared the destruction of Communism as one of his primary objectives.

On 01 September 1939, 1.8 million German troops invaded Poland on three fronts; East Prussia in the north, Germany in the west and Slovakia in the south. The Germans employed exceptional rapid maneuver or "blitzkrieg", or lightning war, coupled with the bombing of towns and refugees, that had never been seen before by the world. This operational practice allowed the Germans to capture much territory early in the war. Both, Great Britain and France attempted to intervene and warned Hitler of the grave consequences of his actions. When Hitler ignored their

The 1st Cavalry Division participates in field exercises held during the first 3rd Army Maneuvers at Camp Polk, Louisiana. (13 to 24 August 1940)

German troops parade through Warsaw, Poland. (September 1939)

warnings, they declared war against Germany on 03 September 1939. This set the stage for an eventual worldwide conflagration involving the two hemispheres. The Polish army, with a million men but little modern equipment, soon fell prey to the blitzkrieg and was defeated in a little over a month.

From 07 to 30 October 1939, against the background of the invasion of Poland, the 1st Cavalry Division conducted their fifth divisional readiness maneuvers in the barren desert terrain around Balmorhea and Toyahvale, Texas. These exercises, made even more memorable and intense by their timing, were observed by many international military personnel. Armed with the knowledge of the power that was employed by the most modern and deadly military force of its time and facing the fact that the United States would soon be drawn into an ultimate war, the US Army began planning for a series of its own major field maneuver operations that would provide realistic wartime training.

In April, May, and from 13 to 24 August 1940, the 1st Cavalry Division participated in the first 3rd Army maneuvers held in the Carvens-Pitkin Maneuver Area at Camp Polk, Louisiana. More evidence of the global nature of the war began to materialize when on 27 September 1940, in Berlin, the governments of Germany, Italy and Japan signed a Three-Power Pact. They agreed to stand by and co-operate with one another in regard to their efforts in greater East Asia and regions of Europe respectively. Their prime purpose was to ostensibly establish and maintain a new order destined to promote the mutual prosperity and welfare of the peoples concerned. Additionally the pact was not to impact any previous political agreements which existed, such as that between Germany and the USSR.

In the wake of the German military successes, the United States accelerated preparations for its own buildup of all military forces to wartime strengths. The overall mobilization of the country represented a transitional phase which blended the increased manpower with the growing industrial output of material and weapons. Returning from the maneuvers, the Division undertook the assignment of constructing barracks for 20,000 anti-aircraft troops at Ft. Bliss, Texas and developing the adjacent Biggs Army Air Base at El Paso. An orderly expansion of the Division was underway by the reactivation of "C" and "G" Troops for all the regiments. In the fall of 1940 the 56th Cavalry Brigade, Texas National Guard, was federalized and integrated into the training programs of the Division.

Early in July 1941 the entire Division was assembled at Ft. Bliss for the first time and an era of intensive wartime maneuvers was inaugurated in preparation for possible war. These maneuvers provided the Division a first hand opportunity to participate in the early tactical evaluations of the military use of light aircraft for artillery fire control and troop reconnaissance. Successful field results with the light planes gained new supporters for their continued use in future maneuvers, two being the commander of the 1st Cavalry Division, Major General Innis Palmer Swift and his Chief of Staff, Colonel Joseph M. Swing. The strength of the Division grew by the separation of the Service Troops from the Headquarters Troop and the activation of the 61st Field Artillery Battalion as the first medium artillery support unit. The authorized manpower strength of the Division was increased to 11,676.

From 05 August to 08 October, the 1st Cavalry Division, then staffed to approximately seventy percent of the authorized strength, participated in the second 3rd Army

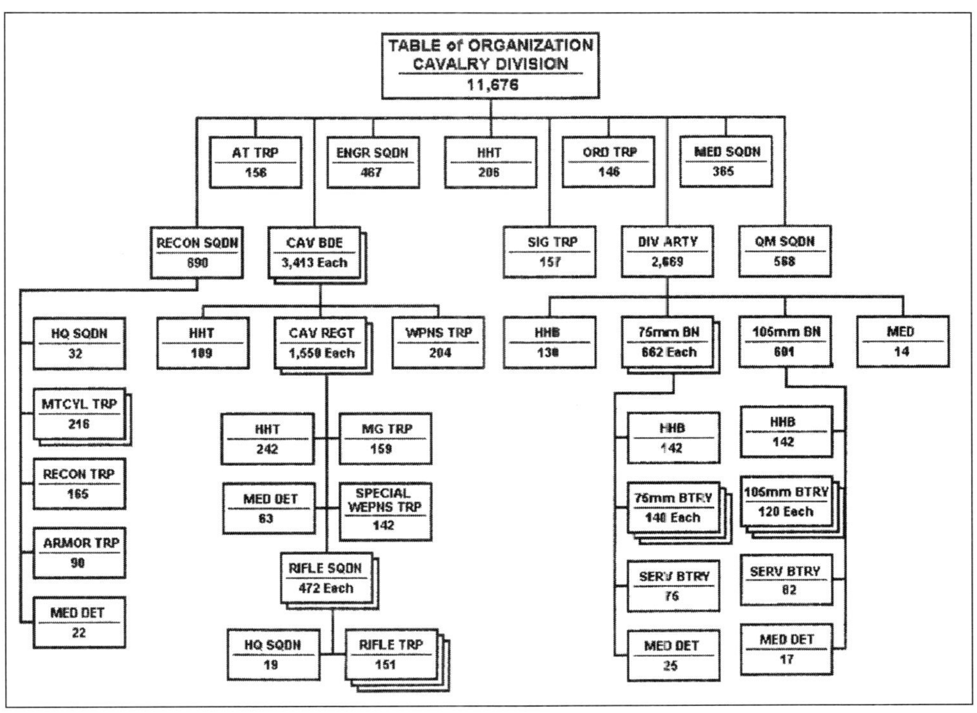

Cavalry Division Organization (1940)

Machine Gun Troop, 12th Cavalry firing 30 Caliber machine gun at Castner Range. (May 1941)

field readiness maneuvers that were held in the vicinity of Leesville, Louisiana. The Blue and the Red Armies that were selected for this set of LOUISIANA MANEUVERS were the finest and best equipped this country could then field. They represented a small part of the US Army Military Establishment of about a quarter million, in total, who would soon form the strong back bone of a mighty army in excess of thirteen million under arms, so called today as "The Greatest Generation,"

For the initial movement of the 1st Cavalry Division, railroad cars were assembled from all over west Texas at El Paso, Texas. In anticipation of extended marches over gravel roads in rural Louisiana, each horse was provided with a spare set of shoes. Material and horses were shipped by rail while the men and their personal gear were transported by motor convoy to the maneuver areas. The maneuvers were a severe test for the men and their horses. The Division covered approximately 900 miles in the maneuver area in the 60 day period.

It was in the 1941 LOUISIANA MANEUVERS that the War Department approved the trial use of light planes for "control of troop movements, scout, patrol, drop bombs, ferry personnel, carry messages, and observe artillery fire". At his own expense, William T. Piper Sr., supplied eight new J-3 Cub airplanes equipped with radios, and a contingent of factory pilots and mechanics. This civilian fleet, called "Grasshoppers", operating through the trying days of the summer and fall maneuvers, proved the flexibility and capability of the light planes to be much more effective than the larger planes used for the same purposes. The after action reports on the maneuvers called for the investigation of their continued use.

On 06 June 1942, following a final series of experiments with organic Army spotter aircraft, the Secretary of War ordered the establishment of organic air observation for field artillery. Subsequently, the adaptation of aerial technology allowed the 1st Cavalry Division to enter World War II with another discipline in its weapon inventory. This capability would be enhanced and improved leading to its use in the organization of the Aviation Brigade, a major maneuvering unit that, today, changes the way that wars are fought.

In the meantime, isolationist politics remained strong in Congress. In spite of this, major priorities were placed on building up the industrial capacity to supply equipment to the Allies in Europe. Many officers and men took leave or returned to civilian life. Other, more dedicated members of the 1st Cavalry Division began to prepare for battle. They had no way of knowing that their first combat engagement would not be for more than one and a half years.

12th Cavalry charging in field exercises at Ft. Seldon, NM. (1941)

Chapter 2: World War II, Pacific

1941-1945

Captured Japanese photograph taken during the attack on Pearl Harbor. In the distance, the smoke rises from Hickam Field. (07 December 1941)

On 07 December 1941, without warning, the Japanese destroyed the American fleet at Pearl Harbor. Although the 1st Cavalry Division was created as a result of a proven need for large horse-mounted formations, by 1940 many thought that the march of progress had left the horse far behind. All doubt was erased with the surprise of the Japanese attack on Pearl Harbor. Immediately, troopers returned to the 1st Cavalry Division from all over the United States. They outfitted their horses and readied their weapons and vehicles in anticipation of the fight against the Axis.

1942 dawned with the Division fated to continue in the role of border patrol which its currently assigned regiments had performed during World War I. Although the Division was anxious for immediate combat, its first wartime mission was to continue border surveillance as a component of the Southern Land Frontier and the Southern Defense Command.

In May 1942, 1,250 1st Cavalry Division Troopers selected from the 5th, 7th, 8th, and 12th Regiments were transferred from Ft. Bliss, Texas as a Non Commissioned Officer (NCO) Cadre group to assist in the organization of the 91st Infantry Division at Camp White, Oregon. In April 1944, the 91st Division went overseas, arriving at Oran, Africa in preparation for the invasion of southern Europe.

By the summer of 1942, most of the commanders were reserve officers who had been in the division for more than a year. These highly motivated officers were the ones that remained with the unit throughout all of WW II. New tactical ideas, including aerial artillery observation, a specialized anti-tank company under division control, and the addition of armored vehicles to the reconnaissance units, were formulated in the planning phases of the LOUISIANA MANEUVERS which were held near Mansfield, Louisiana from 04 August to 19 September. The maneuvers involved more troops than those of 1941. Time passed quickly because everyone was a veteran of large unit maneuvers on horseback. Learning from past years, special alloys of hardened material were welded to the horseshoes and they lasted much longer. The period following the LOUISIANA MANEUVERS became an extension of field exercises and unit testing which continued throughout the winter as rumors of deployment possibilities circulated around the post.

Another element of the 1st Cavalry Division to enter the European Theater was the 91st Reconnaissance Squadron. Formally designated the 1st Armored Car Squadron, the squadron left the 1st Cavalry Division in 1943 to join in the fighting against the Germans and the Italians in the deserts of North Africa. In the same time period, the 62nd Armored Field Artillery and the 161st Engineers left for the European Theater.

The reason that the Army had a continuing interest in horse cavalry operations in 1942 was simply that there was no fully developed strategy for fighting the war in the event of an invasion of the western hemisphere. If the Army, in a tactical situation such as the invasion of France by Germany, could have been guaranteed networks of good roads, ample fuel supplies, gentle familiar terrain and constant air superiority, it might have decided to get along without the horse cavalry operations. However, chances were, that the US defenses might be challenged on less than ideal terrain; perhaps an enemy landing in Western Mexico or along the coast of Brazil, that was friendly to Germany. In such an event mounted soldiers, using sturdy, sure footed-horses, could prove invaluable.

An important operational factor of the horse cavalry was that it could quickly get around and over hazardous terrain such as hills, rocks, trees or desert, but the majority of actual enemy engagement is done fighting dismounted. On such ground, supported by its own artillery and armored units, the 1st Cavalry Division could lick its weight in enemy tanks. At the organizational core of the division were four mounted regiments, each composed of six rifle troops, one machine gun troop, one special weapons troop, and one headquarters troop. Each brigade, composed of two regiments, had another special troop of 37mm anti tank guns and 81mm mortars. In addition, special strong, mobile, mechanized units were being integrated as organic or attached to the Division to support the combat units, quartermaster units, medical corps, and engineer units, that now all rode on wheels.

Near wartime "full strength" levels, the 1st Cavalry Division was equipped with one hundred forty-five armored reconnaissance cars, six hundred trucks of all sizes, and three hundred seventy motorcycles along with thirteen light tanks. Horses combined with the pool of vehicles could pull or carry a complete mobile arsenal of seven hundred .30 caliber machine guns, two hundred sixty five .50 caliber machine guns, five hundred sub-machine guns, sixty seven 37mm anti tank guns, twenty eight 81mm mortars, twenty four 75mm howitzers; four thousand semi-automatic rifles, and ten thousand four hundred .45 caliber pistols. A recent addition to the arsenal included twelve 105mm howitzers.

In February 1943, the entire 1st Cavalry Division was alerted for an overseas assignment. Many of the proud cavalrymen would rather turn their stripes, bars, or stars than trade in their saddles for a seat in any vehicle to become "cushion pounders". Nevertheless, an impatient 1st Cavalry Division was dismounted and the troopers, with mixed emotions, began the process of turning in horses, saddles, and bridles. The troopers continued to feed and water their horses until the Quartermaster assumed control of them. Rather than return the horses to the remount stations, the majority of them were auctioned, at bargain prices, to owners of the large ranches around the El Paso area. For many years following the end of WWII - many of the horses, identifiable by the Preston Brand on their neck, were still serving out their duty to the ranchers.

Well ahead of schedule, the Division began the historic transition and retraining of the troopers in a mobile environment. As planned they began staging for movement to the Southwest Pacific theater as foot solders with the support of mechanized vehicles; tanks, armored cars, trucks, bantam and scout cars. This modern cavalry division had gained tremendous advantages in rapid mobility, extended range and firepower.

The overseas deployment from Camp Stoneman, California to Brisbane, Australia was made in two echelons. The first body, the 5th Cavalry Regiment and the 8th Engineer Combat Squadron, that was attached, would be followed by the remaining units of the main body of the Division. The first echelon traveled by train to the embarkation port of San Francisco, California. On 28 May they boarded the *USS Maui*, a converted troopship leased from the Matson Navigation Company. Traveling under blackout conditions, the sea voyage took seventeen days following the standard procedure of anti-submarine zig-zag maneuvers. Arriving in Brisbane, they moved to Camp Strathpine and began preparations for the arrival of the main body.

By 18 June 1943, the last troops of the division departed Ft. Bliss, Texas for Camp Stoneman, California and later, on 03 July boarded the *USS Montery* and the *USS George Washington*, bound for Australia and their subsequent operations in the Southwest Pacific.

On 26 July, three weeks later, the Division arrived at Brisbane and began a fifteen mile trip to its new temporary home, Camp Strathpine, Queensland, Australia. The 1st Cavalry Division, comprising 15,000 men, totally overwhelmed the tiny locality of Pine Rivers, a population of approximately 4,800 people, who welcomed their arrival. Helped by the troopers' enthusiasm and construction abilities; plus Australian carpenters, woodsmen, road builders, and other experts, Camp Strathpine grew into a modern training operation.

Camp Strathpine became a showcase military facility that was divided into six main areas:

At Camp Strathpine, the majority of the troopers were housed under canvas in tented areas joined by gravel pathways. In contrast, a large number of semi-permanent wood buildings were constructed for use for supply storage, kitchens, mess halls, and recreation facilities.

The Division paraded its strength, new weapons, vehicles and its pride through downtown Brisbane while throngs of appreciative Australians lined the streets or watched from windows of nearby buildings.

Area 1 - Divisional Headquarters located at Joyner, bordered by One Mile Creek, Swift Drive (Samsonvale Road) and Pershing Drive.

Area 2 - The following units were housed in Area 2:
- Military Police Platoon (prison stockade area)
- 302nd Reconnaissance Troop (Mechanized)
- 7th Reconnaissance Squadron
- 603rd Medium Tank Company
- 16th Quartermaster Squadron

Area 3 - Regimental tented camp area for:
- "HHT", 1st Brigade
- 5th Cavalry Regiment
- 12th Cavalry Regiment

Area 4 - Regimental tented camp area for:
- "HHT", 2nd Brigade
- 7th Cavalry Regiment
- 8th Cavalry Regiment

Area 5 - Artillery Area - Now known as French's Forrest Estate at Petrie. The area was used by:
- "HHB", 1st Cavalry Division Artillery
- 61st Field Artillery Battalion (75mm)
- 82nd Field Artillery Battalion (75mm)
- 99th Field Artillery Battalion (105mm)
- 271st Field Artillery Battalion (105mm)

Area 6 - Water pumping station at Young's Crossing and quarters for the Civil Construction Workers who were drafted by the Australian Government to support and carry out building and construction projects.

As soon as all units were established in their quarters, the 8th Engineers moved out, by train, to Waga-Waga, a small town sixty miles west of Sidney and located on a river well suited for stream crossing and swamp terrain training. The artillery units of the 1st Cavalry Division did live firing on a firing range at Flinders on Clear Mountain. Two live practice grenade ranges were located at Cashmere, one to the east of Four Mile Creek next to Area 3 and the other to the north of Winn Road. Two mortar ranges were located at Cashmere, one range being near One Mile Creek, just to the south of today's Ira Buckby Road West and the other across both sides of Winn Road. The target area was immediately to the south of the firing ranges. Both 60mm and 81mm mortars were used.

There were two infantry assault courses in the area, one being near the intersection of Forrest Road and Howze Road west of Area 4. General Infantry training and compass exercises were carried out near Samsonvale township and as far away as Samford, Whiteside and to the north of Petrie. The Division received six months' of intense combat jungle warfare training at Camp Strathpine in the wilds of scenic Queensland and amphibious training at nearby Toorbul Point north of Brisbane Moreton Bay, Port Stephens in northern New South Wales and Camp GanGan, another amphibious training site.

During the fall of 1943, more changes came to the Division. On 11 October, the firepower of the Division was improved by the activation of the 271st Field Artillery. In the reorganization of 04 December, weapons troops "D" and "H" were added to each of the regiments. The 7th Reconnaissance Squadron was reorganized into the 603rd Light Tank Company and the 302nd Reconnaissance Troop (Mech). The 302nd had a specific Table of Organization and Equipment (TO&E) which incorporated a unique radio unit with troops of Lakota and Dakota Indian Tribes who used their ancient tribal Sioux language to communicate with other divisional headquarters troops. This secret organization, formed in the foothills of Australia and later to be known as "The Code Talkers." was recruited at the direction of General MacArthur. The close-knit group of individuals, Phillip Stoney LeBlanc, Edmund St. John, Baptiste Pumkinseed, Eddie Eagle Boy, Guy Rondell, and John Bear King took their task seriously. They saved many American lives using their language as an unbreakable code to fool the Japanese throughout the subsequent Island Campaigns.

As part of the general plan of offense to be carried out in the Southwest Pacific, the Admiralties Islands located some 200 miles north and east of New Guinea had been targeted as one of the main stepping stones in the march toward Japan. Manus was the largest of the island group with Los Negros, the next largest, extended in a rough horseshoe curve to form Seeadler Harbor, the most extensive of several anchorages. This area was designated to became the first of many landing sites of the 1st Cavalry Division on their journey to Tokyo. On 19 December, an advance party of the 2nd Brigade and supporting units departed Camp Strathpine, Australia, arrived in Oro Bay, New Guinea on

Troopers from the 2nd Squadron, 5th Cavalry Regiment transferred to Landing Craft, Medium (LCM) from the Personnel Carrying Destroyers (APD) which had brought them from New Guinea. (29 February 1944)

20 December and opened an Advance CP. On 04 January 1944, the Division established its command post at Cap Sudest, New Guinea, and by 25 February the remainder of the Division had arrived in Oro Bay from Australia. It then began training for operations in the Admiralty Islands.

The front lines were not far away, and a few eager, enterprising troopers found their way into battle - with the US Marines who had recently invaded Cape Gloucester on the Island of New Britain. The fighting was hard and the Leathernecks needed supplies. Many tons of food, ammunition and other essential equipment were loaded on trucks of the 1st Cavalry Division that were driven on Landing Ships, Tank (LST) and moved to Cape Gloucester.

Many of the troopers who had driven the trucks to the supply dumps near the fighting wasted no time when they got there. They grabbed their weapons and moved up to the front lines. The Marines were startled to see the troops taking up positions beside them. They welcomed the help and added firepower. Several of the troopers received decorations from the Marines. However, Headquarters of the 1st Cavalry Division finally - and sternly - quickly put an end to the freelance fighting.

After a period of staging and training in New Guinea, it was time for the 1st Cavalry Division to receive its baptism of fire. On 26 February, with most of the elements of the Division in the vicinity of Camp Borio taking part in the 1st Brigade Amphibious Training Problem, word was received to stop all training. Immediate preparations were made for movement into combat.

At 1400 hours the next day, the 2nd Squadron, 5th Cavalry began to load at Oro Bay. Half of the troops and their equipment were loaded onto three High Speed Transport (APD) destroyers, the *USS Brooks, USS Humphreys*, and *USS Sands*, which had been converted to troop ships. Nine other destroyers, each carrying an average of fifty-seven soldiers, transported the remainder of the landing force. The rest of the 5th Cavalry moved by truck from Camp Borio to Oro Bay, where they embarked on LSTs for their role as a support force if the reconnaissance succeeded.

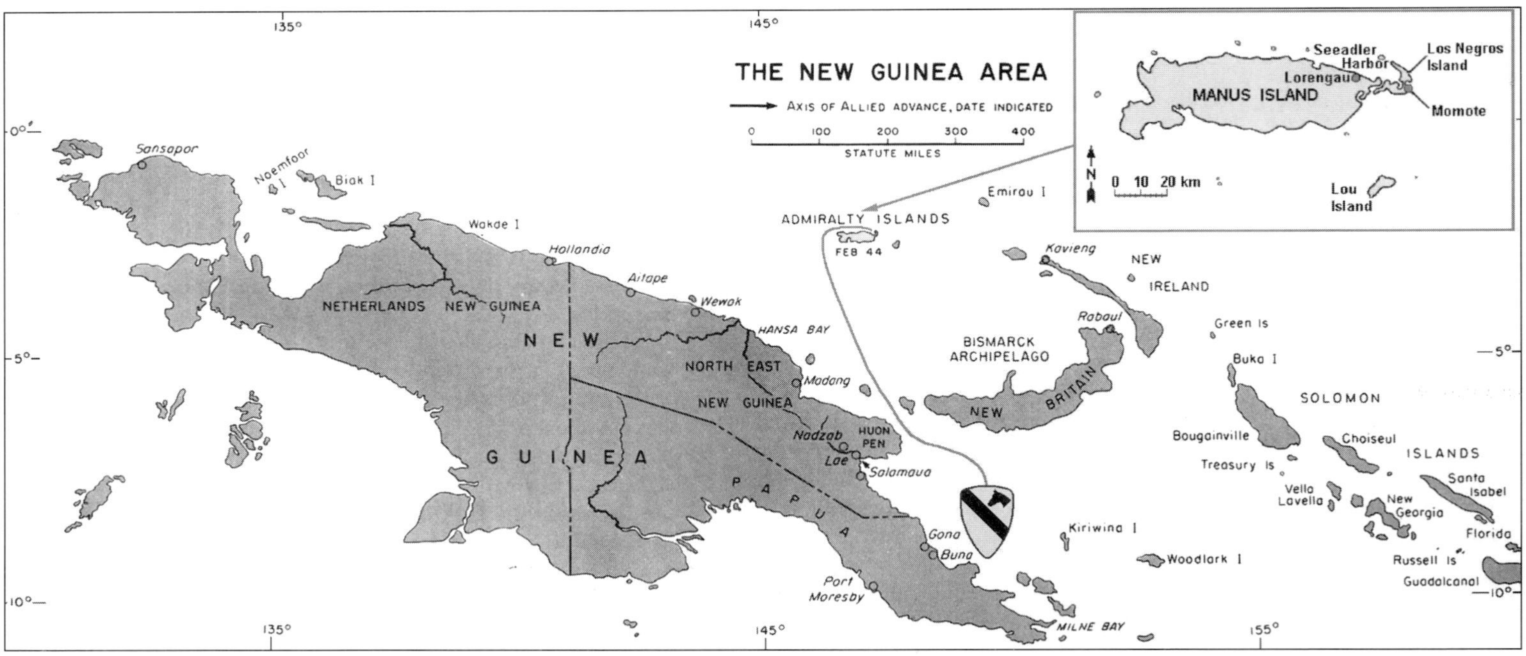

On 13 February General Headquarters, Southwest Pacific Area, issued an order to initiate the first stage of the Island Hopping Campaigns by seizing all of the Admiralties and begin work on a naval base. The major land masses of the Admiralties consisted of Manus and Los Negros Islands separated by a narrow body of water opening into Seeadler Harbor. The airdrome sites of Momote, Los Negros and Lorengau, Manus were designated the objectives with Seeadler Harbor as the invasion zone. The 1st Cavalry Division, the nucleus of the task force assigned to this operation, was already being prepositioned at Camp Borio in the vicinity of Oro Bay, New Guinea, staging area for Operation BREWER.

At 0645 hours on 28 February, 1,026 troopers and their equipment, identified as Task Force BREWER consisting of three High Speed Transport (APD), destroyers; the *USS Brooks, USS Humphreys*, and *USS Sands*, moved out of Cape Sudest, Oro Bay, New Guinea under the command of Brigadier General William C. Chase. They were under the escort of the destroyers *USS Reid, USS Stockton*, and *USS Stevenson*. At 0819 hours, six other destroyers, the *USS Flusser, USS Manhan, USS Drayton, USS Smith, USS Bush* and *USS Welles* joined the Task Force. Their destination was a remote Japanese occupied island of the Admiralties, Los Negros, where they were to make a reconnaissance in force and if feasible, capture Momote Airdrome and secure a beachhead for the reinforcements that would follow.

Even as the Task Force was underway, it received supplementary reconnaissance intelligence from a patrol of Alamo Scouts, a volunteer organization trained to live off the land for weeks at a time and make use of natives familiar with the area and disposition of Japanese troops. Under cover of a diversionary bombing attack, they had landed a mile south of planned invasion area and reported that the enemy was still present in force in a large bivouac area on the southeast part of Los Negros. This information permitted refinements in the Naval Operational Orders to include three separate fire support areas in the supporting bombardment plans.

A rendezvous point, fixed at some twenty miles below Cape Cretin, was reached at 1326 hours. Here the attack group was met by the cruisers *USS Nashville* and *USS Phoenix* and the destroyers *USS Daly, USS Hutchins, USS Beale*, and *USS Bache*, which had come from the Cape Sudest area. General MacArthur and the Commander of the Seventh Fleet, Admiral Kinkaid, were aboard the *USS Phoenix*. The route lay through the Vitiaz Strait, between Long Island and the coast of New Guinea, then into the Bismark Sea. Unchallenged, the convoy arrived at a point about 10 miles south of Los Negros at 0600 hours on D day. The *USS Phoenix, USS Daly*, and *USS Hutchins* led out in column to conduct a reconnaissance approach toward Southeast Point. With the approach of daylight, two observation planes took off from the cruisers.

The units of the 1st Cavalry Division selected to lead the invasion of the Island made final preparations for landing. These units under the command of Brigadier General William C. Chase included:

- 2nd Squadron, 5th Cavalry Regiment
- "B" Battery, 99th Field Artillery Battalion
- 673rd AntiAircraft Machine Gun Battalion (Airborne)
- Headquarters Troop, Reconnaissance Platoon, 1st Cavalry Division
- Headquarters Troop, Communications, 1st Cavalry Division
- 30th Portable Surgical Hospital
- Joining these forces, there were:
- Australian New Guinea Administration Group
- Air Force Supervisory Detachment
- Naval Gunfire Support Party and Air Liaison Party
- "C" Company, 583rd Signal Aircraft Warning Battalion, Air Force

The Support Forces were led by the Commander of the 5th Cavalry Regiment, Colonel Hugh Hoffman. Units assigned to this mission were:

- 5th Cavalry Regiment, less 2nd Squadron
- 99th Field Artillery, less "B" Battery
- 1st Platoon, 8th Engineer Squadron
- 1st Collecting Troop, 1st Medical Squadron
- 1st Platoon, "B" Troop, 1st Medical Squadron
- Signal Detachment, 1st Signal Troop
- "C" Battery, 168th Anti Aircraft Battalion (Gun)
- "A" Battery, 211th Anti Aircraft Battalion (AW)
- 40th Construction Battalion, US Navy
- "E" Company, Shore Battalion, 592 Boat and Shore Regiment

Just after 0800 hours on 29 February, under cloudy skys and a light falling rain, the 1st Cavalry troopers climbed down the nets of the APDs and into the Landing Craft, Medium (LCM) and Landing Craft, Personnel, Ramped (LCPR), the flat bottomed landing craft of the Navy. The landing at Hayane Harbor took the Japanese by surprise. The first three waves of the assault troops from the 2nd

Temporary command posts were setup in support of the many patrols that moved out to the primary objective, the Momote Airstrip.

The 1st Cavalry Division troopers move toward the front lines past a B-25, the first plane to land on the reconditioned airstrip which was put in service exactly 6 days after the initial assault landing. (06 March 1944)

Additional forces arrived at Seeadler Harbor and came ashore at Salami Beach. These forces included the 2nd Brigade, the remainder of the 7th Cavalry Regiment, the 8th Engineer Squadron, the 61st Field Artillery, and a detachment from "A" Company, 69th Signal Battalion.(09 March 1944)

Squadron, 5th Cavalry reached the beach virtually unscathed. The fourth wave was less lucky. By then the Japanese had been able to readjust their guns to fire lower and inflicted many casualties during their landing.

Troops under the command of Lieutenant Colonel William E. Lobit of Galveston, Texas, fanned out and attacked through the rain. They quickly fought their way to the Momote Airdrome and had the entire facility quickly under control in less than two hours. The United Press would hail the Los Negros landing as "one of the most brilliant maneuvers of the war." The Associated Press would call it "a masterful strategic stroke."

Shortly after 1400 hours on "D" day, General MacArthur inspected and praised the cavalry troops' actions and accomplishments. He then ordered General Chase to defend the Momote Airdrome at all costs against Japanese counterattacks. He finally headed back to the beach where he presented the Distinguished Service Cross to Lieutenant Marvin J. Henshaw, 5th Cavalry, of Haskell, Texas. Lieutenant Henshaw had been the first American to land on Los Negros in the first wave, leading his platoon ashore through the narrow ramp of a Higgins boat.

Nightfall was coming when, as it was known, the enemy would counterattack. Early in the morning, around 0200 hours, the enemy came back in force. In the darkness the Japanese made it into the perimeter of the 5th Cavalry. Hand-to-hand fighting broke out near some foxholes. Tough fighting raged the next day and through the night. Japanese pressure on the invasion force remained desperate and intense. The music of the old cavalry charge could virtually be heard when the rest of the 5th Cavalry reinforcements hit the beach in Landing Ship, Tank (LST) and other landing craft. In a coordinated action, the 40th Naval Construction Battalion (Seabees) landed on Los Negros Island in support of the 5th Cavalry. Their mission was to reconstruct the Momote Airdrome. Assigned to defend a large portion of the right flank, the 40th suffered heavy casualties while defending the airfield with the troopers of the 5th Cavalry. Along with the 40th, the consolidated 5th Cavalry Regiment soon secured all of the Momote Airdrome and spent the long night of 02 March, repulsing suicidal attacks, especially against the north and northwest sectors of the perimeter.

After a period of staging in New Guinea, the 12th Cavalry Regiment departed from New Guinea as a part of the combat reinforcements of the Admiralty Campaign. On 02 March, the 12th Cavalry embarked at Cape Sudest, New Guinea and in four LSTs moved to join the forward forces of the 1st Cavalry Division.

The third day on Los Negros, 03 March, was a red letter day for the 1st Cavalry Division. It was the 89th anniversary of the founding of the 5th Regiment. There was little time for celebration; the fresh, well equipped Imperial Marines were counterattacking and the worst was yet to come. Combat raged through the night of 03 March and the morning of 04 March. At one point the Japanese penetrated several hundred yards inside the defense perimeter near "G" Troop. The cavalrymen rallied and they wiped out the attackers. It was during the fierce night fighting that a member of "G" Troop, 5th Cavalry, Staff Sergeant Troy A. McGill of Ada, Oklahoma, with a squad of eight men, occupied a revetment which bore the brunt of a furious attack by approximately two hundred drink crazed enemy troops. Although covered by crossfire from machine guns on the right and left flank, he could receive no support from the remainder of our troops stationed at his rear. All members of the squad were killed or wounded except Sgt. McGill and another man, whom he ordered to return to the next revetment. Courageously resolved to hold his position at all costs, he fired his weapon until it ceased to function. Then, with the enemy only 5 yards away, he charged from his foxhole in the face of certain death and clubbed the enemy with his rifle in hand-to-hand combat until he was killed. At dawn 105 enemy dead were found around his position. For his valiant action, Sergeant McGill received the Medal of Honor.

Personnel of 27th Ordnance Detachment repair a tractor and medium tank off Lorengai Beach, Manus Island, 1200 yards away from enemy fire. The machines had been damaged by enemy land mines, were being put in shape for further use. (26 March 1944)

On 06 March, the 12th Cavalry, along with the 271st Field Artillery Battalion landed on Los Negros Island with minimal resistance. Under cover of the B-25 bombing, they joined up with the 2nd Squadron of the 7th Cavalry Regiment to seize the Salami Plantation and Salami Beach, about three miles north of the Momote Airdrome. The Japanese, expecting an amphibious landing had their guns directed toward the beach. They were surprised by the ground attack from the rear,

The next day after landing, the 12th Cavalry joined with the 2nd Squadron of the 7th Cavalry and the 5th Cavalry and quickly went into action to seize the Papitalai Mission and Lombrum Point before the Japanese could complete building a well fortified defense. In retreat, the enemy left behind large amounts of their food and equipment. On 06 March, the 5th Cavalry went back into action to occupy Porolka and the first American airplane landed on the Momote Airdrome which had been repaired by the Seabees. The next day the 5th Cavalry pushed south and after executing a short amphibious landing assault, overran Papitalai Village.

On 08 March, members of the 12th Cavalry liberated a contingent of sixty nine Sikh solders of the British Empire Forces who had been captured in Singapore in 1942 and moved to Turk, Rabaul and finally to the Momote Airdrome on the Admiralties to be used as forced labor in the construction of defenses for the islands.

On 09 March additional forces arrived at Seeadler Harbor and the same day came ashore at Salami Beach. These forces included the 2nd Brigade, the remainder of the 7th Cavalry Regiment, the 61st Field Artillery, and a detachment from "A" Company, 69th Signal Battalion.

By 11 March, mop up operations were underway all over the northern half of Los Negros and attention was being shifted to a much bigger objective immediately to the west, Manus Island. Successful occupation of Manus Island depended upon control of the area around the coastal village of Lorengau, where the majority of the Japanese defenses were positioned. It was decided that taking the positions at minimum loss would require the establishment of forward artillery positions on some of the small islands west of Seeadler Harbor and north of Lorengau.

Teams from the 302nd Reconnaissance Troop and divisional artillery were sent out to scout three islands in enemy territory. Butjo Island was unoccupied. Bear Point was also unoccupied, but unsuitable as an artillery position. However, Hauwei Island proved to be a "hornet's nest" of Japanese. The 26 man reconnaissance team in a Landing Craft, Vehicle (LCV) was escorted by a Patrol, Torpedo (PT) boat for firepower and protection. Landing and moving inland, they were ambushed and had to engage in fierce fighting resulting in five killed in action, fourteen wounded and three missing in action. They battled back to the beach and reboarded the LCV.

An enemy shell hit the LCV, blowing out one side, and sank it. The PT boat commander had withdrawn without communicating his intentions. The survivors were in the

Troopers of the 12th Cavalry loading into LCVs from the destroyer USS ROE *for the landing on Rambuto Island, southeast of Los Negros, (05 April 1944)*

water about four hours before being spotted by a B-24, which sent another PT boat to rescue them. All of the survivors had been wounded. One of the men tied himself to a tree to shield himself from the friendly fire. He swam and waded from island to island the next day. Another of the men swam to a bell buoy and was recovered the next day.

The reconnaissance mission was costly, yet it revealed that Hauwei, a former coconut plantation, could support two battalions of artillery, providing a level of firepower that would save many American lives during the attack on Lorengau. The assault on Hauwei began on 12 March. Ships anchored inside Seeadler Harbor blasted the island. The 61st Artillery Battalion, firing from Mokerang Plantation near the southern tip of Los Negros, laid down a heavy barrage on the Japanese defenders. The Royal Australian Air Force strafed and bombed the beaches and inland areas shortly before the landing of the 2nd Squadron, 7th Cavalry. By sundown, 13 March, Hauwei Island was cleared of enemy resistance.

At dawn, 15 March, the Manus Island invasion commenced with heavy shelling, naval bombardment and air attacks. Soon afterward, the 2nd Brigade, under the command of Brigadier General Verne D. Mudge, surprised the enemy by swarming ashore at two beaches near the Lugos Mission Plantation. By dusk the 1st Squadron, 8th Cavalry had advanced past snipers and scattered resistance and dug in on the western edge of Lorengau Airdrome, the last airfield controlled by the Japanese. The sixteenth of March was a day of heroes - and casualties - as troopers charged or crawled through heavy machine gun fire to wipe out the enemy positions. Lorengau Airdrome was captured the next day after the 7th Cavalry moved up to relieve the weary 8th Cavalry fighters.

On 18 March, the 2nd Brigade crossed the river in force and drove the enemy from Lorengau Village. The objectives were Rossum, a small village south of Lorengau and Salesia Plantation. By 21 March, the 8th Cavalry had won control of most of the plantation, but the battle for Rossum was slowed by heavy jungle that the Japanese used to their

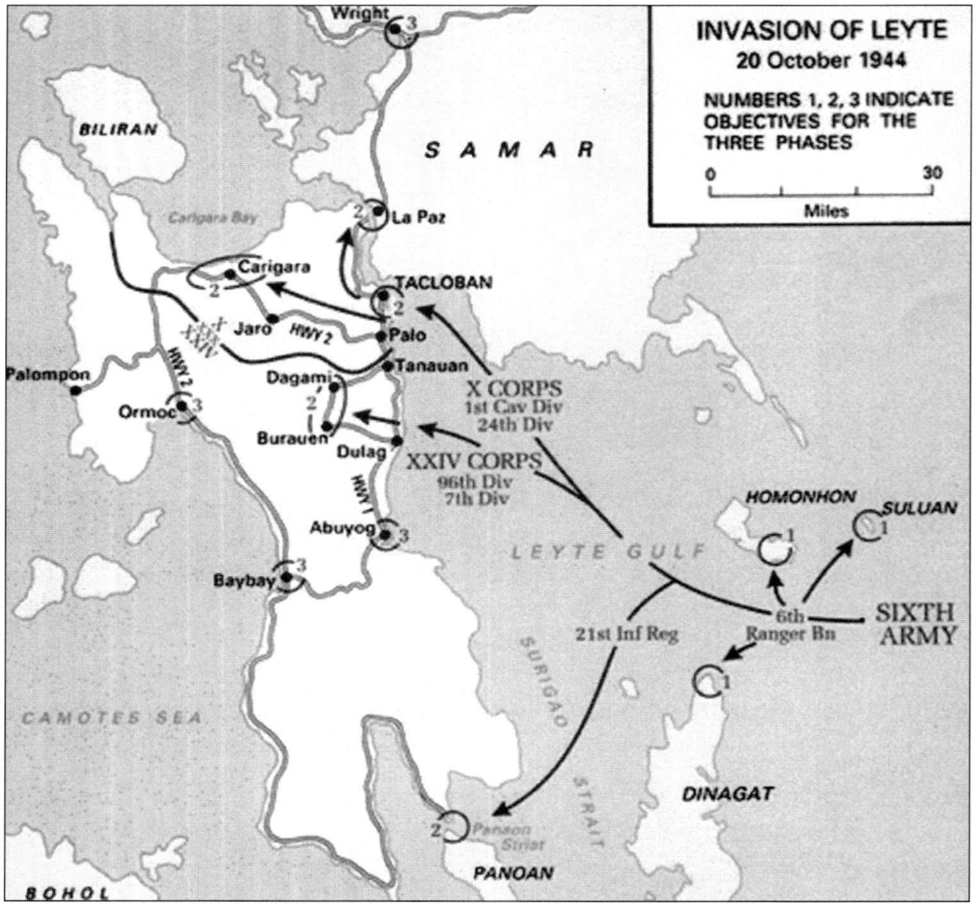

The ship maneuver and invasion plan for Leyte was designed to provide the maximum number of simultaneous attacks over the entire island thus minimizing the opportunity for the Japanese to focus on any single one. (20 October 1944)

advantage. After ninety-six hours of bitter combat the 1st Squadron, 7th Cavalry was relieved by the 1st Squadron, 8th Cavalry. The final push to Rossum was made behind heavy artillery fire and air bombardment.

With attention focused on the opening of new operations at Hauwei Island, the 5th and the 12th Cavalry began working their way south of Papitalai Mission through the rough hills and dense jungles in hand-to-hand combat. Tanks sometimes would give welcome support, but mostly the troopers had to do the dangerous job with small arms and grenades. Two final attacks wiped out the remaining resistance on Los Negros Island. On 22 March, two squadrons from the 5th and 12th Regiments overran enemy positions west of Papitalai Mission. Once again it was tough fighting with the terrain, overgrown with thick canopies of vines, favoring the Japanese. On 24 March, the 5th and 12th Regiments overcame fanatical resistance and pushed through to the north end of the island. On 28 March, the battles for Los Negros and Manus were over, except for mop up operations.

On 31 March, the 1st Squadron, 12th Cavalry moved from Lombrum Point to Mokerang Peninsula and on 01 April launched an attack on Korunist and Ndrilo Islands which lay just off the western tip of Mokerang Peninsula, Los Negros. After land, sea, and air forces had given those islands a hard pounding, the troopers, transported in eighteen native canoes, four captured Japanese collapsible boats, and sixteen engineer half boats landed unopposed. On 03 April, the 2nd Squadron, 12th Cavalry moved to Rambuto Island, southeast of Los Negros, to search out and destroy small bands of enemy soldiers. Because of the numerous coral reefs, the troopers carried their supplies and equipment as they waded ashore in waist-high water. On 07 April, the 1st Squadron, 12th Cavalry was sent on a combat mission to Pak Island where they met light resistance.

On 18 May 1944, the Admiralty Islands campaign officially ended. Japanese casualties stood at 3,317 killed. The losses of the 1st Cavalry Division included 290 dead, 977 wounded, and 4 missing in action. Training, discipline,

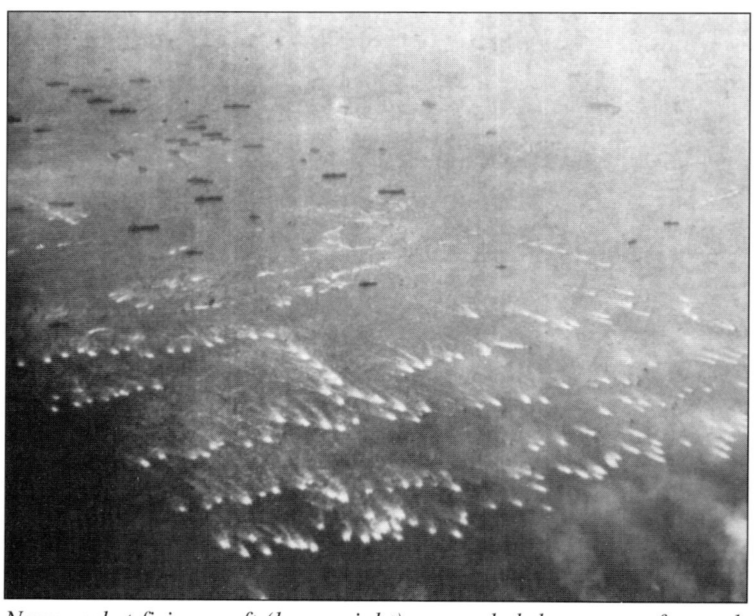

Navy rocket firing craft (lower right) proceeded the waves of assault boats as they moved toward White Beach. (20 October 1944)

"E" Company, 7th Cavalry on White Beach, Leyte, Philippine Islands. (20 October 1944)

General Douglas MacArthur wades ashore during initial landings at Leyte, Philippines Islands (20 October 1944)

determination, and ingenuity had won over suicidal attacks. The troopers of the 1st Cavalry Division were now seasoned veterans.

With the fighting at a lull, the men of the division had time to take stock of what they had just been through. There had been many heroic deeds and many unselfish actions. To preserve the history of accomplishments, traditions and spirit of the 1st Cavalry Division, they decided to organize a 1st Cavalry Division Association which would be open to all members of the division. There on the battlefield, the Association was chartered to promote fellowship, help troopers keep in touch with one another, and hold annual reunions.

General Swift, reassigned to command I Corps, left the division. General Mudge received his second star and along with it, the command of the 1st Cavalry Division. Colonel Hugh F. T. Hoffman assumed command of the 2nd Brigade and was promoted to Brigadier General. Thus the Division command structure had evolved as the farsighted General Swift had arranged. Planning for the next mission, the invasion of the Philippine Island of Leyte, began.

On Columbus Day, 12 October 1944, the 1st Cavalry Division sailed away from its hard earned base in the Admiralties for the Leyte invasion, Operation KING II. Meanwhile, 738 ships of the Navy's Third Fleet sailed toward the Gulf of Leyte. Among the most powerful naval force ever assembled were eighteen aircraft carriers, six battleships, seventeen cruisers and sixty-four destroyers. The invasion force must have appeared awesome as it moved toward the eastern shores of Leyte near Tacloban.

At about 0530 hours, when it began to grow light, on 19 October the invasion forces moved to assigned positions off the landing beaches. The invasion body was composed of two major thrusts; the Southern Attack Group that was divided into two components; Attack Group "Able", carrying the 7th Infantry Division; Attack Group "Baker", carrying the 96th Infantry Division; and the Northern Attack Group that was divided into three components; the San

Military Police Platoon, 1st Cavalry Division directs traffic at the intersection of Routes 1 and 10, outside Tocloban. (26 October 1944)

Ricardo Group, carrying the 1st Cavalry Division, who were designated to land on Beach White, a mile long sandy beach extending southward from the base of Cataisan Peninsula; the Palo Group, carrying the 24th Infantry Division, who were designated to land on Beach Red; and the Panaon Group, carrying the 21st Regimental Combat Team (RCT), who were designated to land on Panaon Island off the southernmost point of Leyte. At 0900 hours, escort ships started giving the enemy at Tacloban a taste of things to come.

At 0645 hours 20 October the flagship *USS Blue Ridge* signaled "Deploy" and the San Rocardo Group, under the command of Admiral W. M. Fechtler and transporting the 1st Cavalry Division, turned to follow the channel between Mariquitdaquit Island and Somar. Designated as TG 78.2, the San Rocardo Group consisted of the attack transports *USS Fremont*, *USS Harris*, *USS Leonard Wood*, *USS Pierce*, transports *USS Herald Of The Morning*, *USS La Salle*, attack cargos *USS Arneb*, *USS Electra*, LSDs *USS Oak Hill* and *USS White Marsh* and was supported by a Destroyer Screen composed of the destroyers *USS Anderson*, *USS Fletcher*, *USS Lavallette*, *USS Jenkins* and thirty-five various support vessels. By 0800 hours the TG had moved to its assigned position of seven miles off the beach transport area.

Following heavy fire support, which began at 0945 hours, landing operations began. Precisely at H-hour, 1000, the first wave of the 1st Cavalry Division hit the beach. The landing, at "Beach White" was between the mouth of the Palo River, to the south, and Tacloban, the capital city of Leyte. By mid afternoon, cargo ships were unloading rapidly. The 7th Cavalry, along with the 44th Tank Battalion, was assigned the mission of securing the Tacloban Airstrip that would be used as a US fighter-bomber base, Troopers of the 5th, 7th, 8th, and 12th Cavalry quickly fanned out across the sands and moved into the shattered jungle against occasional sniper fire.

The fighting near the beaches was still underway when General MacArthur and Philippines President Sergio Osmena waded ashore in ankle deep water. MacArthur soon broadcast his famous message to the Filipinos: "People of the Philippines: I have returned. By the grace of Almighty God, our forces stand again on the Philippine soil - soil consecrated in the blood of our two peoples . . . Rally to me! Rise and strike!" To the Philippine guerilla forces and the seventeen million inhabitants, it was the news they had long awaited.

Even before the smoke of gunpowder could clear, with fighting underway only two miles outside the city of Tobloban, General MacArthur, acting on his own, decided to restore the right to govern the island to civilian hands. On 23 October, in a brief ceremony at the steps of the provincial capital, he formally proclaimed the resumption of constitutional government to the hands of President Sergio Osmena. In a brief speech, MacArthur said "On behalf of my government, I restore to you, a constitutional administration by countrymen of your confidence and choice. As our forces advance, I shall in like manner, restore the other Philippine cities and provinces throughout the entire island."

The Japanese had been able to put an additional 20,000 combat troops ashore on the west side of Leyte shortly after the invasion of the First Team. In a counterattack, Japanese reinforcements had landed at Ormoc Valley, on the other side of the mountain range. They began menacing the X Corps flank from the southwest. Accordingly, the 1st Brigade advanced into the mountains to blunt the threat. The battle through the mountains was the outstanding achievement of the campaign. In record rains, which flooded the island, the supply lines were stretched to the breaking point. The Japanese had dug in on the reverse slopes of the knife-edged ridges, almost immune to artillery fire.

The missions of the 1st Cavalry Division in late October and early November included moving across the northern coast of Leyte, through the rugged mountainous terrain, and deeper into Leyte Valley. After the breakout and securing of the Tacloban Airstrip, the next day, the troops of the 8th Cavalry moved out on their mission to the island of Samar, leaving the 5th, 7th, and 12th Cavalry who moved out to both the swamps and the Japanese to the south, on a two prong approach drive to the north and northwest, clearing the Leyte Valley on the way. On 02 November the 1st Cavalry Division and the 24th Infantry Division attacked Carigara and captured it easily.

As the Japanese continued to pour reinforcements north through the Ormoc Valley, the menace of an infiltration from the mountains separating Leyte and Ormoc became

Filipino volunteers carry supplies into the mountains to reach 1st Cavalry Division troops.

increasingly apparent. To counteract this event, the 12th Cavalry was ordered to push into the high ground west of Leyte Valley and attack the defenses there. On 09 November the Regiment launched its attack at 0900 hours with the support of the 271st Field Artillery. This action began a long and bitter struggle which lasted nearly two months.

While the 12th Cavalry was attacking in the Mt. Pina - Mt. Badian areas, the 5th Cavalry began a probe on the southern flank of the enemy and by 10 November the 1st Squadron had occupied Hill 2926 - Mt. Pina Area. On 11 November the 7th Cavalry took over the defense of the entrance to the Leyte Valley. On 13 November, the 12th Cavalry ran into heavy opposition while pushing to the southwest from Blaud. Two definite Japanese forces were located on high ground above the Naguisan River. On 14 - 15 November the Division continued to secure high ground between Leyte Valley and Ormoc-Pinamapoan Highway.

On 15 November enemy resistance in front of the 7th Cavalry faded after a heavy barrage from the 82nd and 271st Field Artillery Battalions. The 2nd Squadron, 12th Cavalry Regiment got into an intense fight with the Japanese who were well entrenched on Hill 2348, about two miles east of the Ormoc Pinamapoan Highway. The battle for Hill 2348 continued the next day and threatened to be a bloody stalemate. Individual cavalrymen of "G" Troop advanced through heavy machine gun fire and began to silence the Japanese strongholds one by one.

Also on 15th November, the 112th Regimental (Cavalry) Combat Team was attached to the 1st Cavalry Division, and assumed responsibility for the Capocan-Carigara-Barugo area. During the next week the enemy fought defensive delaying actions. On 19 December, the two squadrons of the 12th Cavalry battled their way into the barrio of Lonoy, moving south the next day toward Cananga. The war seemed to speed up as the troopers could use conventional infantry tactics in the open countryside. On 18 November the 1st Squadron, 112th Cavalry relieved the 1st Squadron, 7th Cavalry in the Mt. Minoro area.

On 20 November, the rest of the 12th Cavalry became heavily engaged around Mt. Cabungaan, about three miles south of Hill 2348. The enemy had dug in on the reverse side of sharp slopes. Individual troopers were again faced with the task of searching out and destroying positions in the fog. Throughout the night of 21 - 22 November the 271st Field Artillery kept the Japanese on the northwest side of Mt. Catabaran awake by heavy concentrations of fire. Before the day was over, patrols from the 12th Cavalry had established observation posts within 150 yards of Cananga on Highway 2 in the Ormoc Valley.

On 26 November, both the 12th and 112th Cavalry regiments launched attacks against their immediate opposition. The enemy positions that had given the 112th Cavalry difficulty on the two previous days were seized in the afternoon after a pulverizing barrage from the 82nd and 99th Field Artillery Battalions. On 28 November the 2nd Squadron, 12th Cavalry launched another successful attack on Hill 2348 which took the form of a double envelopment. The 1st Squadron renewed it's attack on positions on Mt. Cabungaan but sharp ridges held up their advance, The 112th Cavalry continued to move toward its objective

Throughout the battle, the supply chain was an example of complex difficulties that were solved. Hauling from the warehouse at Tacloban, motor transport moved cargo to Carigara, a distance of thirty miles, over muddy roads. There LTVs of 826th Amphibious Tractor Battalion picked up the supplies and hauled them westward three miles to Sugud through rice patties churned to a waist deep morass. At Sugud the supplies were man handled into one ton two-wheeled cargo trailers towed by TD-9 tractors that wound their way into the foothills to the supply base established by the 12th Cavalry.

On 01 December the 112th Cavalry engaged the enemy at the ridge south of Limon. On the night of 02 December, the battle for Hill 2348 reached its climax. The 2nd Squadron, 12th Cavalry suffered heavy casualties from the heavy machine gun fire, mortars, and waves of Japanese troops in suicidal attacks. On 04 December, the 2nd Squadron, 12th Cavalry attacked and overcame a position to its front with the enemy fleeing in the confusion. "A" Troop, of the 112th, in a drive to the northwest, made contact with the left flank elements of the 32nd Division. Thus the drive became an unremitting continuous line against the Japanese and enemy elements that were caught behind the line were trapped.

Throughout 07 and 08 December, patrols of the 5th and 12 Cavalries continued mop up operations. The 1st Squadron, 112th Cavalry moved out to locate and cut supply lines of the enemy who were still holding up the advance of the 2nd Squadron. On 09 December, heavy rains brought tactical operations to a near standstill and limited activity to patrol missions. On 10 December, the 2nd Squadron, 7th Cavalry relieved the 2nd squadron, 112th Cavalry southwest of Sinayawan. Enemy action remained at a lull.

Having fought its way to the commanding heights of the mountain ranges after two weeks of laborious action, the Division now had to fight its way down into the Ormoc Valley and seize control of Highway 2. By 14 December the 12th Cavalry was continuing progress to the northeast from Mt. Cabungaan, meeting resistance from small enemy groups determined to initiate a delaying action. The 112th Cavalry continued its push south from Bonbon, along the Leyte River.

On 15 December, while the 5th Cavalry maintained the screen from Mt. Laao, on the south to Mt. Cabungaan on the northwest, patrols from the 7th, 12th and 112th Cavalry mopped up enemy positions that had been overrun on the previous days actions. On 21 December, the 12th Cavalry launched a coordinated attack fighting its way through the mountains and cutting the enemy supply line - Highway 2 and capturing Cananga. The 5th Cavalry followed the path of the 12th, and assembled on high ground overlooking Highway 2.

On 22 December the 7th and 112th Cavalry, with the 32nd Division, swept aside the remains of the once formable "Yamashita Line" and proceeded south. On 23 December the 5th, 7th and 12th Cavalry, supported by the 61st Field Artillery Battalion, began a determined drive toward the west coast of Leyte. The 112th Cavalry continued its mission of cleaning out stragglers and isolated pockets of Japanese in the rear areas.

The Division continued the attack west toward the coast over swamps against scattered resistance. By 29 December the 7th Cavalry had reached the Visayan Sea and initiated action to take the coastal barrio of Villaba. On 31 December after four "Banzai" attacks, each preceded by bugle calls, the small barrio fell. Joined by the 5th Cavalry who assisted in mopup operations, the 7th Cavalry moved on to reach the coast at Tibur. Meanwhile the remainder of the 12th and 112th Cavalry were moved by motor convoy to the Leyte Valley to close in on their respective staging areas. The long wet Leyte/Samar campaign was over except for mop up operations.

Meanwhile on 23 October, in parallel with the above operations, the 8th Cavalry along with the 302nd Reconnaissance Troop had moved out to the neighboring island of Samar, the third largest island in the Philippines. They landed at La Paz and the threat of enemy reinforcements coming from Samar to aid enemy troops Leyte was dispelled. Elements of the 302nd patrolled extensively along the south and southwest coast of Samar, but failed to find any evidence of any enemy strength. At first, the 8th Cavalry advanced slowly north toward Hinabangan which was located on a main east-west road connecting Wright and Taft. By 07 December, Hinabangan was captured. The troopers pushed on and Wright was occupied on the 13th. Taking a left movement, the troopers raced westward toward Catabalogan, the capital of the island, which they secured on the 19th of December. With the aid of guerrillas the Wright-Taft highway was opened. Soon afterward, Taft fell, and the Samar portion of the campaign was over.

By 08 January 1945 the 8th Cavalry had returned from their actions on Samar Island and rejoined the main body of it's Division. On 11 January 1945, when the Leyte-Samar Campaign ended, the Japanese losses were estimated to be 5,937 killed in action and only a handful - three hundred eighty-nine had surrendered. With the last of the Leyte strongholds eliminated, the Division began preparations for movement to Luzon, the main island of the Philippines. Leyte had indeed been the largest campaign in the Pacific War, but the record was about to be shattered by the invasion of Luzon.

On 26 January the 1st Cavalry Division boarded Task Group (TG) 78.8, a LSD Reinforcement Group, under the command of Captain R. W. Cutler. The TG consisting of the *USS Linenwald*, *USS Oak Hill*, *USS Casa Grande*, *USS Epping Forest*, *USS White Marsh* and *USS Shadwell* and escorted by the destroyers *USS Sterett* and *USS Wilson*, formed convoys and departed for the Lingayan Gulf, Luzon Island, the Philippines. On 27 January, under heavy air coverage, the Division came ashore in the Mabilao area of Lingayen Gulf, Luzon without incident, and assembled in the vicinity of Urdaneta.

From 28 to 30 January, the Division relocated thirty-five miles inland and opened its Command Post at Guimba. On 30 January, the 302nd Reconnaissance Troop was given the mission of patrolling to the southeast, along the Pampanga River, to find crossings in order to secure the highway leading south from Cabanatuan. On 31 January the 112th Cavalry, who had supported the Leyte Campaign, was relieved from attachment to the Division. The final planning for the operations in the south and southwest areas of the island were reviewed and updated with the most recent findings of the reconnaissance teams.

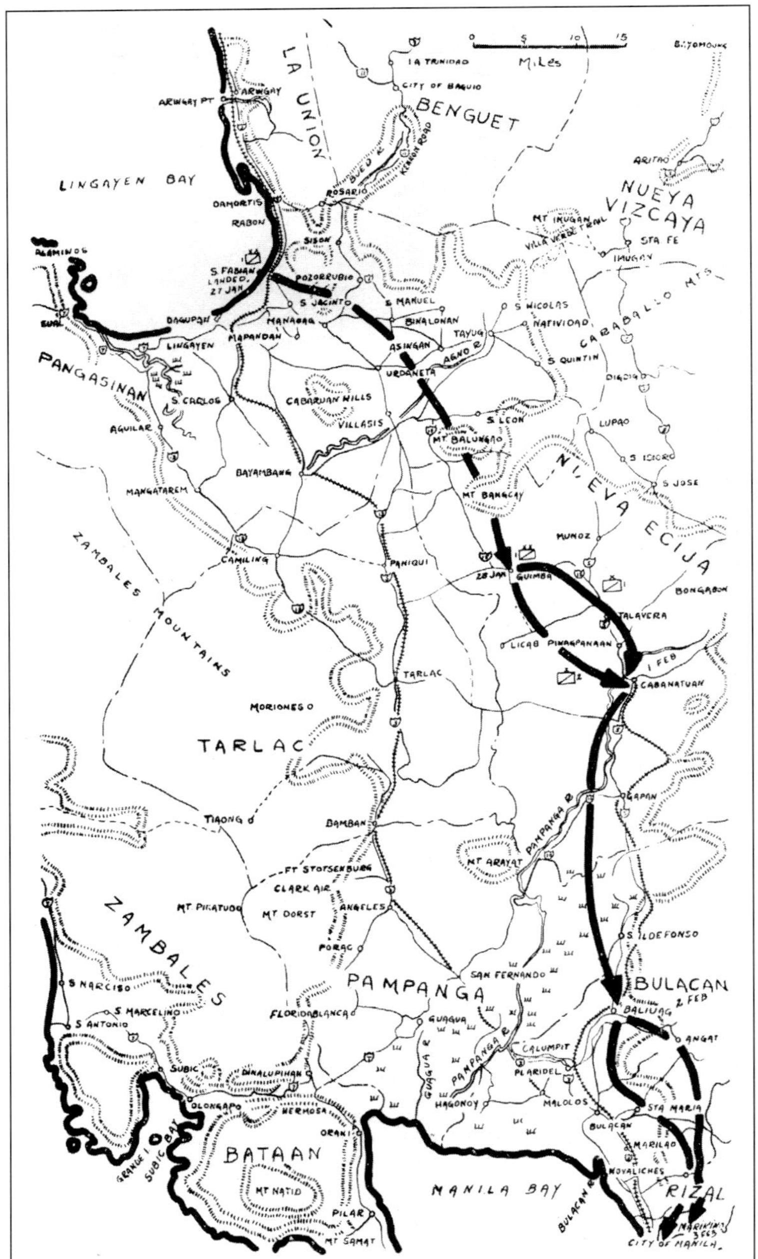

Pushing out at 1 minute past midnight, the "Flying Column" commenced their drive toward Manila. The first objective was Cabanatuan which was captured in a two pronged attack. Dashing southward, through enemy held territory, they led the Division into Manila within 60 hours. (3 February 1945)

The 2nd Squadron, 8th Cavalry Regiment crossing the Angat River, east of Plaridel.

On 31 January, General Douglas MacArthur issued the order "Go to Manila! Go around the Japs, bounce off the Japs, save your men, but get to Manila! Free the internees at Santo Tomas! Take the Malacanan Palace (the presidential palace) and the legislative building!" In an assessment of the situation, the Division's Commander, Major General Verne D. Mudge. decided to attack on a broad front with three mobile tank columns because the positions of the Japanese were vague and the columns had to cross the wide Pampanga River, south of Guimba, en route to Manila. The resulting mission, and the participating units, were dubbed a "Flying Column" by General Mudge.

The rescue operation was divided into three specialized serial elements, The first serial, commanded by Lieutenant. Colonel William E. Lobit, was composed of the Reconnaissance Platoon, 2nd Squadron, 5th Cavalry; Anti-Tank Platoon; Medical Detachment; "A" Battery, 82nd Field Artillery Battalion; "A" Company, 44th Tank Battalion; 3rd Platoon, "A" Troop, 8th Engineers and the 1st Platoon, "A" Troop, 1st Medical Battalion. The second serial, commanded by Lieutenant. Colonel Haskett L. Conner, Jr. was composed of the Reconnaissance Platoon, 2nd Squadron, 8th Cavalry; Anti-Tank Platoon; one section of .50 caliber machine guns; "B" Company, 44th Tank Battalion; "B" Battery, 61st Field Artillery Battalion; 1st Platoon, "C" Troop, 8th Engineers; and the 1st Platoon, "B" Troop, 1st Medical Squadron. The third serial was composed of a Provisional Reconnaissance Squadron, commanded by Lieutenant. Colonel Tom A. Ross, consisting of the 302nd Reconnaissance Troop and the remainder of the 44th Tank Battalion. A Marine Air Control Group, commanded by Captain Samuel H. McAloney, accompanied the serials to provide a communications link to the aircraft from Marine Air Groups 24 and 32 that were used to provide flank and advance cover for the cavalry columns during the mission.

At one minute past midnight, 01 February, the three serials, led by Brigadier General William C. Chase, moved out of Guimba to slice through 100 miles of Japanese held territory. The mission of the 1st Cavalry Division was to dash through the enemy lines and take only force as necessary to get to Manila, not to become embroiled in any large scale battle. At all times during the three day sweep, the nine Marine scout dive bomber patrol, operating as flank guards remained airborne at all times, roaming across the valleys searching every road and trail for signs of enemy movement. Whenever roadblocks were spotted, they reported on the situation and when permission was granted, cleard the area by precision bombing or aerial machine gun screens.

Early in the morning, the 5th Cavalry crossed the Pampanga river and encountered enemy resistance. By 1300 hours, the cavalry forces were locked in a bitter fight with the Japanese near Cabanatuan. The 8th Cavalry crossed the river south of town and turned north to catch the enemy in a pincers movement. By dusk, the 7th and 12th Cavalry had advanced and took over the fight from the lead units.

The Reconnaissance Squadron had swung farther south early on 01 February and approached the town of Gapan at 1330 hours. As the attack moved on to the bridge across the Penaranda River the commander, Lieutenant Colonel Ross, was killed. The commander of the 302nd Reconnaissance Troop, Captain Don Walton, took over the Squadron command immediately and the forces were able to secure the bridge and later with the arrival of "G" Troop, 8th Cavalry, defended it so that the columns could continue their march.

The "race" for Manila was now between the 37th Division and the 1st Cavalry Division, with the cavalry in the lead. In addition to the 302nd Reconnaissance Troop liaison aircraft of the field artillery battalion conducted route reconnaissance and column control. Since the operation began, the reconnaissance units had been fortunate enough to find bridges and fordable crossings almost everywhere they went. The column was able to get around, over, and past each obstacle in its path. The 37th Division, on the other hand, was slowed by difficult crossings which forced

Technical Sergeant R. B. Holland of the Marine Air Group 24, that provided the 1st Cavalry Division a communications link to the Marine aircraft that flew flank and advance cover for the cavalry columns during the mission, patiently waits as he is towed across the Angat River. (February 1945)

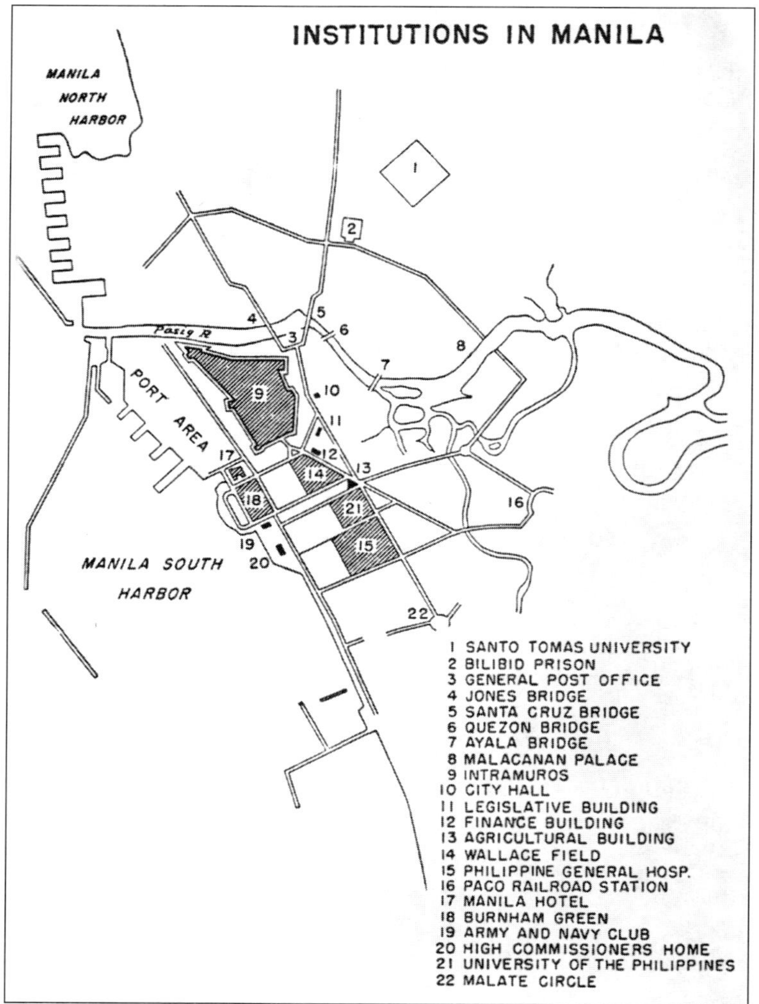

The main established areas of Manila that were entrenched and strongly defended by the Japanese. (03 February 1945)

it to either ferry its artillery and tanks across or wait for the engineers to build bridges. By 02 February, the Flying Column was dashing toward Manila, sometimes at speeds of fifty miles per hour, with individual units competing for the honor of reaching the city first.

During 02 February, the columns pushed south and the 2nd serial leading the way reached Plaridel early in the morning. As the serial crossed the wide Angat River, the 8th Cavalry ran into a Japanese battalion which was dug in on high ground. Grinding out a tough advance, they broke through and continued on. By dusk, the 2nd serial was near Santa Maria. In parallel, the 1st serial crossed the Angat River at Sabang and spent several hours in a bitter fight with a small Japanese force. Following the battle, they turned east on Highway 65 toward Norzagaray.

Early in the morning of 03 February at 0430 hours, the 1st serial moved out. Driving on, they reached Norzagaray at dawn and found the town occupied by Filipino guerrillas. Not stopping, the column turned southwest toward Santa Maria and, fording many streams, reached it at 1500 hours. The 2nd serial, which had crossed the Santa Maria River at noon, moved along Highway 64 to the junction at Highway 52, an outpost manned by the Japanese. After fighting their way through, they left one troop behind to hold the intersection for the units that followed.

By 1630 hours, elements of the 1st Cavalry Division pushed into the town of Novaliches, in the northern outskirts of Manila, with only the steep-sided Tuliahan River separating them from the city proper. A squadron of the 8th Cavalry reached the bridge just moments after Japanese soldiers had finished preparing it for demolition. As the two sides opened fire on one another, the Japanese lit the fuse leading to the carefully placed explosives. Without hesitation, Lieutenant James P. Sutton, a Navy demolitions expert attached to the division, dashed out onto the stone arched bridge and cut the burning fuse. He disregarded the heavy Japanese fire and heaved the other mines and dynamite charges into the gorge under the bridge. The way to Manila was clear.

At 1835 hours, 03 February, the rescue column crossed the city limits of Manila. “F” Troop, 8th Cavalry, under the command of Captain Emery M. Hickman, swept “like lightning” through the heavy Japanese sniper fire to the White House of the Philippines in time to take control of the Malacanan Palace and save it from the torches of the desperate Japanese. As the gates were opened, cheering Filipinos emerged and helped the cavalrymen set up a defense perimeter around the palace grounds.

As the sun set over the ocean behind the advancing Americans, a single tank named *Battling Basic* crashed through the walls surrounding Santo Tomas University, the site of a camp holding almost 4,000 civilian prisoners. The Japanese guards put up little resistance. By 2100 hours, the internment camp at Santo Tomas was liberated and the prisoners, many of whom had been incarcerated for nearly two years, were freed.

The First Team was - “First in Manila”.

Mopping up was in process when it was discovered that the Japanese Guard commander and 70 of his men had taken

Fifth Cavalry, 1st Cavalry Division move on the objective at Harrison Park, strong point of the Japanese defenses in the Southern Section of Manila. This section had to be cleared out in order for our troops to proceed north along Manila Bay toward the central part of Manila for the eventual assault of the Walled City of Intramuros. (February 1945)

The tanks of the 1st Cavalry Division move up into position where the Japanese are holding Rizal Stadium and blocking the south flank of the bay and the route to the Walled City of Intramuros. The tanks have been gassed up and resupplied with ammunition and are moving up one of the side lanes of the Passay District of Manila in order to avoid detection by the Japanese who have a view of the area from the nearby Bay View Hotel. (February 1945)

200 of the internees as hostage and had moved them to the Education Building. With the hostages situation rendering a direct attack on the building impossible, a strong guard was posted and a defense of the area was organized. The next day, under a flag of truce, negotiations for the release of the prisoners began. Late in the afternoon, the Japanese Commander took the position that he and his men, along with their arms and equipment, were to be escorted to a point outside the city or else he would kill all the hostages and make a suicide defense of the building.

Considering the lives of the hostages more important than the capture of the Japanese Commander and his solders, the terms were agreed to. At daybreak, on 05 February, "G" Troop, 5th Cavalry closed in on the door of the Education Building to form an escort for the Japanese who filed out of the building. The small column and their escorts moved to a point near the Pasig River where the two forces parted, the Japanese moving out of sight south and the Troopers returning back to Santo Tomas.

For the next three tough days, troopers of the 1st Cavalry Division battled with flame throwers, bazookas and every weapon on hand to clear the historic Manila Hotel, one of the finest hotels in the Far East and former home of General MacArthur. "A" Troop, 12th Cavalry led the attack, supported by medium tanks and one platoon of heavy weapons. In addition to the maze of tunnels running beneath the building, nearly every room of each floor was heavily fortified with sandbags and automatic weapons. Veteran troopers had to fight "room-to-room" in order to gain control.

Late on the afternoon of 04 February, the 2nd Squadron, 5th Cavalry, was ordered to seize the Quezon Bridge, the only crossing over the Pasig that the Japanese had not destroyed. As the squadron approached the bridge, enemy heavy machine gun fire opened up from a formidable roadblock thrown up across Quezon Boulevard. The Japanese had pounded steel stakes into the pavement, sown the area with mines, and lined up old truck bodies across the road. Unable to advance farther, the cavalry withdrew after nightfall. As they pulled back, the Japanese blew up the bridge.

The next day, 05 February, went more smoothly. Once the 37th Division reached Manila, the northern section of Manila was divided into two sectors, the 37th Division responsible for the western sector and the 1st Cavalry Division responsible for the eastern sector. The 7th and 8th Cavalry fought pitched battles near two water supply installations north of the city. By 10 February, the cavalry had extended its control south of the river. That night, the XIV Corps established, for the first time, separate bridgeheads on both banks of the Pasig River.

Despite initial American euphoria, much fighting remained. Although the approach to the city had been

Troopers of the 8th Cavalry patrol near Santo Thomas, Manila, Luzon, Philippine Islands. (March 1945)

Troopers of the 8th Cavalry, searching for Japanese still hiding in the area, look into a culvert in the battered main street of Santo Thomas, Luzon, Philippines Islands. (24 March 1945)

The fighting around the Kapatalan Sawmill employed a new tactic of cave blasting to drive out the Japanese.

relatively easy, wresting the capital from the Japanese proved far more difficult. Manila, a city of 800,000, was one of the largest in Southeast Asia. While much of it consisted of ramshackle huts, the downtown section boasted massive reinforced concrete buildings built to withstand earthquakes and old Spanish stone fortresses of equal size and strength. Most were located south of the Pasig River which bisects the capital, requiring that the Americans cross over before closing with the Japanese.

On 23 February, during this hard hand-to-hand fighting, a trooper of "E" Troop, 5th Cavalry, Private First Class William J. Grabiarz, E-5 Cavalry, was a scout with a unit that was advancing with tanks along a street in Manila. Without warning, enemy machine gun and rifle fire from concealed positions in the Customs building swept the street, striking down the troop commander and driving his men to cover. As the officer lay in the open road, unable to move and completely exposed to the pointblank enemy fire, Private First Class Grabiarz voluntarily ran from behind a tank to carry him to safety, but was himself wounded in the shoulder. Ignoring both the pain in his injured useless arm and the shouts of his comrades to seek the cover which was only a few yards distant, the valiant rescuer continued his efforts to drag his commander out of range. Finding this impossible, he rejected the opportunity to save himself and deliberately covered the officer with his own body to form a human shield, calling as he did so for a tank to maneuver into position between him and the hostile emplacement. The enemy riddled him with concentrated fire before the tank could interpose itself. Later, the troops found that he had been successful in preventing bullets from striking his leader, who survived. For his valiant action, Private First Class William J. Grabiarz received the Medal of Honor.

One of the final assignments undertaken by the 8th Cavalry was to clear out the Japanese soldiers who had taken cover in the world famous San Miguel Brewery and had provided the earlier heavy sniper fire in the attempt to stop their drive to the Malacanan Palace. It is said that rather than dislodge the occupants using direct artillery fire (which may have also destroyed the inventory of the building), Captain Hickman ordered a fixed bayonet charge which easily put down the Japanese resistance. Soon after the "liberation" of the San Miguel Brewery, General MacArthur joined General Chase and the troopers in a pitcher of the brew before they moved on. By 03 March 1945, organized resistance in Manila was finally wiped out.

The next assignment given to the First Cavalry Division was the difficult task of cracking the Shimbu Line, a few miles east of Manila, and securing a front from Taytay on the North to Antipolo on the south. The goal was to prevent Japanese reinforcements from reaching Manila. Securing the Shimbu Line proved to be a different type of enemy engagement than previously encountered. The division had fought in the jungles of the Admiralties, the mountains and mud of Leyte, open country, and street fighting of Manila. They were now engaged in eliminating the Japanese who were occupying fortified positions in the mountainous environments. The regiments of the division fought abreast as they destroyed the southern flank of the Shimbu Line. From north to south, the units involved were the 5th, 7th, 8th and 12th Cavalry.

On 28 February, General Mudge suffered a serious wound in the abdomen while observing the demolition of a cave that had supposedly been cleared. From deep inside the cave, a diehard Japanese defender managed to lob a grenade that caught General Mudge and his party by surprise.

Brigadier General Hugh F. T. Hoffman immediately assumed command of the Division and carried out the mission. On 11 March, the 1st Brigade Combat Team captured dominating terrain features in and around Antipolo and began mop up operations. The high ground west of

TD-9 Prime Mover in operation over rough terrain pulling an ammunition trailer and a 105mm Howitzer with a full gun crew of "A" Battery, 82nd Field Artillery. (29 June 1945)

Antipolo was secured by the 2nd Brigade Combat Team. Once high ground had been taken, the troopers were relieved by the 43rd Infantry Division and given a weeks rest before taking on a new assignment to help clear southern Luzon of organized Japanese resistance.

On 22 March, elements of the 1st Cavalry Division started their push south into the Batangas Peninsula. On 25 March, on the outskirts of Tanauan, they encountered well organized and mutually supporting enemy bunkers, caves and culverts. In the following 10 days of terrific fighting and under the cover of targeted artillery fire, the barrios of Tanauan, Talisay, Lipa, and San Pablo were taken by elements of the 5th, 7th, and 8th Cavalry Regiments. In the meantime, the 12th Cavalry Regiment was directing an attack on Imoc Hill, which was held by the Japanese.

On 03 April, operations against the final Japanese forces in the Mt. Malepunyo hills began. The 5th Cavalry joined with the 43rd Infantry Division on the eastern shores of Laguna de Bay, severing the main escape route to the north. By the 10th of April, the 1st Squadron, 5th Cavalry, reinforced by a platoon of tanks, entered Mauban of the Pacific to completely cut the Bantangas in half, isolating and trapping the Japanese to the south.

On 12 April, the 5th Cavalry began a drive southeast, down the Bicol Peninsula to clear it of Japanese and link up with the 158th Regimental Combat team. The two forces finally converged at Naga on 29 April, after "B" Troop, 5th Cavalry and a group of engineers made an amphibious assault across the Ragay Gulf at Pasacao.

On 24 April, the 2nd Brigade started its movement north to Siniloan in preparation for relief of the 43rd Infantry Division. Patrols found that the main enemy strength was concentrated around the Kapatailin Sawmill. Next to Antipolo, this position was the most elaborately constructed defense system encountered. On 07 May, after a heavy air bombing, the 7th Cavalry moved toward the sawmill and attacked. Elements of the 8th Cavalry joined in to conduct mop-up operations and consolidate gains. By 09 May, the Sawmill and Nursery areas were secured with only a few scattered remnants of the Japanese to contend with.

On 14 June, the campaign had turned into a rear area skirmish with patrols continuing to mop up stragglers and preventing movement of enemy forces. On 30 June 1945, when the Luzon Campaign was declared finally completed, the 1st Cavalry Division accounted for 14,114 of the enemy killed and 1,199 prisoners of war. At last, the troopers of the 1st Cavalry Division were able to enjoy a longer rest in Lucena, at the southern end of the Tayabas Province.

In July, a new era in the organizational change of the 1st Cavalry Division began with the promotion of Major General William Chase who immediately assumed command of the Division. Planning and training began for a mission that would promise to be more difficult than anything they had previously encountered. This mission was to bring the rages of war home to the Japanese.

On 30 June 1945, General Douglas MacArthur was preparing invasion plans to drive the final bloody strike into the heart of the Japanese homeland. The operation was to be carried out under the overall code name Operation DOWNFALL. In Operation DOWNFALL, scheduled for 01 November 1945, the 1st Cavalry Division was to be among the first American Forces to invade the beaches of the main islands of Imperial Japan, Kyushu. The first phase, Operation OLYMPIC, was the invasion of the southern most principal Japanese Islands, Kyushu, to establish a base for a followup invasion. The second phase, Operation CORONET, the invasion against the main Japanese Island of Honshu and the decisive battle necessary to defeat Japan, was scheduled to be launched in March 1946.

MacArthur selected General Walter Krueger, commanding general of the Sixth Army, to command the Kyushu Invasion Force, consisting of three Army and one amphibious (Marine) corps, totaling 14 combat divisions. The 550,000 man invasion force would be delivered by approximately 3,000 ships, including 66 aircraft carriers and over 2,600 combat aircraft, under the command of the Fifth Fleet. The Third Fleet would

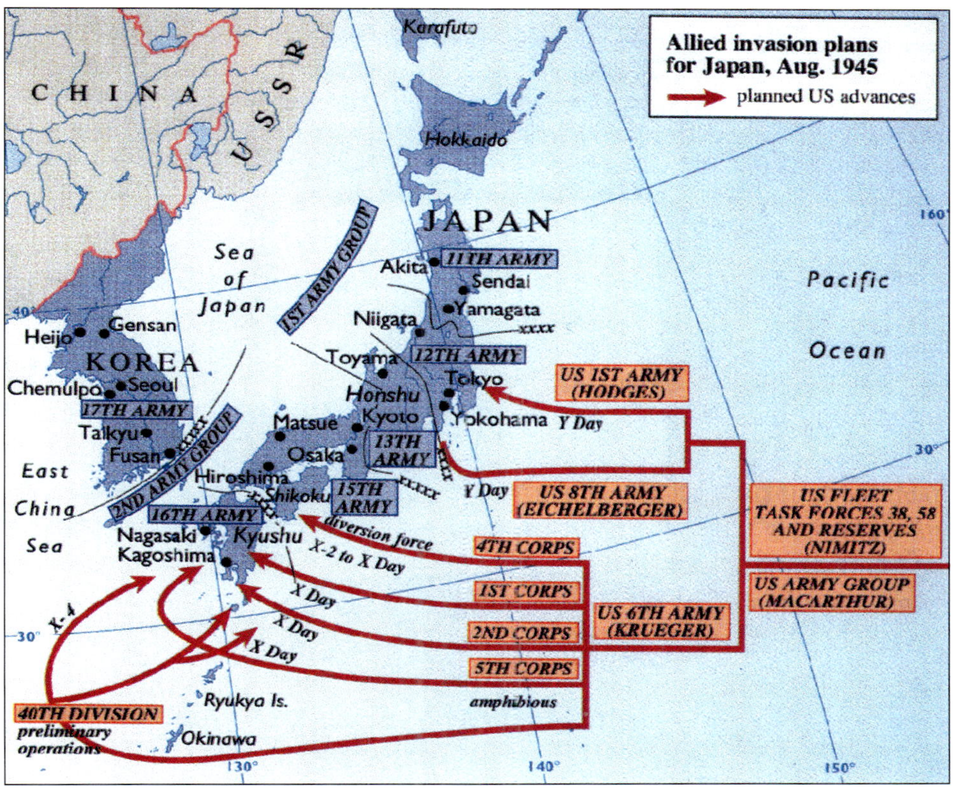

A conventional invasion of the main islands of Japan was planned, under the overall code name Operation DOWNFALL, for the late summer 1945. Based upon the experience of attacking other Japanese held Islands, casualties of the invading forces were estimated at 500,000 while Japanese losses were projected in the millions.

support the invasion by blocking any enemy attempts to reinforce Kyushu from Honshu and Hokkaido.

Army Intelligence estimated that approximately 735,000 enemy troops occupied Kyushu and several small islands to the south. In addition, a force of 5,000 Kamikazes was also ready to descend on the invaders.

The first phase of Operation OLYMPIC was scheduled for 27 October. The 40th Infantry Division and the 158th Regimental Combat team would seize positions in the Koshiki islands southwest of Kyushu, and on Tanega, Make, Take, and Io Islands south of Kyashi, where it was estimated 25,000 Japanese troops were dug in waiting for the invasion.

The 2nd, 3rd, and 5th Marine divisions would invade near Kushikene island, on the southwest coast and split into two elements. One would move toward Sendai and the other to the port city of Kagoshima. Meanwhile, the Eastern Assault Force consisting of the 25th, 33rd, and 41st Infantry Divisions, would land on the southeast coast and capture Miyazzki and the adjacent airfield.

A dense column of smoke rises more than 60,000 feet into the air over the Japanese port of Nagasaki, the result of an atomic bomb, the second ever used in warfare, dropped on the industrial center from a US B-29 Superfortress. (08 August 1945)

In a simultaneous attack, the Southern Assault Force, composed of the 112th Regimental Combat Team and the 43rd Infantry, 1st Cavalry, and Americal Divisions, would land at Ariake Bay and drive inland to capture Shibushi and Kancya.

Three days after the main invasion, the 77th, 81st, and 98th Infantry Divisions, would assault the southern coast of the island west of Kiaman-Dake and attempt to move north and west, bottling up the enemy forces on the southern tip of Kyushu. The 11th Airborne Division, would stand in reserve to support the assault divisions. If not needed as reserves, the 11th Airborne would also land at Kiaman-Dake and link up with the Marine Forces at Kagoshima.

Depending on the success of Operation OLYMPIC, The massive undertaking of Operation CORONET, a 4 month operation, would be launched on 01 March 1946. As many as 28 divisions were earmarked for this phase, including all six Marine divisions and the remainder of all US Naval forces in the Pacific. In all, nearly 5 million men would participate in the operation.

Estimates of the casualties in the battle for the main islands of Japan were addressed. The early stages of the landings would have been particularly bloody with both sides suffering a combined death rate of some 1000 men an hour according to one estimate. Recent intelligence estimates indicate that if Operations OLYMPIC and CORONET had been executed as planned, it would have been the largest bloodbath in American history. These recent estimates set losses at more than one million with a death toll exceeding the number suffered in both World Wars, Korea, Vietnam, and the Persian Gulf. Although American forces had superior fire power and were better trained and equipped than the Japanese soldier, the close-in, fanatical combat between infantrymen would have been mutually devastating.

The little enthusiasm of President Truman for a ground invasion of the Japanese home islands and the accompanying casualties played a major factor in his decision to approve the use the atomic bomb. The dropping of the atomic bombs on Hiroshima (06 August), Nagasaki (09 August) and the subsequent announcement of the surrender of Japan made all the invasion preparation superfluous. Operations OLYMPIC and CORONET, along with the training details and mission they encompassed, were soon forgotten along with speculation about a Pacific Theater experience that never came to pass.

On 13 August 1945, the 1st Cavalry Division was alerted that they had been selected to accompany General Douglas MacArthur to Tokyo and would be part of the 8th Army in the occupation of Japan. Preparation for loading out began immediately and continued night and day. A staging area was set up at Llipa and an advanced command post was set up at Batangas Bay. Soon, ships of the Navy Transport Division 16 began arriving.

Movement of the 1st Cavalry Division to Yokohama, Japan was under control of Navy Task Force (TF) 33, commanded by Rear Admiral John L. Hall. On 18 August attack transports *USS Brisco*, *USS Cecil*, *USS Highlands*, *USS Missoula*, *USS Rutland*, *USS St Mary's*, *USS Talladega* and the attack cargo ships *USS Yancy* and *USS Whiteside* began loading 1st Cavalry Division equipment and personnel for the last leg of their WW II journey through the Western Pacific Ocean. The loading was completed on the 23rd and as part of TF 33, they weighed anchor on the 25th. However, TF 33 had to turn back because of a tropical storm in the vicinity. The typhoon delayed the TF for only a day, as they weathered the fringes of the storm at Subic Bay before getting underway again and headed north along the east coast of Luzon through the China Sea.

On the morning of 02 September TF 33 entered Tokyo Bay, the day Japan signed the formal articles of surrender on the deck of the battleship *USS Missouri*, anchored there. After the conclusion of those ceremonies, all of the ships of TF 33 hauled down their battle flags and headed into Yokohama harbor and entered the dock areas to start unloading operations.

General Douglas MacArthur signs as Supreme Allied Commander during formal surrender ceremonies on the USS MISSOURI in Tokyo Bay. Behind General MacArthur are Lieutenant General Jonathan Wainwright and Lieutenant General A. E. Percival. (02 September 1945)

Chapter 3: Occupation Of Japan

1945-1950

As the surrender terms are being signed aboard the USS Missouri, assault boats carrying elements of the 1st Cavalry Division pass the breakwaters at the inner harbor of Yokohama

At 1030 hours the leading ships were in place in the inner harbor of Yokohama. The First Team was given the honor of leading the Allied Occupational Army into Tokyo. At that time, a reconnaissance party headed by Colonel Charles A. Shelton, the Division Chief of Staff, went ashore to contact the advance party of Lieutenant Colonel Moyers S. Shore which had arrived by plane five days earlier to reconnoiter and select assembly area locations for the landing parties.

The initial assembly areas were within five blocks of the docks. By nightfall, the troops of the 1st Cavalry Division were occupying staging areas throughout the harbor. The Division Artillery was quartered at Hara-Machida. The 8th Cavalry was passed to the control of XI Corps with the mission of occupying Yokohama and was stationed at Kawasaki. The 12th Cavalry was assigned the mission of seizing five airfields in the vicinity of Tachikawa which was completed by 04 September. The next day, long lines of vehicles moved west to the designated bivouac area in Hara-Machida District, where Japanese Signal and Ordnance schools were located. At noon on 05 September, a reconnaissance party headed by Colonel Charles A. Sheldon entered Tokyo. This embarkment was the first official movement of American personnel into the capital of the Japanese Empire.

At 0800 hours on 08 September, a history making convoy left Hara-Machida with Tokyo as its destination. Headed by Major General William C. Chase, Commanding General of the 1st Cavalry Division, the party included a veteran from each troop of the division. Passing through Hachioji, Fuchu and Chofu, the convey halted briefly at the Tokyo city limits. General Chase stepped across the line thereby putting the American Occupational Army officially in Tokyo and adding another "First" to its name.

- "First in Tokyo"

The honor of the first enlisted man to enter the city officially went to a member of "D" Troop, 12th Cavalry, Private First Class Paul Davis of Fairland, Ottawa County, Oklahoma. On the same morning, General MacArthur made his official entry into Tokyo, escorted by the 2nd Squadron, 7th Cavalry, the 302nd Reconnaissance Troop, and the Division Band carrying the Division standards and colors. The official flag raising ceremony took place at 1700 hours.

Soon after the surrender terms were signed, troopers disembarked from the transport USS Cecil *at the port facilities of Yokohama*

The first semi-permanent camp in Tokyo was set up at Yoyogi Parade Grounds near adjacent to the Meiji Inner Shrine.

Alongside the main gateway road to Tokyo, a billboard boastfully displays an expression of "Pride of Accomplishments"

1st Cavalry troopers formed the first honor guard for the American Embassy in Tokyo. There, the colors of all the units of the Division were massed for the flag raising ceremony.

The flag used in the ceremony was the flag that flew over the White House on Pearl Harbor Day and had been flown over the Battleship Missouri while the surrender terms were being signed.

Following MacArthur's entrance into Tokyo, the 7th Cavalry moved by truck and rail from Hara-Machida to Tokyo. The bivouac area was on the Yoyogi Parade Grounds adjacent to the Meiji Inner Shrine. Japanese Army uniforms were much in evidence but the insignia was removed from them in compliance to the surrender terms.

The first mission of the Division was to assume control of the city of Tokyo. "F" Troop, 2nd Squadron, 7th Cavalry Regiment was selected as honor guard for General MacArthur's Headquarters at the Imperial Hotel. Other troops of the 2nd Squadron, 7th Cavalry were posted in and around the American Embassy, where MacArthur had taken up residence. On 10 September, the 8th Cavalry was relieved from the XI Corps and moved to Kawaski were its command post was set up at Keio University. Except for the Division Artillery which remained at Hara-Machida, the remainder of the Division was concentrated at Yoyogi Parade Grounds.

During the next week, most of the units moved to the Meiji Shrine Park in Tokyo. The first units to move into more permanent quarters, the 12th Cavalry and the artillery battalions, were located at the Japanese Military Preparatory Academy at Asaka, fourteen miles northwest of the center of Tokyo. The 1st Brigade Headquarters command post and 5th Cavalry were located at Camp McGill at Otawa, approximately twenty miles south of Yokohama. The 2nd Brigade Headquarters had its command post at the Imperial Guard Barrack Buildings in Tokyo along with the 8th Cavalry Regiment. The 7th Cavalry was situated at the Merchant Marine School. Division Headquarters and other units were stationed at Camp Drake near Tokyo.

On 11 September, medical members of the 1st Cavalry Division assisted in frustrating the suicide attempt of Hideki Tojo, war-time premier of Japan whose name had been placed first on the list of Japanese war criminals. As the Military Police entered his residence to arrest him, Tojo

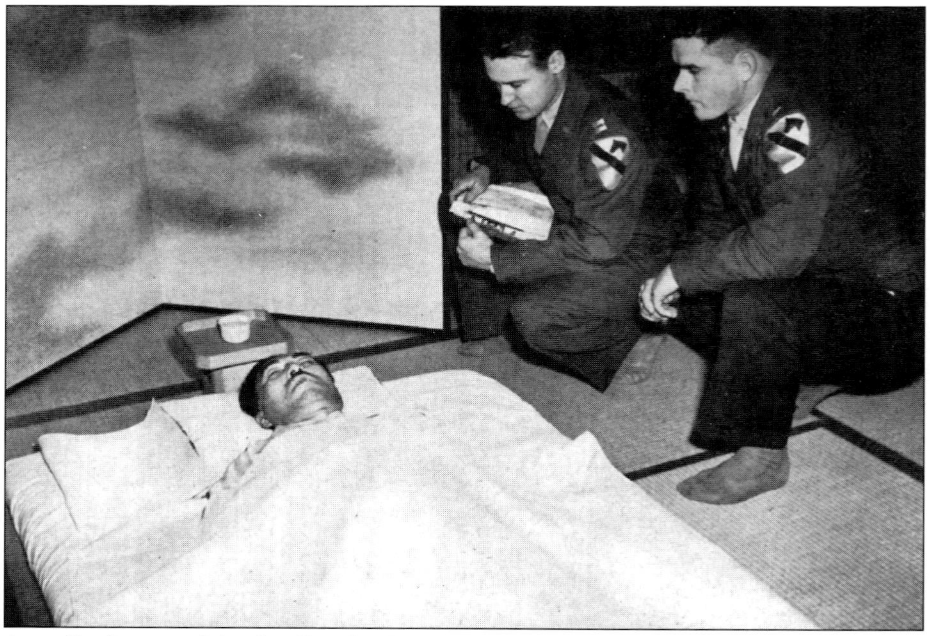

A medical team of the 1st Cavalry Division work feverishly over Hideki Tojo, former wartime premier, who attempted to commit suicide while being arrested for war crimes.

shot himself with a small pistol, but before he could bleed to death he was evacuated in a 1st Cavalry Division ambulance and given medical treatment so that he survived to stand trial by the International Tribunal.

Now that the war had ended and it became apparent that the occupation of Japan would not require a tremendous force of arms, the redeployment of veterans with long service records went forward on a large scale. This meant that the 1st Cavalry Division, one of the units selected to remain overseas, would be charged with an increasingly larger zone of occupation. On 16 September, the 1st Cavalry Division was given responsibility for occupying the entire city of Tokyo and the adjacent parts of Tokyo and Saitama Prefectures.

In October, the 1st Cavalry Division commenced the seizure of all stocks, precious metals, jewelry, foreign currency and narcotics which were in the possession of Japanese business firms and industrial plants. By the end of 1947, more than 76,000,000 grams of gold, a billion grams of silver, 7,000,000 grams of platinum and a large quantity of diamonds had been confiscated by the division and impounded in the vaults of the Bank of Japan. On 11 October, patrols from the 1st Cavalry Division officiated at the release of thirty-four political prisoners who had been confined by the Japanese Government for periods ranging up to eighteen years.

On 06 November, the division took over the operation of the repatriation center at the port of Urga, south of Yokohama, on Tokyo Bay. Japanese nationals, including army, navy, civil service and diplomatic personnel returning from overseas, were processed and sent to their homes in Japan.

The beginning of 1946 was welcomed as a new dawning of peace for the 1st Cavalry Division. The days of privation, hardship, suffering, and death were over for the first time since 07 December 1941. That year found the 1st Cavalry Division in control of Tokyo and vicinity, the capital of the war-built Japanese Empire. Early 1946 also saw the Military Police Platoon smash a large black market ring with the arrest of two large blackmarket operators in Tokyo. On 01 March, the 1st Cavalry Division was given the control and occupational responsibility of an area that embraced five thousand square miles with a population of twenty million including the following prefectures: Tokyo, Kanagawa, Chiba, Yamanashi, Saitama, Nagano, Gumma, Tochigi Niigata, Fukushima, and Ibaraki. The Division now had coast-to-coast control of Central Honshu.

At the end of April, increased unrest among labor groups and signs of possible disturbances in the metropolitan areas brought the Division to an alert status which lasted until 02 May. The demilitarization of war plants and factories and the seizure of illegally held weapons and war making equipment continued on a lesser scale. By the end of October, very few caches of arms were being found and nearly all plants had been converted to peace-time pursuits. The huge 1st Tokyo Military Arsenal was finally cleared of all arms and ordnance supplies.

The year of 1946 came to a close with a December quota of 5,253 replacements from the states. Their basic training varied from three weeks to completion of the basic course.

Permanent Division Headquarters were set up at Camp Drake at the outskirts of Tokyo, The facility was named in honor of Colonel Royce A. Drake who lost his life while commanding the 5th Cavalry Regiment at Leyte.

Recovered precious metals were escorted by armored guards to the Bank of Japan where the 1st Cavalry Division was assigned to stand guard over the recovered assets while it was inventoried and later stored in the vaults.

The 1st Cavalry Division participates in the Army Day Parade in downtown Tokyo - the first one held after the end World War II. (06 April 1946)

They were immediately integrated into the division military training program. and by the end of the year the trained replacements brought the Division to authorized strength. By the end of December, all division units had completed their training programs.

The 1st Cavalry Division began 1947 with the continuation of its occupation of the heart and nerve center of the Japanese Empire. Although there was no change in occupational policy, there had been vast changes among the troopers themselves. Over sixty percent of the combat veterans of the Division had been replaced by new arrivals from the states. By this time, the yearly turn-over of personnel averaged close to one hundred percent. Their time was spent in receiving advanced training, guard duty, patrolling, and specialist assignments. By March, the 1st Cavalry Division was deep in the throes of its intensive Mobilization Training Program.

In June, many relocations of organizational units began in order to accomplish training and occupational missions. Units of the 2nd Brigade rotated between Tokyo and Camp Palmer for marksmanship training on individual arms. Other movements saw "C" Troop, 12th Cavalry replace "B" Troop, 12th Cavalry at Hiratsuka; "G" Troop, 8th Cavalry took over duties of "F" Troop at Fuchu; "C" Battery, 82nd Field Artillery Battalion supplanted "B" Battery at Kanuizawa and "E" Troop, 7th Cavalry was attached to the 71st Quartermaster Depot.

The month of September proved to be one of the most eventful periods of the occupation history of the 1st Cavalry Division. The Division experienced an acceleration in training and ceremonies in addition to the numerous, unusual incidents of typhoons and floods that tested its resourcefulness and ability to perform under emergency conditions. On 30 September, in one of the more welcome events of month, the 95th Light Tank Company was attached to the division by the Eighth Army Headquarters.

As 1948 opened, the influence of the occupation was everywhere. Japan had been converted into a peaceful nation with a framework of government under its new constitution that would make it a lasting democracy. The reduction of troops continued throughout the year. The 5th Cavalry, split between Camp McGill and Camp Drake, had seriously reduced ranks. The 8th Cavalry Regimental Headquarters were collocated at the Marine Guard Barracks in Tokyo. The 12th Cavalry, reduced to cadre strength, had been relieved of its occupational duties.

On 01 May, the entire Division was placed on an alert status in preparation for any outbreaks of violence that might occur as a result of the May Day festivities and Korean elections. The alert remained in effect from 01 to 12 May, with surveillance patrols in all areas of responsibility, which could very well account for the fact that no outbreaks of a subversive nature occurred in their area of responsibility. On 09 September, coupled with replacement training and continuous emphasis on field training, all divisional units participated in a Command Post Exercise (CPX). During this CPX, all headquarters down to all regimental and squadron levels, moved into the field for a two-day exercise.

All ranks looked forward toward the new year of 1949 in anticipation of bringing the Division up to standards of combat efficiency and morale for which it stood. Personnel increased approximately 70 percent over the previous levels.

On 25 March, the 1st Cavalry Division was redesignated as 1st Cavalry Division (Infantry) and organized as a regular, triangular infantry division. The 12th Cavalry was deactivated and most of its members were transferred to the 32nd Infantry Regiment. This move reduced the strength of the Division to approximately 12,000.

The Division continued to carry out the dual mission of occupation (maintain guard detachments at the Bank of Japan, Civil Property Custodian Warehouse, Tokyo Quartermaster Depot, and escort war criminals from Sugamo Prison to Eighth Army Headquarters for trial) while emphasizing training. By the end of the year, combat effectiveness had risen sharply over that of the previous year. The emphasis on training conducted by the Division had netted the desired result.

Training - Keystone of the future for the First Team.

The beginning of 1950 called for an increased level of training to improve the increasing combat effectiveness of the Division. Unknown at the time, the real combat effectiveness of the 1st Cavalry Division was to be tested later in the year. Many of the troopers who had joined the Division in the later years of the occupation would soon earn their spurs across the Sea of Japan in a fight against a new foe - Communism.

They would learn the full and bitter meaning of a new concept of "limited war". They would find that massive battles could be waged without Congressional declarations of war, without maximum efforts to supply ample manpower, support or materials to the front lines, and surprisingly, a growing trend, without the complete support of the American people.

Chapter 4: Korean War

1950-1951

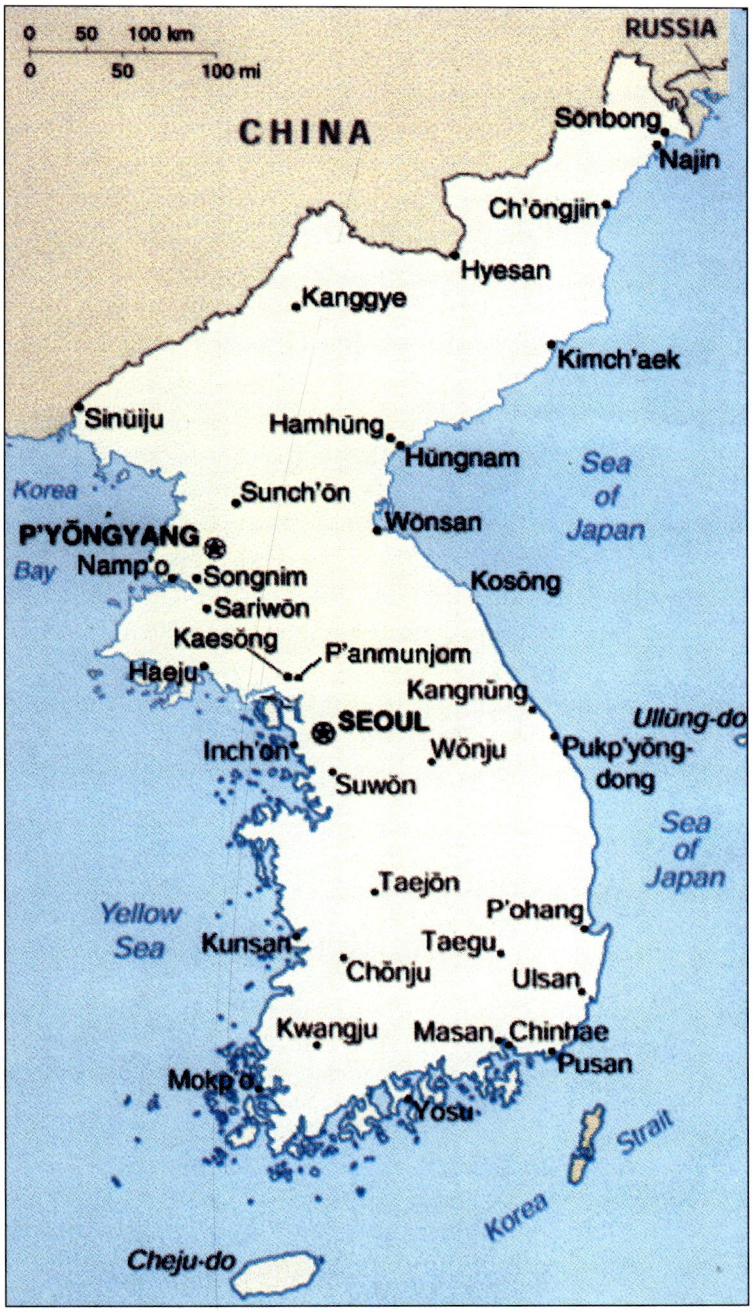

Korean Peninsula - Following the surrender of Japan in World War II, Soviet troops occupied the Korean Peninsula north of the 38th parallel while the US troops controlled the south. In 1948 the division of Korea was formalized with the declaration of North and South Korea entities, however, both factions claimed sovereignty over the entire peninsula and after the US and Soviet troops were withdrawn, armed clashes escalated.

It happened before dawn on 25 June 1950, a Sunday morning that began with a gentle rain and then exploded with cannon and automatic rifle fire from North Korean soldiers marching toward Seoul, South Korea. Less than 5 years after the terrible devastations of World War II, a new war had broken out in a distant country whose name means "The Land of Morning Calm".

The decision of the United States to send immediate aid to South Korea came two days after the fast moving North Korean Army broke through the Republic of Korea (ROK) defenses and sent tanks into the capital city of Seoul. In addition to the US Air Force, Navy and Marines, a 500 man battalion from the 24th Infantry Division, including many specialists and noncommissioned officers transferred from the 1st Cavalry Division, arrived 1 July. More help was on the way. "A" Company of the 71st Heavy Tank Battalion, previously assigned to the 1st Cavalry Division, arrived in Korea early in July and was immediately attached to the 24th Infantry Division and experienced its first combat at Taejon.

On 06 July, General MacArthur called Major General Hobart Gay, Commanding General 1st Cavalry Division, and informed him to make plans for the 1st Cavalry Division to make an amphibious landing at Inchon. In a questionable state of readiness, the 1st Cavalry Division had been weakened by the earlier transfer of approximately 750 noncommissioned officers to the 24th and 25th Divisions to strengthen their combat capabilities in Korea.

In the early stages of establishing a defensive position of air cover, the Navy and Marine aircraft operated off carriers stationed in the Sea of Japan. However the Air Force operated at considerable disadvantage at this time. There were only two dirt airstrips in South Korea suitable for operational use by F-51 and C-47 type aircraft, K-2 at Taegu and K-3 P'ohang Airdrome (also referred to as the Pohang-dong or Yonil Air Base) at Yonil (N 35.99 E129.34), on the east coast of South Korea.

As part of an advance party of the four US Army divisions committed to Korea, the 1st Battalion, 7th Cavalry Regiment was the first unit of the 1st Cavalry Division to be deployed. On 09 July, this 1st Battalion arrived on the east coast of South Korea with the mission to provide a security force for the Air Force at the Yonil Airfield, the main airport just six miles below Pohang-dong, and fire support for the ROK 23rd Regiment. As soon as the airfield was secure, all of the liaison and artillery aircraft and critical supporting maintenance personnel of the Air Section were self deployed to South Korea where they set up maintenance operations and prepared for a protracted period of combat operations of courier and mail transport, supply support, and artillery observation.

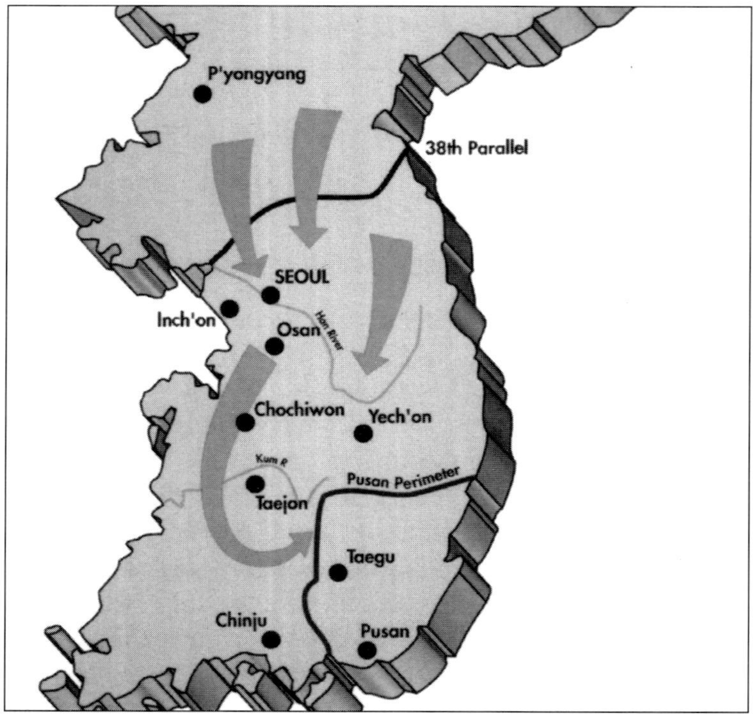

Unexpectedly the North Korean People's Army (NKPA) supported by Soviet made T-34 tanks invaded South Korea in force. Finally, the United Nations (UN) forces halted the advance along a front which came to be known as the Pusan Perimeter. (June to September 1950)

Between 12 and 14 July, the Division loaded on ships in the Yokohama area. However, at that time, the steady enemy successes south of the Han River had changed the objective of a landing in the rear of the enemy at Inchon to a landing on the east coast of Korea at Pohangdong, a fishing village sixty miles northeast of Pusan. The date of the landing was rescheduled to 18 July. The new mission of the division was to reinforce the faltering 24th Division. From Pohangdong the 1st Cavalry Division could move promptly to the Taejon area to provide direct support to the 24th Division.

The organization of the 1st Cavalry Division deployment followed standard amphibious practice. The landing force, commanded by Major General Hobart Gay, consisted of the 1st and 2nd Battalions, 5th Cavalry, 2nd Battalion of the 7th Cavalry, 1st and 2nd Battalions, 8th Cavalry Regiments, an artillery group of four battalions, a combat engineer battalion, special troop units, along with quartermaster support, administrative units and equipment. These were to be moved to the Pohangdong area by a naval transport group designated as "Task Force 90". The amphibious transport group consisted of the command flagship *USS Mount McKinley*, the transport *USS Cavalier*; three cargo ships - *USS Oglethorp, USS Titania, USS Union*, the landing ship tank *USS LST-611*; fifteen LSTs assigned from the Japanese Shipping Control Administration; two fleet tugs - the *USS Cree* and *USS Lipan*; one salvage ship - the *USS Conserver*; six landing utility ships (LSU); six minesweepers - the *USS Pledge, USS Chatterer, USS Kile, USS Redhead, USS Mockingbird, USS Osprey*, and *USS Partridge*; a high speed transport ship - the *USS Diachenko*; a gunfire support group of *USS Juneau, USS Coller, USS Higbee, USS James B. Kayes*, and the Australian Navy Destroyer *HMAS Bataan*; an underwater demolition team of UDT-3 and units assigned for reconnaissance, control purposes at the objective, administration of the beaches, and the like. Deep air support was the responsibility of the Air Force, which by this time had a fighter squadron on the Pohangdong air strip. Close air support at the objective would be provided by the Seventh Fleet, which was coming up from Okinawa for the occasion.

On 14 July, as the minesweepers started sweeping the waters of Yongil Man Bay - the designated landing site. The tractor group of LSTs, towing the LSUs and with two fleet tugs as escort, sailed from Tokyo Bay. On 15 July, "Task Force 90" command ship *USS Mt. McKinley*, under the command of Rear Admiral J. R. Doyle and final elements of the transport convoy followed a route south along the coast of Japan, then north by the Bungo Strait. Turning westward through the Inland Sea, the force steamed past Shimonoseki and into the Korean Strait. On the way, their progress was monitored by Russian submarines. Occasionally one of the accompanying destroyers would intercept their path, driving them away for a while. Early in the morning of 18 July, tractor and transport groups joined, and the ships moved into the Yongil Man Bay. Fighting had been reported only a few miles north of Pohangdong, but the ROK 3rd Division still held the road, and at 0559 hours Admiral Doyle signaled the Task Force to "Land the Landing Force in accordance with the plan for an unopposed operation."

On 18 July, although peaceful, the scene at Pohangdong was a busy one. From the ships of the transport group at anchor in Yongil Man, troops and vehicles were shuttled

USS Mount McKinley - The landing Task Force 90, commanded by Rear Admiral J. R. Doyle was christened with the code name "Bluehearts". The first sizable planned naval operation of the war had drawn the attention of the press, and 26 correspondents embarked on the command ship, USS Mount McKinley, *to accompany the 1st Cavalry Division. (15 July 1950)*

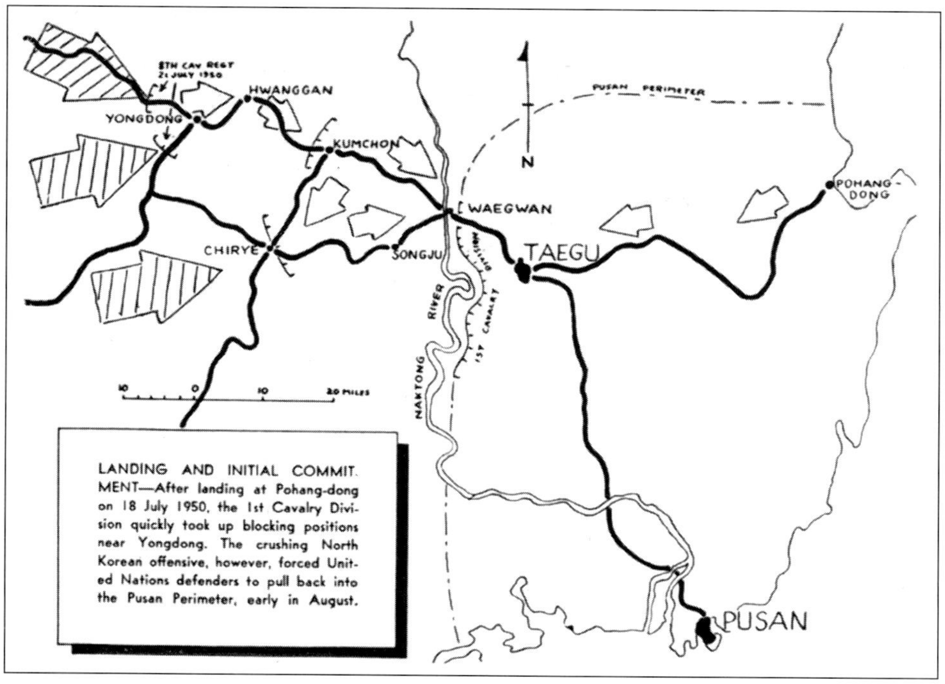

LANDING AND INITIAL COMMITMENT—After landing at Pohang-dong on 18 July 1950, the 1st Cavalry Division quickly took up blocking positions near Yongdong. The crushing North Korean offensive, however, forced United Nations defenders to pull back into the Pusan Perimeter, early in August.

ashore. Nine of the LSTs disgorged their cargo along the jetty wall and on the beaches of the Yongil Man Bay, along with the smaller landing craft. Seven were ordered out to Kuryongpo around the point to unload vehicles. Troop landing was begun at 0715 hours with vehicle and general cargo unloading commencing at 0930 hours.

Lead elements of the 8th Cavalry Regiment landed soon after daylight in the early morning of 18 July, successfully carrying out the first amphibious landing of the Korean War. The 13th Signal Company quickly followed behind the last wave of the 8th Cavalry. The first troops of the 5th Cavalry Regiment came ashore at approximately 1630 hours. Only a small combat air patrol from the carrier Valley Forge was retained overhead to protect the ships and landing forces. All major ships had been emptied by midnight, while the LSTs had discharged all personnel, all vehicles, and more than half their bulk cargo. More than 10,000 troops and 2,000 vehicles, and almost 3,000 tons of cargo had been put ashore.

The North Koreans (NK) were 25 miles away when elements of the 1st Cavalry Division came ashore. At noon on 19 July, General Gay assumed command ashore and the 5th Cavalry started toward Taejon. In the afternoon, with unloading completed, ships of the Attack Force shifted to heavy weather anchorages as Helene, the first typhoon of the season, was reported heading for the Korea Strait. The next day, the 8th Cavalry followed and closed in on an assembly area east of Yongdong by evening, unaware that the strength of Typhoon Helene, which had swept the eastern coast of Korea, would prevent the 2nd Battalion, 7th Cavalry Regiment and 82nd Field Artillery Battalion from landing until 22 July. By the end of 22 July, all regiments were deployed in battle positions, in itself a remarkable logistical achievement in the face of the heavy typhoon that had pounded the Korean coastline.

On 22 July, the 1st Cavalry Division assumed responsibility for blocking the enemy along the main Taejon-Taegu corridor. Concurrently the 8th Cavalry relieved the 21st Infantry Regiment, 24th Division, in its position at Yongdong. The 1st Battalion, 8th Cavalry was deployed north of Taegu, now the temporary capital of South Korea and astride the direct line of enemy advance. The 2nd Battalion was deployed to cover the road from Muju to the southwest. The next morning the 1st Battalion, 8th Cavalry, in front of Yongdong, reported that it had destroyed three enemy T34 tanks with 3.5-inch rocket launchers in the first use of the weapon. The enemy began to close on the 1st Cavalry Division for the battle at Yongdong. In the meantime the 2nd Battalion, 8th Cavalry was hit by heavy artillery fire and a mortar barrage, and North Korean infantrymen swarmed toward their entrenched positions.

The 7th and 9th Regiments. NK 3rd Division began their attack on the Yongdong positions. They made their first penetration southwest of Yongdong, establishing a roadblock a mile and a half behind the 2nd Battalion, 8th Cavalry. At the same time other enemy units heavily engaged the 1st Battalion, 8th Cavalry northwest of Yongdong in a frontal attack.

On 23 July four different attempts by three light tanks failed to dislodge the enemy behind the 2nd Battalion. As a reinforcement, the 1st Battalion, 5th Cavalry, and the 16th Reconnaissance Company moved toward the cutoff battalion. By noon, enemy troops were attacking the 61st and 99th Field Artillery Battalions which were supporting the 2nd Battalion, 8th Cavalry, indicating that the infiltration had been extensive.

On the other approach road, northwest of Yongdong, heavy automatic fire from quad-50's, 37-mm. fire from "A" Battery of the 92nd Antiaircraft Artillery Battalion, and artillery fire from the 77th Field Artillery Battalion helped the 1st Battalion to repel enemy attacks.

The large numbers of Korean refugees crowding the Yongdong area undoubtedly helped the enemy infiltrate the 1st Cavalry Division positions. For example, on 24 July, a man dressed in white carrying a heavy pack and accompanied by a woman appearing to be pregnant, came under suspicion. The couple was searched and the woman's assumed pregnancy proved to be a small radio hidden under her clothes. She had used this radio for reporting American positions.

By the morning of 25 July enemy forces had infiltrated the positions of the 1st Cavalry Division so thoroughly that they were forced to withdraw. The 1st Battalion, 8th Cavalry executed an orderly and efficient withdrawal, covered by the fire of the Heavy Mortar Company and the two batteries of the 77th Field Artillery Battalion. The mortar men finally lost their mortars and fought as infantry in the withdrawal.

Meanwhile, the situation worsened on the road southwest of Yongdong. The artillery support was so

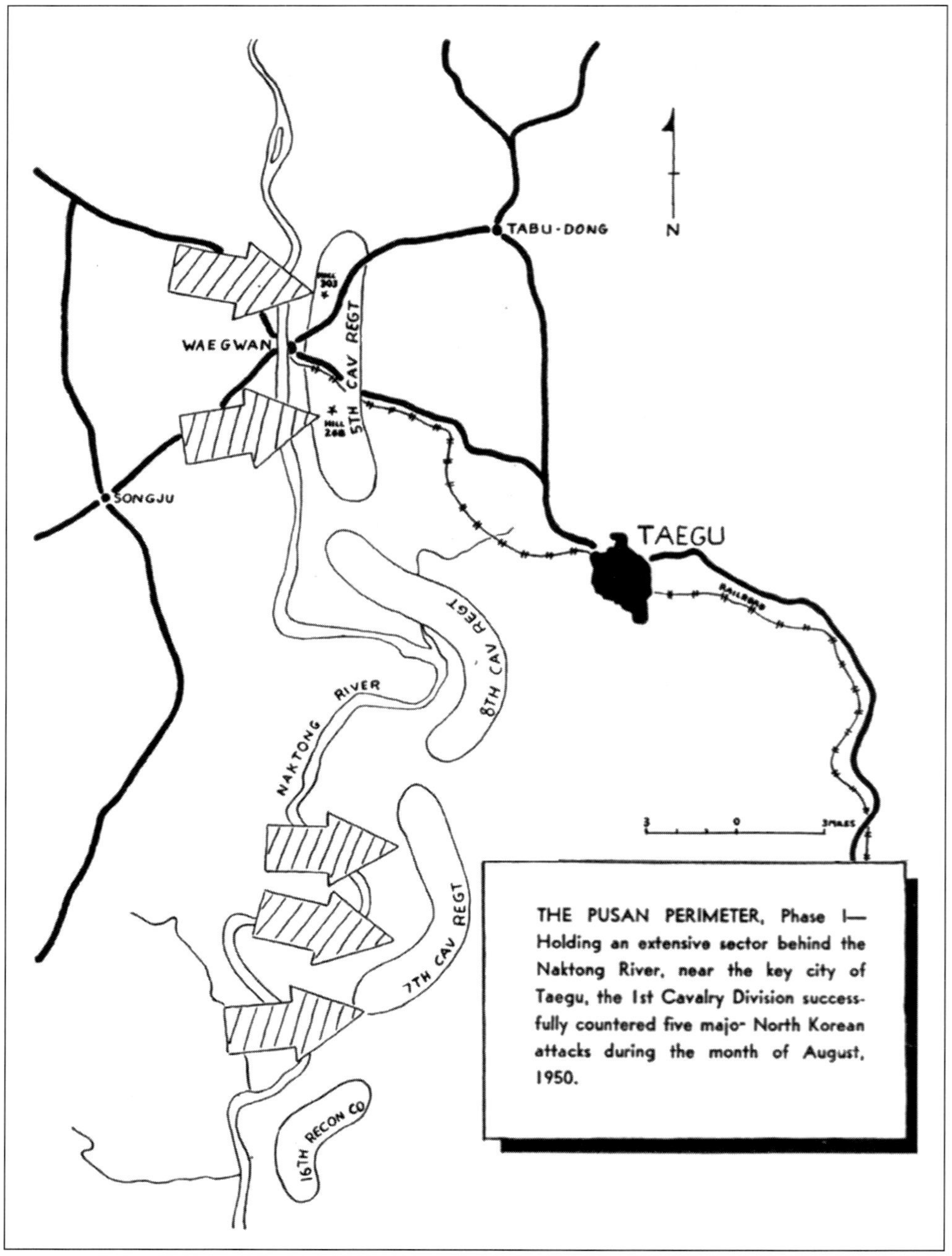

THE PUSAN PERIMETER, Phase I—Holding an extensive sector behind the Naktong River, near the key city of Taegu, the 1st Cavalry Division successfully countered five majo- North Korean attacks during the month of August, 1950.

concentrated that shells falling close to the positions of the 2nd Battalion southwest of Yongdong wounded four men. Together with an attack by the battalion, the enemy roadblock briefly opened and the bulk of the battalion escaped to Yongdong with the exception of "F" Company, 8th Cavalry; the 16th Reconnaissance Company; and the 1st Platoon, "A" Company, 71st Tank Battalion, at the rear of the column that were cut off. Only four of eleven light tanks broke through the enemy positions. Crews abandoned the other seven tanks and walked over the hills as part of a group of 219 men, most of them from "F" Company.

As the space between the battalions became increasingly threatened, the 1st Battalion, 5th Cavalry moved into the gap to absorb some of the pressure. Closer to Yongdong, the 2d Battalion, 5th Cavalry moved to assist the cutoff units of the 1st Battalion, 8th Cavalry on its right. En route, "F" Company ran into trouble, encountering an overwhelming concentration of North Korean Infantry soldiers. Only twenty-six men from the relief units managed to escape and return to friendly territory. Altogether, the 5th Cavalry Regiment suffered 275 casualties that day.

The 7th Cavalry, initially held in the 8th Army defenses at Pohangdong, was released to the 1st Cavalry Division on 25 July, and began moving up to join the 5th Cavalry Regiment. During the next few days a defensive line was formed at Hwanggan with the 7th Cavalry moving east and the 5th Cavalry replacing the 25th Infantry Regiment.

The Korean War was chaotic and difficult for the artillery. Classical front lines disappeared. To make up for their own lack of artillery, the North Koreans made battery positions their prime targets. Batteries had to fight off invaders in close combat and still fire their guns in support of the combat operations. During the first few weeks, the division artillerymen were fighting with small arms alongside their thundering artillery pieces. Artillery units often found themselves surrounded and artillerymen were called upon to fight side-by-side with the infantry. One cannoneer suggested that the crossed cannons of the artillery be changed to one cannon and one rifle.

On 29 July, in order to preclude being flanked by the enemy forces moving south toward Chirye, the 1st Cavalry Division took up new defensive positions at Kumchon, an important road center 30 miles northwest of Taegu. The 5th Cavalry blocked the Kumsong-Chirye road. The 8th Cavalry went to a position astride the Sangju road north of Kumchon. The 7th Cavalry remained at its Hwanggan position until adjacent units had been withdrawn, and then it fell back to a new position on the Yongdong road about six miles northwest of Kumchon.

Following the establishment of the defensive positions, the 16th Reconnaissance Company, the lead reconnaissance unit of the 1st Cavalry Division, conducted a patrol toward and beyond Chirye. They encountered strong enemy fire. With their return route blocked by the North Koreans, the troopers moved ahead until their vehicles were knocked out, then moved on foot to friendly lines. Out of the original roster of 57 men, 44 made it safely back by crossing 35 miles of difficult mountain terrain.

On 05 August, "A" Company, 71st Heavy Tank Battalion was reorganized as "A" Company, 71st Tank Battalion and reassigned to the 1st Cavalry Division. By mid September, "A" Company had lost twenty of its original issue of twenty-two M-24 light tanks because their 75mm guns could not penetrate the armor on the Russian built T-24 tanks. After they lost their tanks in combat, there were enough survivors to form a machine-gun platoon and they spent the next thirty days on the line fighting as infantry. On 16 October, the unit was deactivated and relieved from assignment to the 1st Cavalry Division because of its heavy losses.

On 07 August 1950, the 70th Heavy Tank Battalion arrived at the port of Pusan on the transport *USS General Brewster* and on 12 August was attached to the 1st Cavalry Division. Soon after their attachment, they joined the Division in the launching of a major offensive of probing and striking attacks in multiple directions in the Taegu area in defense of the Pusan Perimeter. In carrying out the probes, the 5th Cavalry Regiment, with "A" Company of the 70th Tank Battalion, captured several strategic points along the Naktong River. The 8th Cavalry Regiment, with "B" Company of the 70th Tank Battalion, halted the advance of the North Koreans west of Taegu. The 7th Cavalry Regiment, with "C" Company of the 70th Tank Battalion, launched a series counterattacks. Throughout its remaining campaigns in Korea, the 70th Tank Battalion remained employed as the armor support to the 1st Cavalry Division.

The 1st Cavalry Division was assigned to defend a 35 mile sector along the Naktong River, extending from three miles north of Waegwan, south to the area defended by the 24th Infantry Division. The rugged, mountainous terrain of Korea and the lack of developed transportation and communications systems, created significant challenges to the 8th Engineer Battalion. Most of the initial engineer work involved demolition of bridges and important facilities in an attempt to delay the North Korean advance to the south. In the Pusan Perimeter the 8th Engineer Battalion not only worked on standard defensive and construction projects, but also manned the front lines when the enemy threatened to penetrate the perimeter.

On 09 August, the enemy hurled five full divisions and parts of a sixth at the Naktong defenders near Taegu. The 1st Battalion, 5th Cavalry bore the brunt of this attack. The North Koreans gained some high ground - but not for long. The 1st Battalion, 7th Cavalry moved against them, hitting their flanks with a coordinated artillery strike by the 61st Field Artillery. At noon, the 7th Cavalry attacked with orders to continue to Hill 154, closer to the river. Hill 268 was covered with thick brush and it was difficult to maneuver. The attack was repulsed, but the next day, an air strike softened the objective and the 1st Battalion, 7th Cavalry pushed the remaining enemy back across the river. In seizing Hill 268, known as "Triangulation Hill", the troopers accounted for 400 enemy dead. The First Team pulled back from some of its positions and tightened its defenses.

On 12 August, the 2nd Battalion, 7th Cavalry was attacked by units from the 10th NK Division with the objective to gain the high ground east of Youngpo. With good artillery support from the 77th Field Artillery Battalion, the enemy was pushed back across the river. On 14 August, a second powerful punch was delivered to the 2nd Battalion, 7th Cavalry, but timely assistance from the 1st Battalion, ordered out of reserve, the 8th Engineers, the 16th Reconnaissance Company, and the artillery caused the termination of the attack which had penetrated to Samuni-dong, about 12 miles from Taegu. Out of 1,700 enemy attackers, some 1,500 died.

Simultaneously, on 14 August, a second enemy attack was launched at the boundary between the 1st ROK Division and the 1st Cavalry Division at Waegwan. "G" Company, 5th Cavalry situated on Hill 303, the right flank anchor of the US Eighth Army Command, began receiving small arms fire from the ROK side of the boundary. On 15 August, "F" Company withdrew, creating a situation in which "G" Company and a supporting platoon of "H" Company motarmen were surrounded. "B" Company and a platoon of tanks tried to break through to "G" Company, but were driven back.

On 16 August, another unsuccessful rescue attempt was made. Finally, during the night, "G" Company was able to elude the enemy and get off Hill 303. On 17 August, the 5th Cavalry, with support by "A" Company, 70th Tank Battalion, massed another attack on Hill 303. At 1400 hours, under the cover of a napalm and bombing attack, they were able to retake the hill. In retaking the hill, the regiment came upon a painful scene; the bodies of "H" Company troopers, with hands tied in back, who had been sprayed with burp gun fire.

5th Cavalry Light Machine Gun (LMG) position in Korea.

The Pusan Perimeter continued to hold. With added reinforcements, Pusan became a staging ground and depot for United Nations supplies and soldiers from all around the world. The gallant Greek Battalion was the first of the United Nations forces to become attached to the Division. The rapid buildup was an encouraging factor to an army that might possibly face another "Dunkirk". The defenders now outnumbered the attackers and had the equipment and firepower to go on the offensive.

Meanwhile, back in the states, priority was given to a rapid build up of military manpower necessary to counter the North Korean threat. A mobilization of all available active duty and reserve Army personnel initiated a marshalling of forces at Ft. Lawton, Washington, Ft. Ord California and Camp Stoneman, California for rapid embarkment to Korea. On 11 August, the soldiers of the units that were destined to be redesignated the 3rd battalions of the 5th, 7th, and 8th Cavalry Regiments, 1st Cavalry Division, boarded the *USS General John Pope (AP-110)* at Oakland, California, and sailed out into the bay and under the Golden Gate Bridge for movement to Korea. Following a fourteen day routine of automatic weapons training, map reading, and calisthenics, they arrived in Pusan and disembarked on 25 August. On 26 August, the Eighth Army cut orders assigning the three new Battalions to the 1st Cavalry Division. The availability of these units allowed the fill out of the TO&E of the 1st Cavalry Division, which in 1949, was authorized a 3rd battalion for each of its regiments.

The 3rd Battalion, 14th Infantry RCT, 10th Mountain Division, from Camp Carson, Colorado was redesignated as the 3rd Battalion, 5th Cavalry Regiment, the 2nd Battalion, 30th Infantry RCT, 3rd Division, from Ft. Benning, Georgia was redesignated as the 3rd Battalion, 7th Cavalry Regiment and the 3rd Battalion, 7th Infantry RCT, 3rd Infantry Division, from Ft. Devens, Massachusetts was redesignated as the 3rd Battalion, 8th Cavalry Regiment. The next day the battalions moved out by rail to join their respective regiments. By the end of August, the four US divisions in Korea had finally been built up to their authorized level of three battalions per regiment.

On 29 August, the 1st Cavalry Division sector of coverage was shifted to the north and northwest mountainous areas. The Division took up defensive positions northwest of Taegu along a line eight miles long. The 5th Cavalry was positioned on the west, the 7th Cavalry was positioned in the center and the 8th Cavalry was positioned on the northeast of the 7th. Critical terrain consisted of Hills 346, 490, and 518 in the center and Hills 188, 302, and 303 east of Waegwan and a corridor known as "The Bowling Alley" running north and south. From Taegu thru Tabu-dong the terrain of the corridor was flat, with irrigation ditches and rice paddies. On either side were rugged mountains. To reduce the attack options of the North Koreans around Waegwan, the dominating terrain linking Waegwan and Tabu-dong, just twelve miles northwest of Taegu, Hill 518 and Hill 346 would have to be reduced.

On 02 September, following artillery bombardments and air strikes, in a coordinated effort, the 7th Cavalry attacked Hill 518 and the 8th Cavalry advanced on the right flank and attacked Hill 490. Massive artillery support, four fifths of the division firepower, was used to support the advancing troopers. However, after three days of hard fighting the Division Commander ordered a withdrawl to shorten and strengthen defensive lines. The withdrawal movement

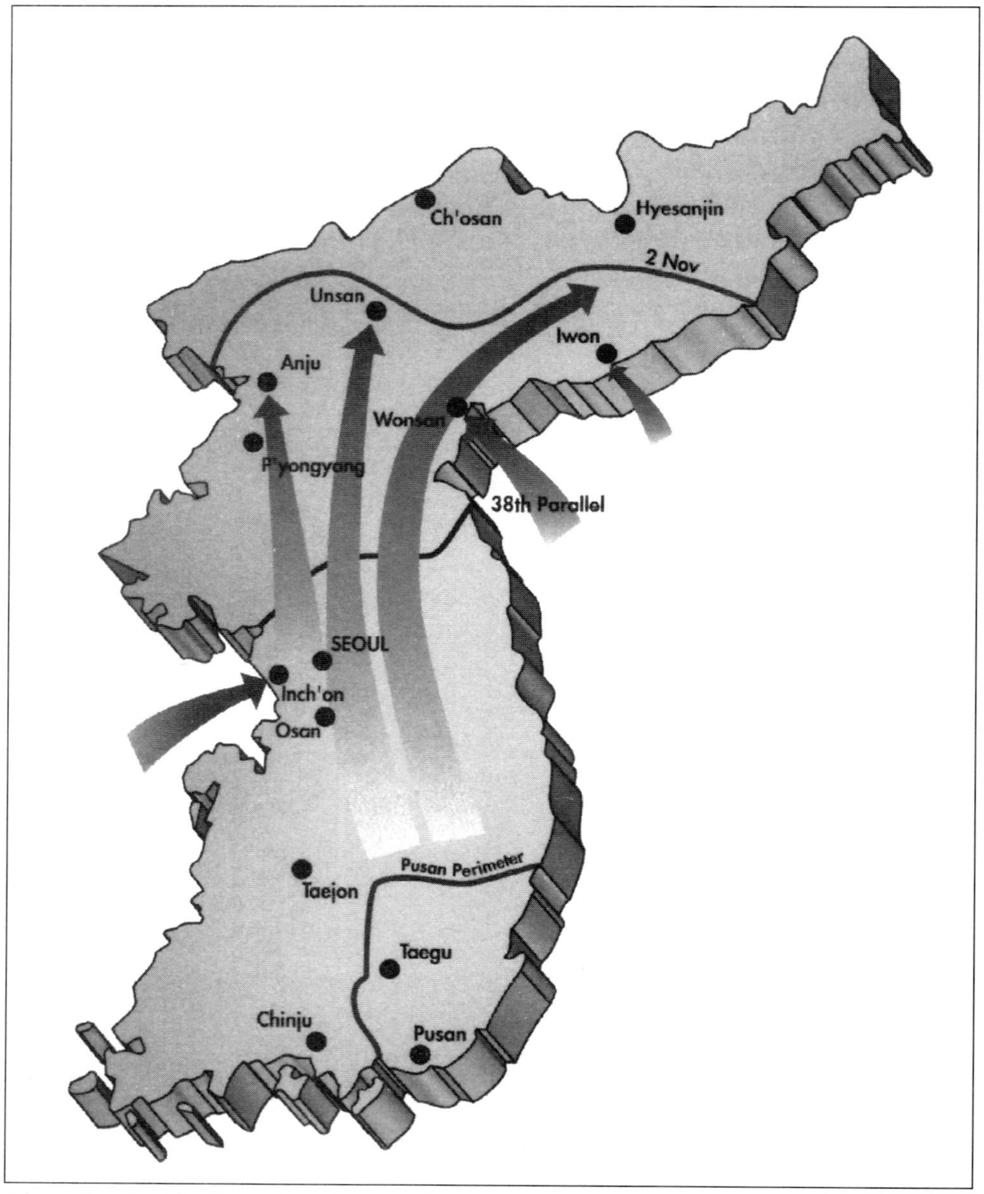

Phase II - The UN forces break out of the Pusan Perimeter, cross the 38th Parallel, fan out across the expanding front and pursue the NKPA forces to the northern boundaries of the China - North Korean border. (September to November 1950)

began on the right with the 8th Cavalry, then the 7th Cavalry in the Hill 518 area and finally the 5th Cavalry in the Waegwan area. The key to the withdrawal was Hill 464, behind the 2nd Battalion, 7th Cavalry, that dominated the Waegwan - Tabu-dong road. The mud created by the heavy rains which fell 05 to 06 September, slowed all wheeled and tracked vehicles.

East of Tabu-dong the enemy had begun to exert pressure on an area known as the "Walled City." It was located among a series of high mountain ranges and huge granite boulders positioned to prevent landslides. On the highest mountain was an ancient Buddhist Shrine, the key to the defense of Taegu, which overlooked the straight corridor leading to the city - just twelve miles away.

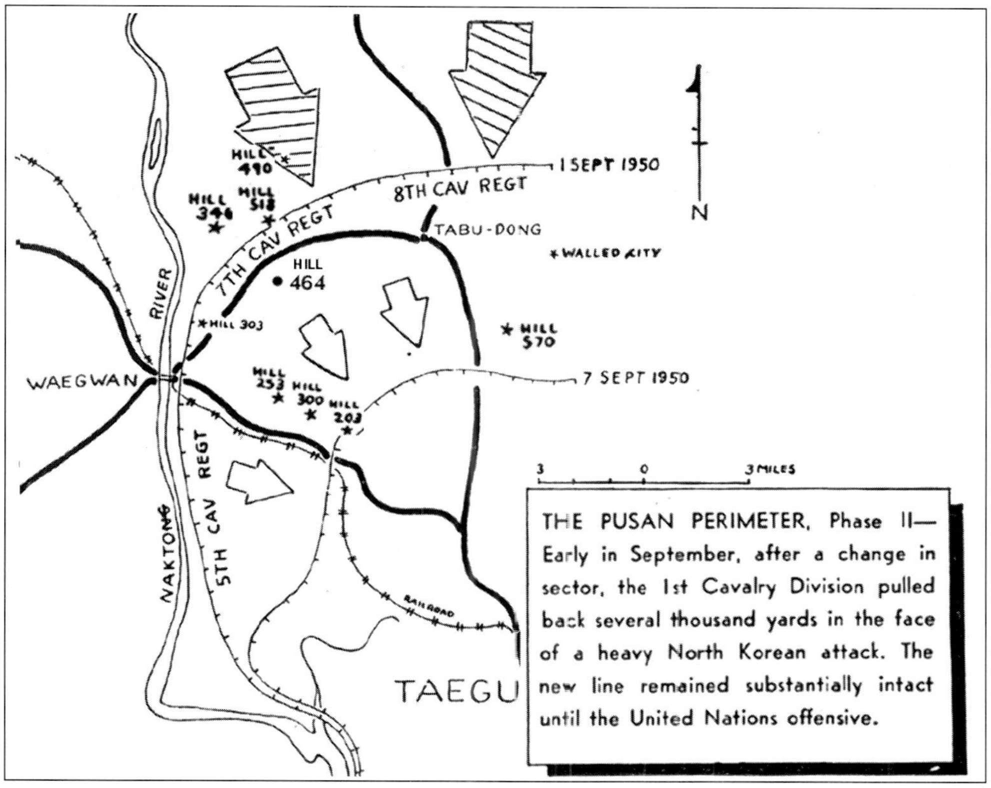

THE PUSAN PERIMETER, Phase II—Early in September, after a change in sector, the 1st Cavalry Division pulled back several thousand yards in the face of a heavy North Korean attack. The new line remained substantially intact until the United Nations offensive.

By 04 September, Division headquarters, at Taegu, was alive with activity. Brigadier General Frank A. Allen, the Assistant Division Commander was busy organizing "Task Force Allen", including of two provisional infantry battalions composed of bandsmen, technical, and miscellaneous troops. The 8th Combat Engineer Battalion was pressed into service as infantry. "D" Company was given the mission to secure the "Walled City of Kasan". The engineers fought their way to their destination and held it.

It was in this battle of repelling counterattacks on 04 September, that Private First Class Melvin L. Brown, "D" Company, 8th Engineer Combat Battalion, distinguished himself by conspicuous gallantry and intrepidity above and beyond the call of duty in action against the enemy. While his platoon was securing Hill 755 (the Walled City), the enemy, using heavy automatic weapons and small arms, counterattacked. Taking a position on a fifty foot-high wall, Private First Class Brown delivered heavy rifle fire on the enemy. His ammunition was soon expended and although wounded, he remained at his post and threw his few grenades into the attackers causing many casualties. When his supply of grenades was exhausted his comrades from nearby foxholes tossed others to him and he left his position, braving a hail of fire, to retrieve and throw them at the enemy. The attackers continued to assault his position and PFC Brown weaponless, drew his entrenching tool from his pack and calmly waited until they each peered over the wall, delivering each a crushing blow to the head. Knocking ten or twelve enemy from the wall, his daring action so inspired his platoon that they repelled the attack and held their position. For his valiant action, Private First Class Brown received the Medal of Honor.

On 06 September, at 0300 hours, the 2nd Battalion, 7th Cavalry disengaged from the enemy on Hill 464 and fought its way to the east. The 5th Cavalry, occupying positions on Hill 303, came under heavy fire and was driven from key terrain, however, they were able to recapture the lost ground with the aid of the 70th Tank Battalion elements. During the next few days, the situation was very tenuous. The enemy had gained major footholds east of Naktong and south to within about 8 miles of Taegu in the vicinity of Hills 314 and 570.

On 10 September, during an advance on a strategic enemy held hill near Kasan, Corporal Gordon M. Craig, 16th Reconnaissance Company and his company were subjected to intense hostile grenade, mortar, and small-arms fire. Corporal Craig and four comrades moved forward to eliminate an enemy machine gun nest that was hampering the advance of the company. At that instant an enemy machine gunner hurled a hand grenade at the advancing men. Without hesitating or attempting to seek cover for himself, Corporal Craig threw himself on the grenade and smothered its burst with his body. His intrepid and selfless act, in which he unhesitantly gave his life for his comrades, inspired them to attack with such ferocity that they annihilated the enemy machine gun crew, enabling the company to continue its attack. For his valiant action, Corporal Gordon M. Craig received the Medal of Honor.

On 12 September, the 3rd Battalion, 7th Cavalry was assigned to retake Hill 314 just north of Taegu. Many accounts of heroism and professionalism occurred in this successful engagement and the Battalion was awarded the Presidential Unit Citation (PUC). Subsequently, the 2nd Battalion, 8th Cavalry attacked and after a hard fought battle, regained Hill 570 and also was awarded the PUC. The North Korean drive ground to a halt on 13 September, seven miles short of Taegu. The momentum began to change and plans were laid for an all-out offensive.

The turning point in this desperate struggle came on 15 September 1950, when MacArthur unleashed his plan, Operation CHROMITE, an amphibious landing at Inchon, far behind the North Korean lines. In spite of the many objections by critics of the plan, the Inchon landing was an immediate success allowing the 1st Cavalry Division to break out of the perimeter and link up with the Inchon Landing. The routes north were heavily mined. Rather than have the engineer battalion methodically screen and dig up the mines, 17 tanks of "A" Company, 70th Tank Battalion drove into the minefield to rapidly clear the mines along the routes.

On 22 September at 0800 hours "Task Force Lynch" attacked north along the "bowling alley" and made good progress with the close air support and a gutsy move to breach minefields with M-4 tanks. In twenty two hours since leaving Tabu-dong the Task Force had advanced 35 miles, captured 5 tanks, 50 trucks and secured a crossing over the Naktong River. Orders from above sent the Task Force north to another crossing site where they crossed the river and assembled in Poun.

On 26 September General Walker granted authority for the Task Force to proceed north and link up with the 31st Infantry near Osan. Led by four M-4 tanks from "C" Company, 70th Tank Battalion and followed by the 3rd Battalion, 7th Cavalry, the 77th Field Artillery Battalion, the 3rd Platoon, Heavy Mortar Company and the regimental I&R Platoon it covered 106.4 miles in less than 21 hours.

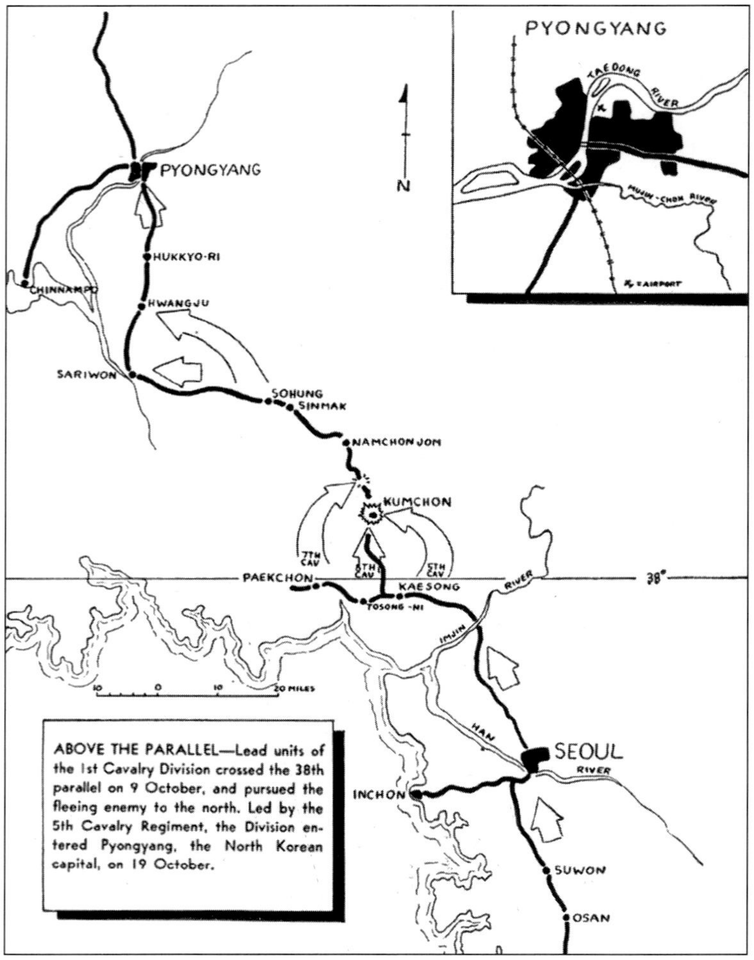

ABOVE THE PARALLEL—Lead units of the 1st Cavalry Division crossed the 38th parallel on 9 October, and pursued the fleeing enemy to the north. Led by the 5th Cavalry Regiment, the Division entered Pyongyang, the North Korean capital, on 19 October.

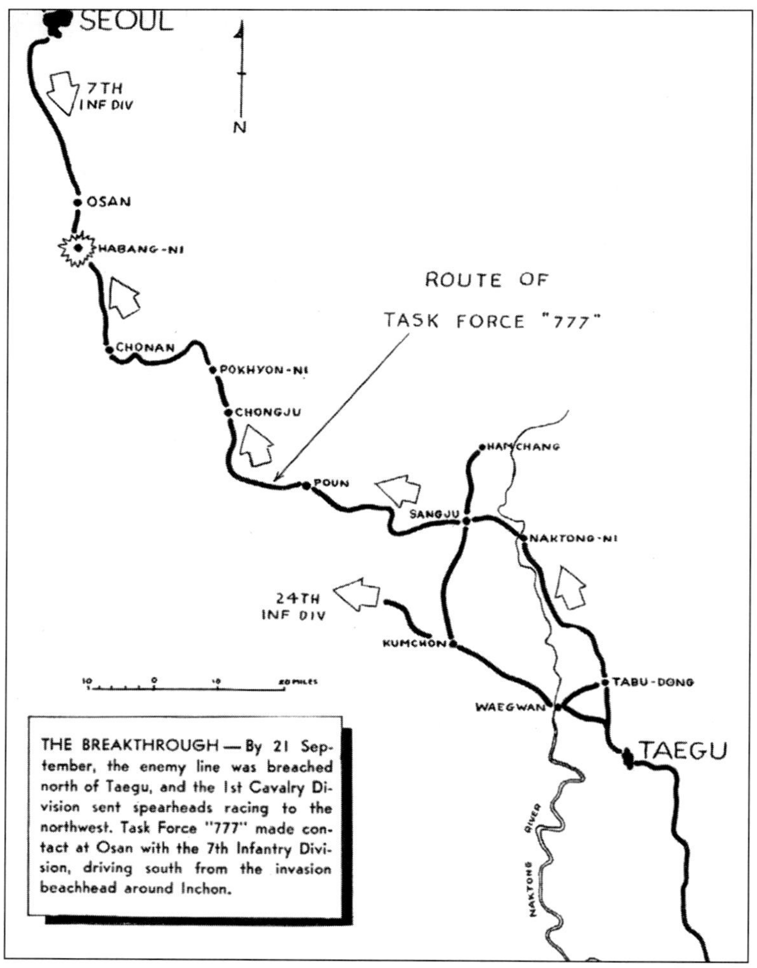

THE BREAKTHROUGH—By 21 September, the enemy line was breached north of Taegu, and the 1st Cavalry Division sent spearheads racing to the northwest. Task Force "777" made contact at Osan with the 7th Infantry Division, driving south from the invasion beachhead around Inchon.

On 27 September, north of Osan at a small bridge, "L" Company, 7th Cavalry, met elements of "H" Company, 31st Infantry, 7th Division. In this rapid advance to Osan, the 1st Cavalry Division cut off elements of the North Korean 105th Armored Division in the Ansong and P'yongt'aek area and miscellaneous units in the Taejon area. On 28 September, elements of "C" Company, 70th Tank Battalion, and "K" Company, 7th Cavalry, with the strong assistance of fighter-bombers, destroyed at least seven of ten T-34's in the Pyongtaek area, five by air strikes. Elements of the 16th Reconnaissance Company barely escaped destruction by these enemy tanks, and did suffer casualties.

From 28 September to 03 October, major efforts concentrated on mopping up operations in the large sector assigned to the 1st Cavalry Division. By 04 October, the division had resumed the offensive to the north. On 05 October, the 1st Cavalry Division advanced north of Seoul for the purpose of securing the US I Corps assembly area near the 38th Parallel. Led by "I" Company, the 5th Cavalry crossed to the north side of the Imjin River at Munsan-ni. On 07 October, the 16th Reconnaissance Company entered Kaesong, and that evening elements of the 1st Battalion, 8th Cavalry arrived there. By the evening of 08 October, the 7th and 8th Cavalry had secured the I Corps assembly area in the vicinity of Kaesong. Some of the troops were within small arms range of the 38th Parallel. On 09 October, the 1st Cavalry Division crossed the 38th Parallel.

On 09 October, Private First Class Robert H. Young, a trooper of "E" Company, 8th Cavalry was deep in enemy territory. His company, spearheading a battalion drive, suddenly came under a devastating barrage of enemy mortar and automatic weapons crossfire that inflicted heavy casualties among his comrades and wounded him in the face and shoulder. Refusing to be evacuated, Private First Class Young remained in position and continued to fire at the enemy until wounded a second time. As he awaited first aid near the company command post the enemy attempted an enveloping movement. Disregarding medical treatment, he took an exposed position and firing with deadly accuracy killed five of the enemy. During this action he was again hit by hostile fire that knocked him to the ground and destroyed his helmet. Later when supporting tanks moved forward, Private First Class Young, his wounds still unattended, directed tank fire which destroyed 3 enemy gun positions and enabled the company to advance. Wounded again by an enemy mortar burst, and while aiding several of his injured comrades, he demanded that all others be evacuated first. Throughout the course of this action the leadership and combative instinct displayed by Private First Class Young exerted a profound influence on the conduct of the company. His aggressive example affected the whole course of the action and was responsible for its success. For his valiant action, Private First Class Robert H. Young received the Medal of Honor.

On 10 October, the 89th Tank Battalion was attached to the 1st Cavalry Division to strengthen the armor support for the northern offensive.

On 12 October, 1st Lieutenant Samuel Coursen, a member of "C" Company, 5th Cavalry, was engaged in an all out offensive of Hill 174. While "C" Company was under heavy enemy small-arms fire, the platoon of Lieutenant Coursen received enemy fire from close range. The platoon returned the fire and continued to advance. During this phase one of his men moved into a well camouflaged emplacement, which was thought to be unoccupied, and was wounded by the enemy who were hidden within the emplacement. Seeing the soldier in difficulty Lieutenant Coursen rushed to the aid the man and, without regard for his personal safety, engaged the enemy in hand-to-hand combat in an effort to protect his wounded comrade until he himself was killed. When his body was recovered after the battle seven enemy dead were found in the emplacement. As the result of 1st Lieutenant Coursen's violent struggle several of the heads of the enemy had been crushed with his rifle. His aggressive and intrepid actions saved the life of the wounded man, eliminated the main position of the enemy roadblock, and greatly inspired the men in his command. For his valiant action, 1st Lieutenant Samuel Coursen received the Medal of Honor.

On 13 October the 1st Cavalry Division began to close in on one of the main arsenals of the NVA, the Kumchon Pocket, an enemy strongpoint heavily defended with tanks, self propelled guns, and anti aircraft guns. With the 7th Cavalry blocking the Hanpo-ri bridge on the road north of Kumchon, the decisive action now rested with the 5th and 8th Cavalry, which were trying to compress the pocket from the south and the east. Moving west from the Sibyon-ni road, the 5th Cavalry encountered an almost continuous mine field in its approach to Kumchon, and it also had to fight and disperse an enemy force of over 300 NK soldiers. Overcoming these difficulties, the 5th Cavalry pressed ahead and by the evening, it was approaching Kumchon.

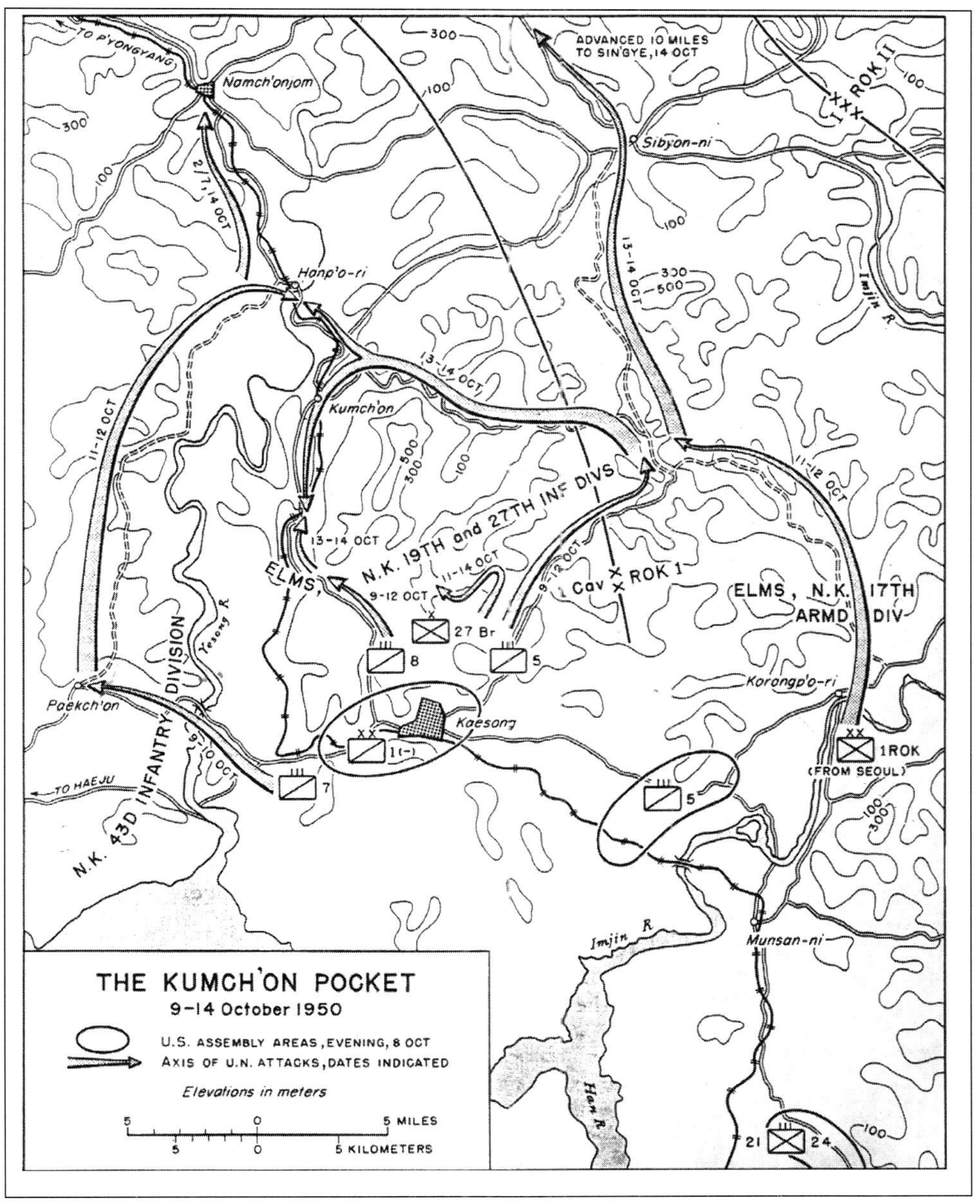

The Kumchon Pocket, an enemy strongpoint heavily stocked with tanks, self propelled guns, and anti aircraft guns, was strongly defended for five days by the North Koreans. (09 to 14 October 1950)

Strong opposition confronted the 8th Cavalry moving north on the main highway where the enemy apparently had concentrated most of his available forces and weapons. There, on the morning of the 13th, an artillery preparation employing proximity fuze air bursts blanketed the North Korean positions. While the enemy force south of Kumchon fought desperately and successfully to prevent the 8th Cavalry from closing in on the town from the south, a large enemy column of trucks with an estimated 1,000 soldiers moved northward out of Kumchon toward Namchonjom. At the Hanpo-ri bridge they ran directly into the 7th Cavalry roadblock. In the ensuing action, the 7th Cavalry, aided by air strikes, killed over 500 and captured 201 of this force.

At midnight of the 13th, the 2nd Battalion, 5th Cavalry, resumed its attack on Kumchon from the east. After dispersing an enemy force near the edge of town, the battalion then entered and seized the northern part of it. The 3rd Battalion, 5th Cavalry, following, seized the southern sector of Kumchon. Advancing northwest, the 2nd Battalion, 5th Cavalry joined elements of the 7th Cavalry above Hanpo-ri at noon on the 14th. An enemy force, of some 2,400 men, which had been stopped by the 7th Cavalry roadblock at Hanpo-ri, escaped into the hills as the 2nd Battalion approached from the south. Meanwhile, attacking south from Kumchon, the 3rd Battalion, 5th Cavalry neared a special task force of the 8th Cavalry Regiment which had attacked north during the morning and already had lost two tanks to enemy action. The two columns, the 3rd Battalion, 5th Cavalry, and the special 8th Cavalry task force met just after noon about four miles south of Kumchon. By the close of 14 October, the day Kumchon fell to the 1st Cavalry Division, the enemy positions between the 38th Parallel and the North Korean capital of Pyongyan, and enemy front lines as such had ceased to exist. The NK forces were in a state of utter confusion.

On 14 October, the Korean War took a grim new turn when the first element of the Chinese Communist Forces (CCF), the 334th Regiment, 119th Division, of the 15th Chinese Field Army crossed the Yalu at Andong. By moving only at night, they were able to penetrate the area and move undetected into North Korea in great numbers. Only scout units moved during daylight to determine routes for the next night's march. They were ordered, under penalty of death, to freeze motionless if they heard aircraft. Their only heavy weapons were mortars, but they came in increasingly vast numbers.

Trained and battle hardened in guerilla warfare, the CCF carried none of the baggage of a modern army. Masters of concealment, they moved and fought best by night. They wore thick, padded, green or white uniforms, caps with a red star, and carried a personal weapon, 80 rounds of ammunition, a few stick grenades, spare foot rags, a sewing kit, and a week's rations of fish, rice and tea.

Meanwhile, on 15 October, after moderate resistance from enemy positions, the 2nd Battalion, 7th Cavalry and "C" Company, 70th Tank Battalion secured the city of Namchonjam. On 17 October, they made a flanking movement to the right of the main highway to Pyongyang. As the 7th advanced toward Pyongyang, they were stopped by strong enemy resistance near the town of Hukkyo-ri. General Gay ordered a flanking movement, but when he realized that the troopers were fatigued from constant movement, he ordered the 5th Cavalry to bypass the roadblocks and go forward to Pyongyang.

On 19 October at 1100 hours, the 2nd Battalion, 5th Cavalry reached the southwestern edge of Pyongyang, the capital city of North Korea, at the same time that the ROK 1st Division came in from the east. This event marked the third "First" for the division -

"First in Pyongyang"

On 25 October 1950, serious fighting began in the sector of the ROK 6th Division. The sudden intervention of Communist Chinese forces dashed hopes of a quick end to the war. In spite of urgent reports that the Chinese were preparing to enter the battle in force, MacArthur and other high commanders remained convinced that these new troops were Chinese volunteers of Korean descent, numbering no more than 30,000, who had joined North Koreans as replacements.

On 28 October 1950, orders came from I Corps to saddle up the rest of the division and move north. The Korean war seemed to be nearing a conclusion. The

Yalu River, Andong - The CCF moved great numbers of battle-hardened veterans of the 334th Regiment, 119th Division, of Fifteenth Chinese Field Army across the Yalu. into North Korea, undetected. (14 October 1950)

North Korean forces were being squeezed into a shrinking perimeter along the Yalu and the borders of Red China and Manchuria. By now, more than 135,000 Red troops had been captured and the North Korean Army was nearly destroyed.

By 29 October, the 8th Cavalry Regiment along with its supporting units advanced north from Pyongyang to Sukchon, Sinanju and to the vicinity of Unsan, with the mission of relieving ROK elements of the I Corps in the area. Later that day, the 8th Cavalry received orders to extend the attack all the way to the Yalu River.

On the morning of 30 October, the 5th Cavalry Regiment, under the command of Lt. Col. Harold K. Johnson, arrived at Yongsan-dong. The mission of the 5th Cavalry was to protect the rear of the 8th Cavalry, which had continued on north to Unsan where it was to relieve part of the ROK 1st Division. The 1st Battalion, 8th Cavalry, under the command of Maj. John Millikin, Jr., arrived at Unsan that afternoon. In conferring with the United States Army Advisory Group, Korea (KMAG) officers attached to the ROK 12th Regiment, Millikin and his company commanders learned that the ROK line, about 8,000 yards north of Unsan, was under attack and being pushed back.

On 31 October, the 2nd and 3rd Battalions, 8th Cavalry, relieved the ROK 12th Regiment. But on the right an enemy attack during the night had driven back the ROK 2nd Battalion more than a mile. Its commander wanted his troops to regain the lost ground before they were relieved. Millikin's 1st Battalion, however, moved into a defensive position behind the ROK 2nd Battalion line north of Unsan. That afternoon, General Milburn, US I Corps commander, visited the 8th Cavalry regimental command post and was advised that everything was all right.

By 01 November, the 8th Cavalry and its supporting units had advanced to within 50 miles of the Chinese border and the three battalions had moved up to relieve portions of the ROK 1st Division. The arrival of the US 8th Cavalry Regiment at Unsan had set in motion a redeployment of the ROK 1st Division. Upon being relieved west of Unsan, the ROK 11th Regiment had shifted southeast to establish contact with the ROK 8th Division on the Corps boundary. The ROK 12th Regiment moved to a reserve assembly area at Ipsok south of the Kuryong River, six air miles from Unsan. Still engaged in the battle at Unsan, the ROK 15th Regiment was desperately trying to hold its position across the Samt'an River east of the 8th Cavalry Regiment. In short, the US 8th Cavalry was to the north, west, and southwest of Unsan; the ROK 1st Division to the northeast, east, and southeast of the town.

Later in the morning of 01 November, patrols from the 1st and 2nd Battalions, 8th Cavalry, clashed with soldiers clearly identified as Red Chinese. Contact with the Chinese had begun increasing that afternoon, starting in the sector of the 1st Battalion, north of Unsan, then spreading west into the sector covered by the 2nd Battalion. By 1200 hours 01 November, the Chinese had cut and blocked the main

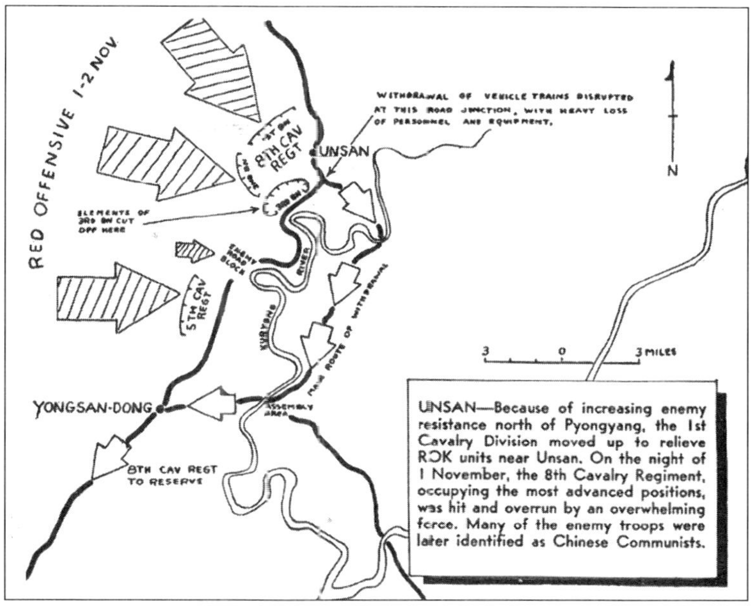

UNSAN—Because of increasing enemy resistance north of Pyongyang, the 1st Cavalry Division moved up to relieve ROK units near Unsan. On the night of 1 November, the 8th Cavalry Regiment, occupying the most advanced positions, was hit and overrun by an overwhelming force. Many of the enemy troops were later identified as Chinese Communists.

road six air miles southwest of Unsan with sufficient strength to turn back two rifle companies which had been strongly supported by air strikes during daylight hours. The 7th Cavalry at Chinnampo was recalled.

The CCF had set the stage for an attack that night against the 8th Cavalry Regiment and the ROK 15th Regiment. In the afternoon of 01 November, the CCF attack north of Unsan had gained strength against the ROK 15th Regiment on the east, and gradually it extended west into the zone of the 8th Cavalry Regiment. At 1700 hours, the first probing attacks, accompanied by mortar barrages, came against their right flank units, "A" and "B" Companies, 1st Battalion. There was also something new in the enemy fire, support-rockets fired from trucks.

When dusk fell that evening enemy soldiers were on three sides of the 8th Cavalry - the north, west, and south. Only the ground to the east, held by the ROK 15th Regiment, was not in Chinese possession. At 2330 hours, the CCF launched an all out attack on the positions of the 1st Battalion, 8th Cavalry. As the battle grew, the attack of the CCF, well planned and executed in strength, broke through the ROK 15th Regiment. Following the issue of warning alerts of an impending withdrawal and armed with the most recent intelligence data, Colonel Holmes, Chief of Staff, 1st Cavalry Division, issued a final order for the 8th Cavalry Regiment to withdraw at 2400 hours. Soon afterwards, at about 0100 hours 02 November, the CCF cut the withdrawal route of the 1st and 2nd Battalions.

Major Milikin, 1st Battalion Commanding Officer, telephoned Colonel William Walton, 2nd Battalion Commanding Officer, that he would try to hold Unsan until the 2nd Battalion cleared the road junction south of it. Then he would withdraw. The 3rd Battalion, southwest of Unsan, was to bring up the regimental rear.

The 1st Battalion, 8th Cavalry had expended its basic ammunition as well as the reserve which had been sent down from the Regiment. "A" Company had engaged in "hand-to-hand" combat on both its flanks. The 1st Battalion Commanding Officer, Major Millikin requested additional

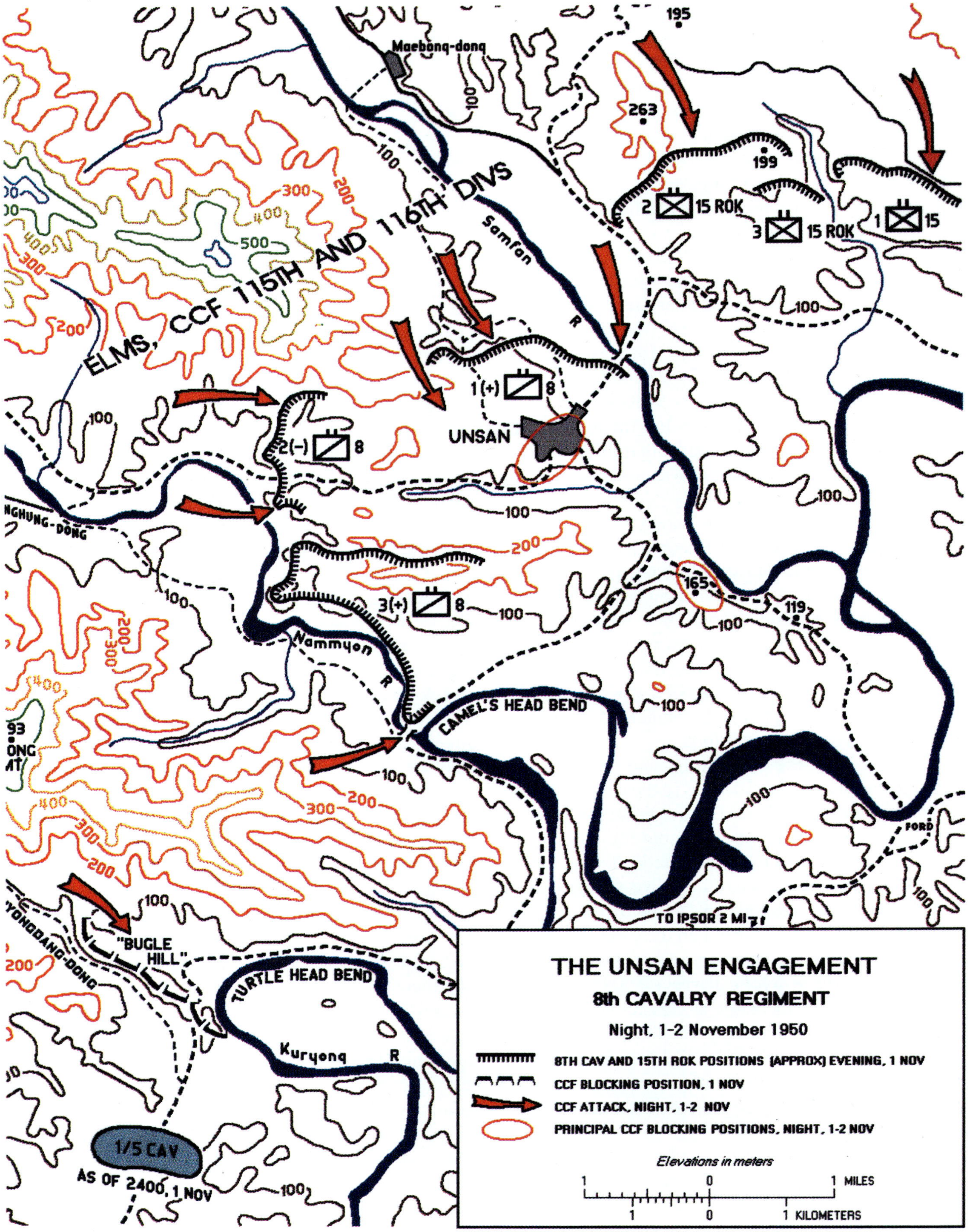

By 01 November, the 8th Cavalry and its supporting units had advanced to within 50 miles of the Chinese border and the three battalions had moved up to relieve portions of the ROK 1st Division.

issues of ammunition. Receiving the division withdrawal order at midnight, with the 1st and 2nd Battalions, 8th Cavalry in heavy contact, the Regimental Commander, Colonel Palmer, ordered a withdrawal to the south because the main southwest road was blocked. The plan was for the 3rd Battalion to cover the withdrawal.

The entire rear areas were swarming with the CCF. With heavy close-in fighting, the convoys of the 8th Cavalry trains managed to withdraw under fire and to break through the CCF lines. Mostly, the men withdrew in scattered groups or as individuals. Many of the groups were lost as well as critical equipment needed to support the withdrawal.

By 0200 hours, 02 November, the Chinese had blocked the last remaining road for a possible retreat overland. Southwest of Unsan, the 3rd Battalion had dug in just north of the Nammyon River. By dawn it was completely surrounded. The bulk of the battalion, trapped by the Chinese, formed two islands of resistance. All day long fighter aircraft and bombers pounded the enemy positions. The battalion took heavy losses in its officers and enlisted men. The Commanding Officer, Major Ormond, was badly wounded and the staff were all wounded or missing in action.

The troopers used the daylight respite gained from the air cover to dig an elaborate series of trenches and retrieve rations and ammunition from the vehicles that had escaped destruction. An L-5 plane flew over and dropped a mail bag of morphine and bandages. At dusk, a helicopter also appeared and hovered momentarily a few feet above the 3rd Battalion, intending to land and evacuate the more seriously wounded, but enemy fire hit it and it departed without landing. The battalion group was able to communicate with the pilot of a Mosquito plane overhead who said a relief column was on its way

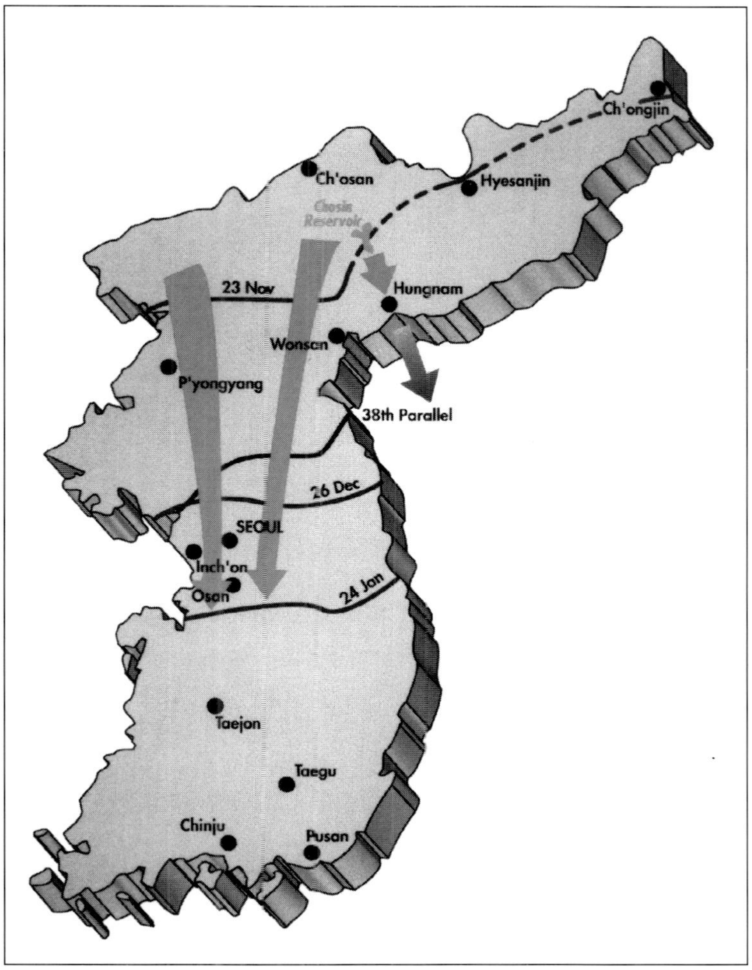

The Chinese Communist Forces (CFF) cross the China - North Korean border and launch a surprise attack with 300,000 soldiers which began the longest retreat in US Military history. Again Seoul fell to the enemy. (November 1950 - January 1951)

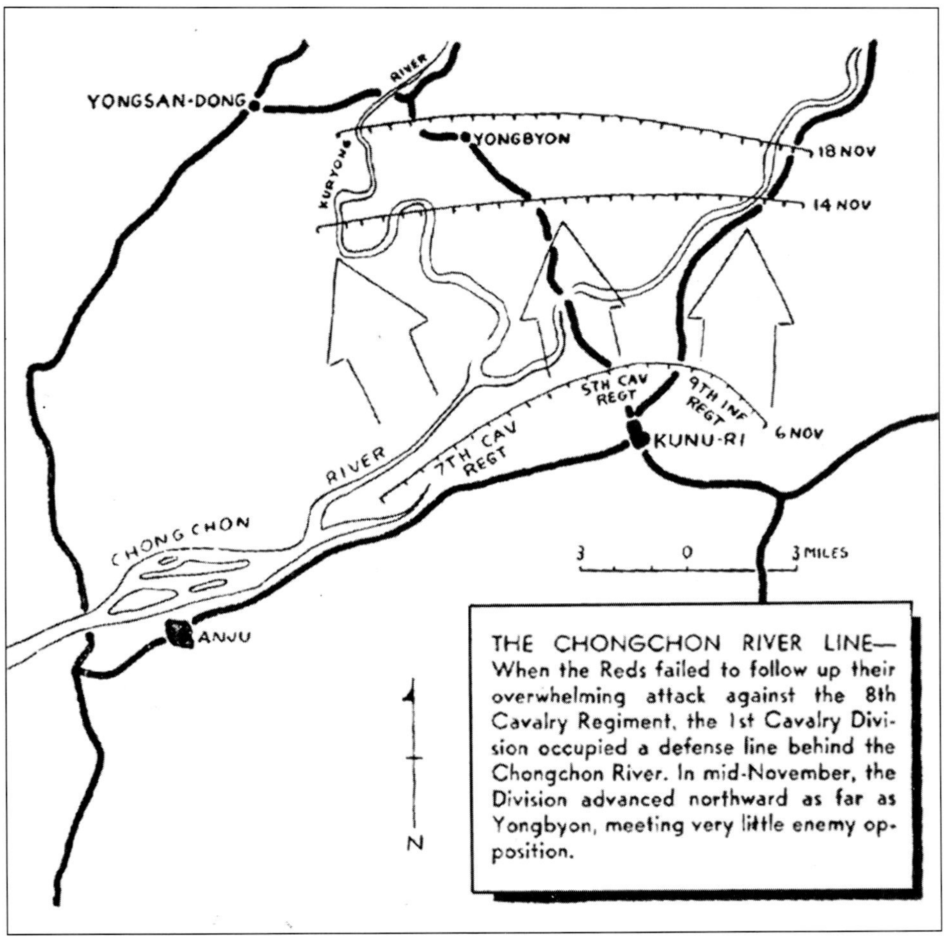

The 1st and 2nd Battalions of the 5th Cavalry attempted a break through from the southwest, but the CCF on "Bugle Hill" could not be dislodged from their defensive positions. The 5th Cavalry, after receiving more than 350 casualties, pulled back.

Just after dark, a plane dropped a message to the 3rd Battalion with orders that they are to begin an orderly withdrawal. The withdrawal route indicated was the only one possible, east from the road fork south of Unsan, across the the Kuryong River, and then by the main supply route of the ROK 1st Division to Ipsok and Yongbyon.

After examining all the options, the remaining men of the 3rd Battalion decided to stand and fight even though they faced a full division of the CCF. The night brought a heavy bombardment of 120mm mortar fire and a mass attack by the CCF. During the lull that followed, large numbers of enemy died outside the perimeter. With their own ammunition nearly spent the men searched the battlefield around the perimeter to retrieve weapons and ammunition from the enemy dead.

On the morning of 03 November a three-man patrol went to the former battalion command post dugout and discovered that during the night the Chinese had taken out some of the wounded. That day there was no air support. Remaining rations were given to the wounded. Enemy fire kept everyone under cover. The night of 03 November was a repetition of the preceding one, another barrage followed by a mass attack, with the Chinese working closer all the time. Again with their own ammunition almost gone, after each enemy attack had been driven back, the men would crawl out and retrieve weapons and ammunition from the enemy dead.

The morning of 04 November disclosed that there were about 200 men left able to fight. Casualties had risen to about 250 men. A discussion of the situation brought the decision that those still physically able to make the attempt should try to escape. Captain Anderson, the battalion surgeon, volunteered to stay with the wounded. Lietenants Mayo and Peterson left the perimeter to scout a way out. They found a route to the east and about 200 escaped that night, however, the next day they were surrounded by Chinese and had to break up into small groups. Eventually many were captured or killed. In all, less than 200 enlisted men and ten officers escaped to rejoin their regiment.

On 05 November, the Eighth Army announced that "as a result of an ambush" the 1st Cavalry Division would receive all the new replacements until further notice. In the next twelve days, The Eighth Army assigned 22 officers and 616 enlisted men as replacements to the 1st Cavalry Division. Nearly all of them went to the 8th Cavalry Regiment.

This event would be the most painful chapter in the proud history of the 1st Cavalry Division. At approximately 1600 hours on the afternoon of 06 November, the action of the 3rd Battalion, 8th Cavalry, as an organized force came to an end. It died gallantly. At first, more than 1,000 men of the 8th Cavalry Regiment were missing in action, but as the days passed, some of them returned to friendly lines along the Ch'ongch'on. Eventually the estimate was revised to a count of more than 600 officers and men that were lost at Unsan, most of them from the 3rd Battalion.

The heroic 3rd Battalion commander, Major Ormond, was among the wounded captured by the CCF in the perimeter beside the Kuryong. He subsequently died of his wounds and, according to some reports of surviving prisoners, was buried beside the road about five miles north of Unsan. Of his immediate staff, the battalion S-2 and S-4 also lost their lives in the Unsan action. There were a few others who escaped later, some from captivity, and were given the status of recovered allied personnel.

On 06 November, the 5th and 7th Cavalry Regiments began to set up a defense line along the south bank at the Chongchon River, which runs from northeast to southeast, with the mission of preventing any enemy movement across the natural barrier of the river. During the next five days, continued active patrolling made little contact with the enemy forces. However, the 61st Field Artillery Battalion met with different circumstances. They were assigned as direct support to the British Brigade that was defending a sector along the Ch'ongch bridgehead. A Chinese battalion slipped by the Argyll infantry and attacked the 61st FA. Immediately the battery commanders placed all men except gun crews on a perimeter around the guns and fought off the oncoming Chinese. The batteries fired some 1400 rounds point-blank at a range of 50 yards. Brigadier Coad said, "It was up to the very highest tradition of any artillery regiment." The 61st received the Presidential Unit Citation for this action

Two weeks after the Unsan action, tank patrols were still bringing in men wounded at Unsan and fortunate enough to have been sheltered and cared for by friendly Koreans. On 22 November, the Chinese themselves, in a propaganda move, turned free 27 men who had been prisoners for two weeks or longer, 19 of them captured from the 8th Cavalry Regiment at Unsan.

In order to execute their battle plan, the Chinese and the nearly beaten North Korean forces had a trio of powerful allies located half way around the world. Three Britons, two working in the British Embassy in Washington, DC and a third heading the American Department in London, were Soviet

60mm Mortar Crew in Korea.

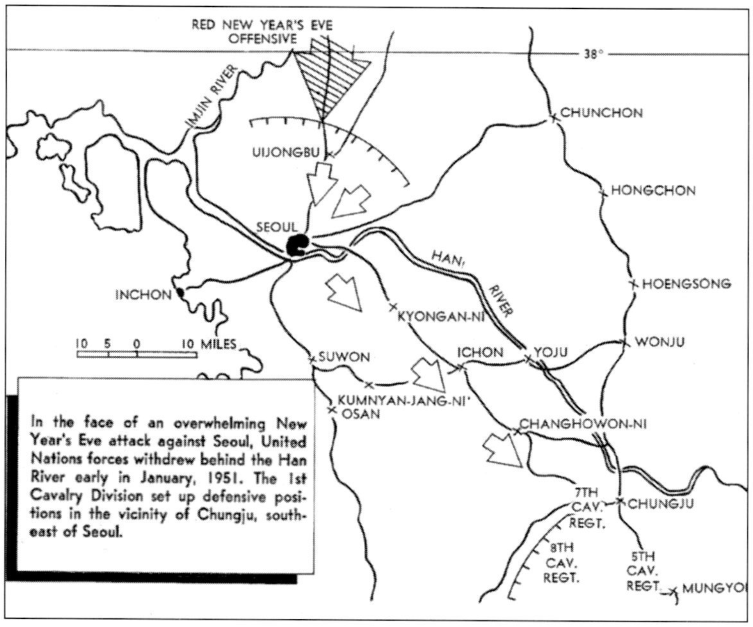

In the face of an overwhelming New Year's Eve attack against Seoul, United Nations forces withdrew behind the Han River early in January, 1951. The 1st Cavalry Division set up defensive positions in the vicinity of Chungju, southeast of Seoul.

agents. The three spies; H.A.R. "Kim" Philby, Guy Burgess and Donald Maclean, had access to communications between MacArthur and the Pentagon because Great Britain had sent its Commonwealth Brigade to be part of the UN military forces in Korea. Copies of communications relative to military planning of UN military organizations were sent directly to Moscow and relayed to Peking.

A massive confrontation with the Chinese seemed inevitable. But the Chinese did the unexpected; they drew back into the frozen hills from which they had suddenly materialized. On 24 November, General MacArthur launched a counterattack of 100,000 UN troops. Taking a chance, General MacArthur believed it was necessary to push the Chinese back across the border. On 25 November, the 1st Cavalry Division moved up to the Taedong River, positioning itself behind the front lines.

On 26/27 November, the enemy shook off heavy casualties and threw great waves of troops at two battle-weary ROK divisions. The ROK II Corps folded quickly, leaving the First Team astride the Chinese penetration. The 5th Cavalry was ordered to move south from Kunu-ri and join the division defenses at Sinchang-ni. On 28 November, The lead element - the I&R platoon, was ambushed and lost all but three of its troopers. Two battalions of the 5th Cavalry attacked the roadblock and cleared a passage for the remainder of the Regiment.

On 29 November, the 7th Cavalry fell back to Sinchang-ni, where at about midnight, the Chinese launched a strong attack. Although they were repulsed in a counter attack, they were able to infiltrate with a small unit that attacked the battalion command post before being dispersed. The counterattack gave the UN time to set up new defensive lines and begin an orderly withdrawal from North Korea.

On 15 December, the 1st Cavalry Division moved northeast of Seoul to the vicinity of Uijong-bu and assumed a defensive position. By 28 December, the true extent of the enemy buildup had become clear. At least 20 Red Chinese divisions were poised for a drive on Seoul. Now there were almost a million and a half Chinese and North Korean troops on the Korean peninsula. The UN Command had less than 250,000 seasoned soldiers to repulse this juggernaut.

When the new year of 1951 began, the First Team defenders readied their weapons, shored up their defenses, and waited in the bitter cold. This time there was no surprise when the Chinese artillery began pounding the UN lines in

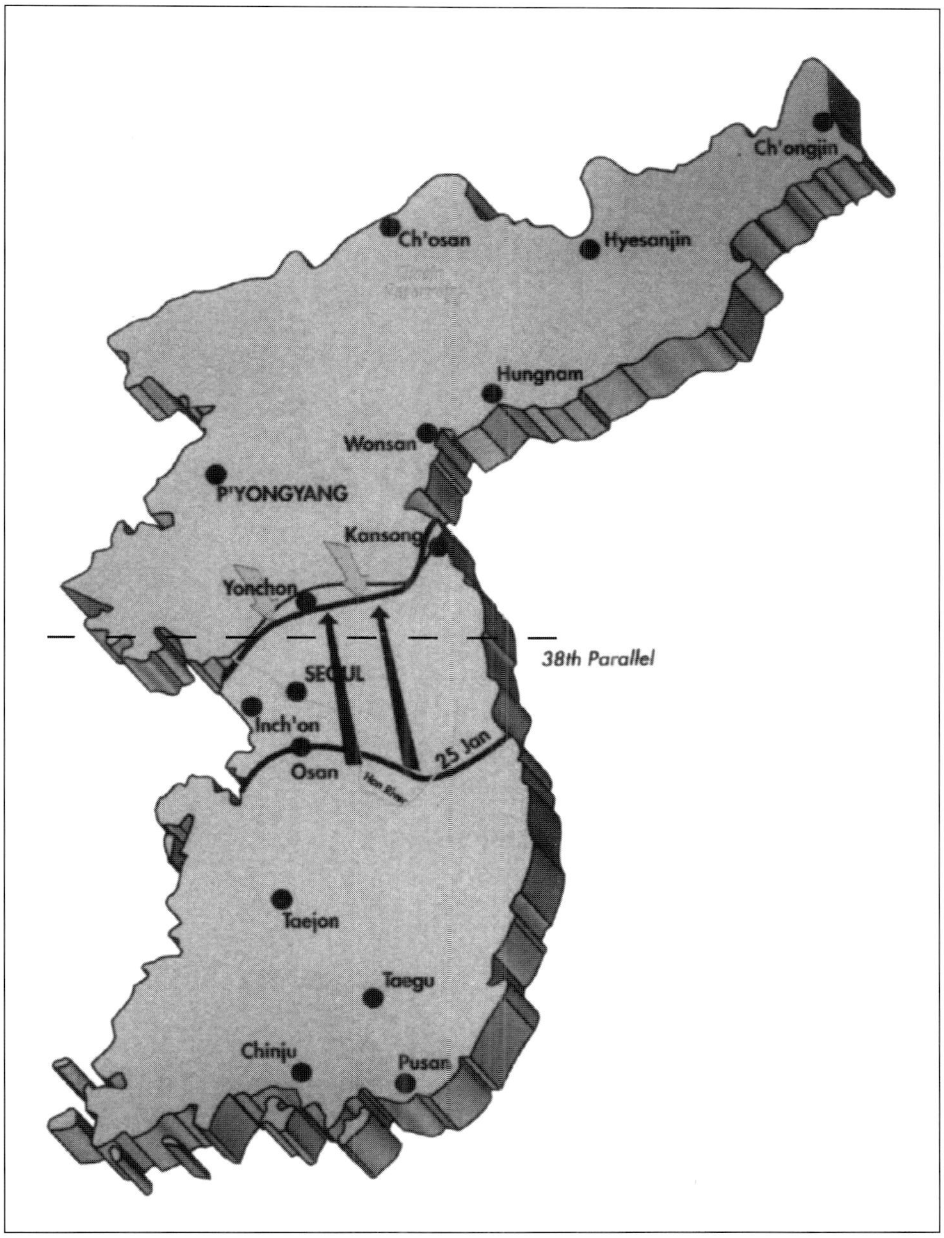

Halting the CCF advances, the UN forces initiated a series of well executed and pin pointed attacks recovering Seoul and the surrounding areas. Following a series of counter attacks, which began in July, an armistice was eventually negotiated dividing the two territories along a Demilitarized Zone located north of the 38th Parallel (January 1951 - July 1953)

As an elderly Korean with a child on his back watches, tankers of the 70th Tank Battalion move up on the north bank of the Pukhan River, above Chunchon in a drive to decimate fleeing enemy forces. (01 April 1951)

the first few minutes of 1951. The units forward of the 38th Parallel were hit by the Chinese crossing the frozen Imjin River. Ignoring heavy losses, the Chinese crawled through mine fields and barbed wire. The UN Forces abandoned Seoul and fell back to the Han River. The Chinese drive lost its momentum when it crossed the Han and a lull fell over the front.

During this pause, one of the most remarkable turnarounds in military history began to take shape. Lieutenant General Matthew B. Ridgway had been put in command of UN Ground Forces, replacing General Walton Walker who had died 23 December when his jeep collided with a South Korean truck. General Ridgway, arriving amid the chaos, immediately went to work restoring order and confidence among the officers and men of the Eighth army. He ordered a change in tactics. The UN Forces would now fight a "war of maneuver", with more emphasis on inflicting enemy casualties and capitalizing on their inability to carry enough supplies to sustain drives longer than a week.

On 22 January 1951, the First Team, joined by the revitalized 3rd Battalion, 8th Cavalry rebounding from its tragedy at Unsan, moved back into action. The movement, designated as "Task Force Johnson" began as a reconnaissance in force. Its mission was to assess the enemy situation in the area, disrupt enemy attack preparations, and destroy maximum enemy personnel and material. Elements comprising "Task Force Johnson" were the Headquarters Command Group, 3rd Battalion, 1st Platoon, "C" Company, 3rd Platoon, Heavy Mortar Company of the 8th Cavalry, Medical Company of the 8th Engineer Battalion, "C" Battery of the 99th Field Artillery, "A", and "B" Companies and the Reconnaissance Platoon of the 70th Tank Battalion. "B" Company, 70th Tank Battalion had been partially restaffed by tankers from the deactivated "A" Company of the 71st Tank Battalion. In addition the force was assisted by organic aircraft from the Division and a flight of tactical air support aircraft.

In the IX Corps sector, along route 17 towards Kyong-ni, the 1st Cavalry Division attacked with the 8th Cavalry on the left and the 7th Cavalry, with the Greek Battalion, on the right. Just past the first control Line "A", the Greek Battalion on Hill 381 was counterattacked by a large enemy force. The battle began before dawn and raged on for the rest of the day. By afternoon, the Chinese had had enough and retreated leaving 800 dead and a proud battalion of United Nations Command (UNC) soldiers.

In the western sector, the 8th Cavalry had met with resistance and forward progress was slowed. The 5th Cavalry was ordered to go around the 8th and seize Hill 312. It was taken in desperate hand to hand combat between the troopers of the 1st Battalion, 5th Cavalry and the strongly dug in Chinese. For a time, the outcome hung in the balance, then the third platoon, "A" Company came charging up the hill with fixed bayonets. The enemy positions were overwhelmed and, although small hand to hand engagements continued for a while, the hill belonged to the 5th Cavalry.

The platoon leader was 1st Lieutenant Robert M. McGovern who led his men up the reverse slope and as his unit got near the enemy on the crest of the hill, they came under heavy machine gun and rifle fire. Despite a wound sustained in this initial burst of withering fire, 1st Lieutenant McGovern assured the men of his ability to continue on and urged them forward. Forging up the rocky incline, he fearlessly led the platoon to within several yards of its objective when the ruthless foe threw and rolled a vicious barrage of hand grenades on the group and halted the advance. Enemy fire increased in volume and intensity and 1st Lieutenant McGovern, realizing that casualties were rapidly increasing and the morale of his men was badly shaken, hurled back several grenades before they exploded. Then, disregarding his painful wounds and weakened condition, he charged a machine gun emplacement which was raking his position with flanking fire. When he was within ten yards of the position a burst of fire ripped the

"L" Company, 5th Cavalry .50 caliber machine gun position, (02 June 1951)

carbine from his hands, but, undaunted, he continued his lone-man assault and, firing his pistol and throwing grenades, killed seven hostile soldiers before falling mortally wounded in front of the gun he had silenced. The incredible display of valor by 1st Lieutenant McGovern imbued his men with indomitable resolution to avenge his death. Fixing bayonets and throwing grenades, they charged with such ferocity that hostile positions were overrun and the enemy routed from the hill. For his valiant action, 1st Lieutenant Robert M. McGovern received the Medal of Honor.

In the counterattack, the Eighth Army moved slowly and methodically, ridge by ridge, phase line by phase line, wiping out each pocket of resistance before moving farther north. The advance covered 2 miles a day, despite heavy blinding snowstorms and sub-zero temperatures.

By 12 February, the 5th Cavalry reached Line "E" and was relieved by the 28th British Brigade and sent to Yoju. The 7th Cavalry was not as fortunate. On 14 February, heavy fighting erupted around an objective known as Hill 578, which was finally taken by the 7th Cavalry after overcoming stiff Chinese resistance. During this action General MacArthur paid a welcome visit to the First Team.

Not far away, at the town of Chipyong-ni, the 23rd Regimental Combat Team and a French Army Battalion were surrounded by five Chinese divisions. In desperate fighting, the two units killed thousands of Chinese but were unable to break out.

Hearing of their situation. the 5th Cavalry formed a rescue force, called "Task Force Crombez" to counterattack

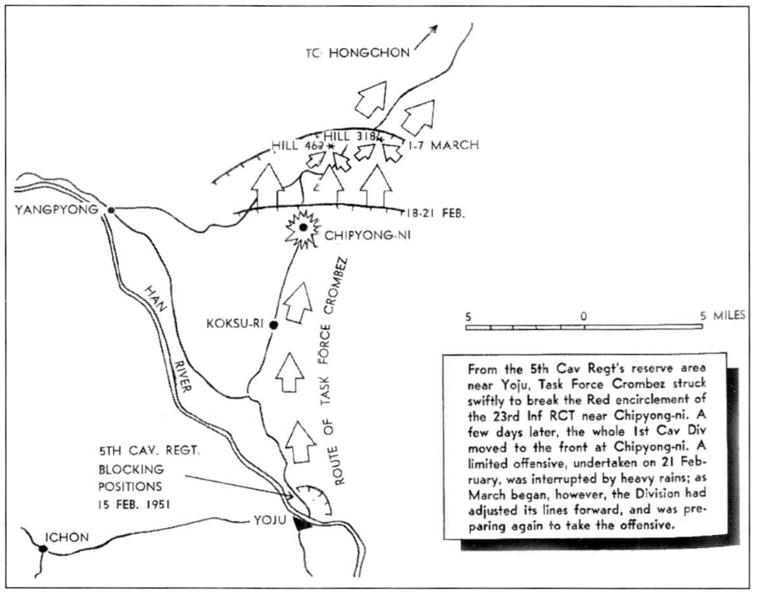

From the 5th Cav Regt's reserve area near Yoju, Task Force Crombez struck swiftly to break the Red encirclement of the 23rd Inf RCT near Chipyong-ni. A few days later, the whole 1st Cav Div moved to the front at Chipyong-ni. A limited offensive, undertaken on 21 February, was interrupted by heavy rains; as March began, however, the Division had adjusted its lines forward, and was preparing again to take the offensive.

along a road running from Yoju to Chipyong-ni via Koksu-ri, a distance of 15 miles. The troopers had painted tiger stripes on their armored tanks to give them a psychological advantage. The sight of the tiger-striped M-4A3 and M-46 tanks sent many of the Chinese running from their entrenched positions. As the fleeing Chinese raced through open ground, they were cut down by heavy fire from the tanks and escorting troopers of "L" Company, who had taken heavy casualties in their mission of tank protection enroute to Chipyong-ni. On 15 February, Task Force Crombez broke through the perimeter of Chipyong-ni ending the standoff. The victory at Chipyong-ni marked the first time in the Korean War that the Chinese had been dealt a major defeat.

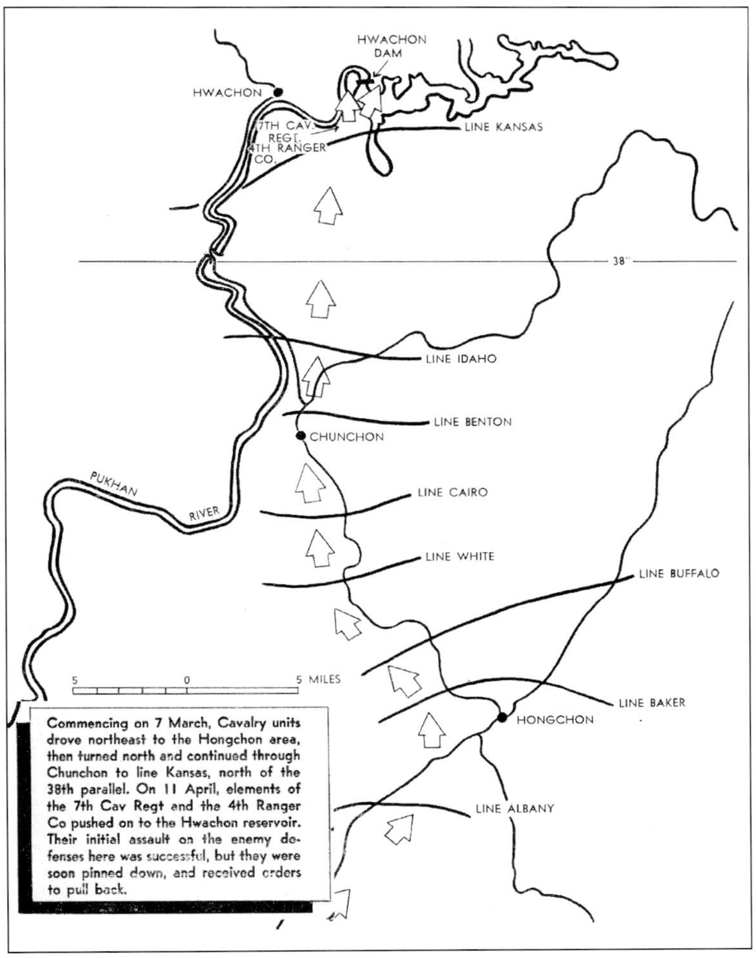

Commencing on 7 March, Cavalry units drove northeast to the Hongchon area, then turned north and continued through Chunchon to line Kansas, north of the 38th parallel. On 11 April, elements of the 7th Cav Regt and the 4th Ranger Co pushed on to the Hwachon reservoir. Their initial assault on the enemy defenses here was successful, but they were soon pinned down, and received orders to pull back.

The 1st Cavalry Division slowly advanced though snow and later, when it became warm, through torrential rains. The Chinese Army was slowly but firmly, being pushed back. On 14 March, the 3rd Battalion, 8th Cavalry had crossed the Hangchon River. The Division, along with the 1st Marine Division conducted a double envelopment to capture the town of Hongchon. On the 15th, Seoul was recaptured by elements of the Eighth Army. New objectives were established to keep the Chinese from rebuilding and resupplying their forces and to advance to Line "Kansas", which roughly followed the 38th Parallel and the winding Imjin River.

By 09 April, the 1st Cavalry Division was approaching the Line "Kansas" with the Hwachon Reservoir on its eastern flank. There was some apprehension that the enemy would open the gates and flood the Pukhan River destroying all bridges that crossed the river and bring havoc to X Corps. The dam was on the northwest side of the reservoir in a rather inaccessible location from a southern approach. The 7th Cavalry was ordered to capture the dam which would eliminate the possibility of enemy destruction. The approaches to the dam severely restricted vehicular movement and the artillery could not be brought into range to support an attack. In an alternate approach, the 4th Ranger Company, attached to the 7th Cavalry, was to cross the reservoir by boat and attack from the east as the 2nd

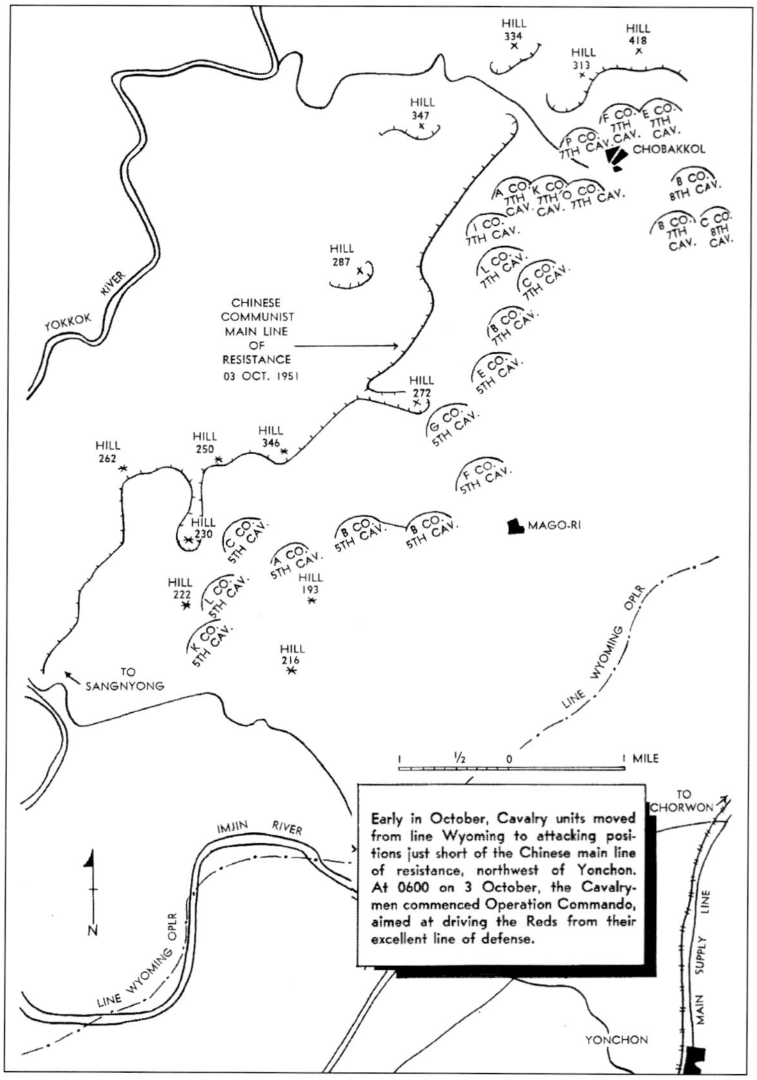

Early in October, Cavalry units moved from line Wyoming to attacking positions just short of the Chinese main line of resistance, northwest of Yonchon. At 0600 on 3 October, the Cavalrymen commenced Operation Commando, aimed at driving the Reds from their excellent line of defense.

Battalion, 7th Cavalry attacked from the southwest. The Americans made three separate assaults on the defenders, but none was successful in evicting them. Before another attempt could be organized, the troopers were pulled out for another fight. On 17 April, ROK marines crossed the reservoir and found that the Chinese had abandoned the dam and left it undefended.

On 22 April, 21 Chinese and 9 North Korean divisions slammed into Line "Kansas". Their main objective was to recapture Seoul. At the beginning of the Communist attack, the balance of the 1st Cavalry Division remained in reserve until the complete collapse of the ROK Division in the IX Corps sector had left the Seoul-Chunchon axis open to the enemy. The 1st Cavalry Division joined in the defense line and the bitter battle to keep the Reds out of the South Korean Capital. On 25 April, elements of the 5th Cavalry, with "A" Company, 70th Tank Battalion, closed in on the Kapyong area to relieve the hard pressed 27th British Commonwealth Brigade. On 28 April, the Division occupied Line Golden, north of Soul astride the main highways from Uijong-bu and Munsan-ni. With adequate reserves, fortified positions, and a narrower front that allowed concentration of artillery fire, I Corps was in the strongest position that it occupied since the beginning of the offensive.

Stopped at Seoul on 15 May, the Chinese shifted three armies to the east and assisted by rain and fog, attempted a go around maneuver in the dark and attacked the center UNC forces. The 8th Army pushed them back to Line "Kansas". Later the First Team moved deeper into North Korea to distract the enemy from its central offensive, reaching the towns of Yonchon and Chorwon, below the "Iron Triangle", the Chinese supply base.

After a week of hard fighting, the line was stabilized and X Corps assumed the offensive which gave the 1st Cavalry Division the opportunity to resume its movement north towards Sochon-ni just north of the 38th parallel. On 24 May, the Division crossed the Line "Topeka" and two days later reached Sochon-ni which was located on Line "Kansas". Patrol bases established north of Line "Kansas" marked the third time that the First Team crossed the 38th parallel.

June was a period in which each regiment established up to battalion size patrol bases which was a new tactic designed especially for the terrain and battle conditions of Korea. Such bases utilized perimeter defenses supported by mortar and artillery concentrations. The bases were situated on prominent terrain features and utilized this height advantage to inflict a terrible toll on the enemy who employed "human wave" tactics. Improving on the defenses of Line "Kansas", a new Line "Wyoming" was established approximately 35 miles north of Line "Kansas".

By 01 August, following a "reserve" status of 10 days, all regiments were on Line "Wyoming" and resumed their patrolling from bases established along the outpost line of resistance. In mid August, the 5th and 7th Cavalry sustained determined enemy attacks on their bases. The 2nd Battalion, 8th Cavalry, in one of the largest offensives of this period, led an attack on Hill 272 and after the third attempt, seized the objective.

The patrol continued into the early days of September when "K" Company, 5th Cavalry was attacked by two Chinese regiments. "K" Company, at the time, was composed of a majority of replacements, many of whom had never participated in all out combat. By morning, the company had used up four basic loads of ammunition and the situation became critical. A jeep, loaded with ammunition, drove over the heavily mined, four miles of road, from the Main Line of Resistance (MLR) to the perimeter of "K" Company, receiving enemy fire all the way. This type of action occurred daily along the MLR.

On the nights of 21 and 23 September, the 2nd and 3rd Battalions, 7th Cavalry repulsed waves of Red Chinese with hand to hand fighting. But harder work followed when Operation COMMANDO, a mission to push the Chinese out of their winter defense positions south of the Yokkok River, was launched. The objective, the existing Chinese main line of resistance, was designated Line "Jamestown".

To carry out the I Corps broad advance against the entrenched 42nd, 47th, 64th and 65th Chinese Armies, five divisions were used in the execution of Operation

COMMANDO. On the western flank, the ROK 1st Division was to leave Line "Wyoming", cross the Imjin River and move to Kaesong. On the eastern flank the British Commonwealth Brigade was to take the high ground between Samich'on and Kyeho-dong. Still farther east, the 1st Cavalry Division, was to move northeast on an 8 mile front between Kyeho-dong and Kamgol. On the right flank, the 3rd Division was to advance and capture Hills 281, 324, and 373 northwest of Choron to join the 25th Division to take the terrain northeast of Chorwon where the Hantan and Namdae Rivers came together.

On 03 October, the day that Operation COMMANDO was to begin, the main line of enemy resistance was directly in front of the 1st Cavalry Division. Elements of the 139th and 141st Divisions of the CCF 47th Army facing the 1st Cavalry Division had constructed defenses similar to those encountered on Heartbreak Ridge - strong bunkers supporting each other with automatic weapons fire with heavy concentration of artillery and mortars interdicting the approach routes to the hills and ridges. Barbed wire aprons and mines guarded the trenches and bunkers which were well stocked with ammunition.

Task Force MAC, consisting of the 70th Tank Battalion and the 16th Reconnaissance Company, on the left flank of the 1st Cavalry Division had the mission of advancing along the east bank of the Imjin toward Kyeho-dong and move west with the British Commonwealth Brigade, protecting the left flank of the 5th Cavalry Regiment. The 5th and 7th Cavalry were to attack abreast across the division front. The 8th Cavalry was to remain in reserve. All of the artillery units of the division were to participate in the attack. The 61st and 82nd Field Artillery supported the 5th Cavalry and the 77th and 99th Field Artillery supported the 7th Cavalry. Additional artillery units, the 936th, "A" Battery of the 17th and "A" and "B" Batteries of the 204th, were assigned from I Corps and positioned along the main line of resistance, 4 to 6 miles from Line "Jamestown".

An hour before the attack was launched, a heavy barrage of artillery fire began to soften the enemy positions. Then at 0600 hours, the five divisions moved out. The reaction of the enemy in front of the 1st Cavalry Division was immediate and violent. Task Force MAC, on the left, encountered heavy mine field concentrations coupled with strong artillery and mortar fire. The 5th Cavalry assaulted the four immediate objectives, Hills 222, 272, 287 and 346. The Chinese refused to give any ground, directing artillery fire at the three battalions of the 5th as they pushed up the hills. Six attempts by the 3rd Battalion won a foothold on Hill 272, but enemy pressure forced a withdrawal later in the day. The 3rd Battalion could only get a lasting success against Hill 222. After another frontal attack, the Chinese abandoned the hill and fell back to the north.

The situation of the 7th Cavalry was similar. Attacking with the Greek, 2nd and 3rd Battalions abreast, they stormed Hills 313 and 418 along the ridge. Both, the Greek and 2nd Battalion fought their way to the ridge line, but suffering heavy casualties, neither could hold the ground. Despite heavy fighting on 4 October, there was little forward progress. Elements of the 8th Cavalry reinforced the 7th Cavalry on the right and assaulted the ridges west of Hill 418, but the enemy clung tenaciously to its positions. During the day, elements of the CCF 140th Division moved up to reinforce the CCF 139th Division which had been hit hard by the constant battering from the 1st Cavalry Division. The 1st Cavalry Division, now had to contend with the bulk of the elite CCF 47th Army.

The first crack in the Chinese defense came on 05 October, when the 1st Battalion, 8th Cavalry discovered that, during the night, the enemy had withdrawn the majority of their forces from Hill 418 which was the anchor for Line "Jamestown". By afternoon, the 1st Battalion had cleared the ridge 1,400 yards to the northeast. The 2nd Battalion, 7th Cavalry then moved up the ridge southwest of Hill 418 and occupied Hill 313 without opposition.

On the following day, 06 October, the 2nd Battalion, 8th Cavalry launched an attack on Hill 334, and after two attempts, seized the objective. Heavy enemy resistance, during the day and later at night, was beaten down. At Hill 287, over 4,000 yards southwest of Hill 334, the 1st Battalion, 7th Cavalry, fought its way to the crest and held on to part of the hill at nightfall. The enemy prisoners taken indicated that the many of the enemy units were decimated in the opening days of the battle and was falling back to prepared defense lines to the northwest.

On 07 October, the 7th Cavalry completed the seizure of Hill 287 and sent the 3rd Battalion forward two miles southwest to take Hill 347. Attacking from the south, the 3rd Battalion began to clear the hill at the end of the day. The fall of Hill 347 meant that the 1st Cavalry Division now dominated the high ground comprising Line "Jamestown" and the northeastern half of the divisional sector.

The breach of the northeast had little immediate effect on the Chinese defense of the terrain directly across from

Hill 346 - "Old Baldy - Chemical defoliation was not necessary to obtain a clear view of the enemy. By now, the rugged terrain of Hill 346, showed the effects of heavy artillery barrages that had stripped all vegetation and trees leaving twisted, charred stumps.

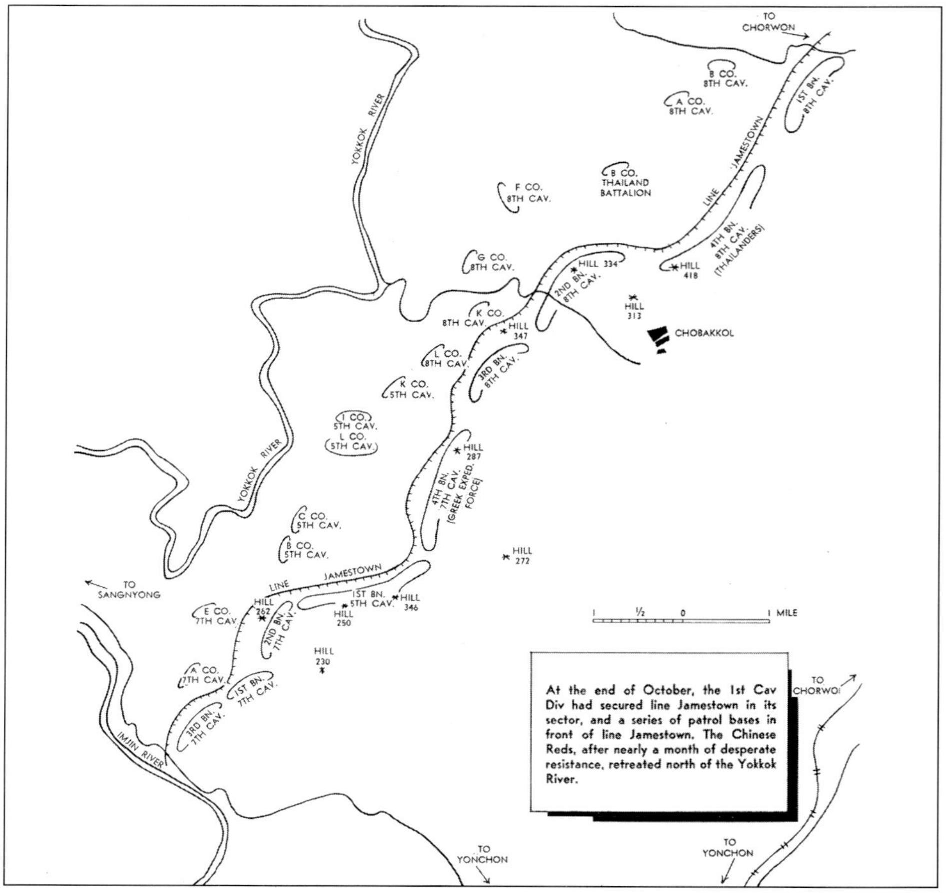

At the end of October, the 1st Cav Div had secured line Jamestown in its sector, and a series of patrol bases in front of line Jamestown. The Chinese Reds, after nearly a month of desperate resistance, retreated north of the Yokkok River.

the 5th Cavalry. After eight days of intense pressure against Hills 230, 272 and 346, the Chinese still refused to give any ground, but the punishment they had absorbed and the drain of manpower and ammunition stocks were beginning to show. On the night of 14 October, the Chinese abandoned Hill 272 allowing the 8th Cavalry take possession without contact. The control of Hill 272 opened the eastern approach to the key hill of the remaining defense line - Hill 346.

On 15 October, a new Operational plan. called POLECHARGE, was put in effect. The 5th Cavalry was reinforced with the Belgian Battalion from the US 3rd Division and given the mission of taking Hill 346, and then push on to Line "Jamestown" with the support of the 8th Cavalry. On 16 October at 0300 hours, the planned assault got underway, but the heavy fire power of the enemy stopped the advance of the 5th Cavalry. The 8th Cavalry began their drive northeast of Hill 346 and made progress. However, they could not flank the objective. For the next two days the 5th and 8th Cavalry sustained the pressure on Hill 346 without success.

On 18 October, the 1st Battalion 5th Cavalry and took Hill 346 against stiff resistance. Later in the day, the 3rd Battalion met similar resistance on Hill 230, but the 2nd Battalion managed to seize Hill 340, just left of Hill 346. The 8th Cavalry pushed well beyond Hill 287 and took over control of the area north of Hill 346. That night the Chinese gave up and retreated. By 19 October the 1st Cavalry Division had seized the last of its objectives on Line "Jamestown" as the enemy retreated north of Yokkok-chon to his next line of defense. With these final gains, the mission of Operation COMMANDO was complete.

The cost to the enemy had been high. Estimates of enemy losses during the 3 to 19 October period placed the total well over 21,000, including 300 prisoners. Nearly 16,000 casualties had been inflicted upon the enemy by the 1st Cavalry Division alone, as it reduced the crack CCF 47th Army to half strength. Later data, gained from intelligence reports, pointed out that Chinese Commanders may have had a lack of interest in the fate of front line regiments which had been ordered to resist to the end rather than ordering up reserves in a heavy counterattack to attempt the retaking of lost territory .

On 23 October, Operation STONEWALL was launched. Its objective was to strengthen Line "Jamestown" by building a wall of defenses which would prevent possible enemy counterattacks. The battle for establishing the outposts on the rugged high ground just south of the Yokkok River proved to be extremely hard. The enemy was unwilling to give up the high ground which would give the 1st Cavalry Division a better command position. The 5th Cavalry Regiment experienced the greatest difficulty in securing their assigned areas. On the first day, only "K" Company was able to secure its objective. The next day "A" and "C" Companies moved forward to successfully secure their bases, but following a night counterattack, they had to fall back to the main line. Resistance was similar elsewhere, initial efforts were often without success. Heavy bombardment increased the intensity of the repeated assaults and objectives were taken. but not without losses.

On 28 October, in action against the enemy near Chong-dong, 1st Lieutenant Lloyd L. Burke, "G" Company, 5th Cavalry observed that intense enemy fire had pinned down leading elements of his company committed to secure commanding ground. 1st Lieutenant Burke left the command post to rally and urge the men to follow him toward three bunkers impeding the advance. Dashing to an exposed vantage point he threw several grenades at the bunkers, then, returning for an Ml rifle and adapter, he made a lone assault, wiping out the position and killing the crew. Closing on the center bunker he lobbed grenades through the opening and, with his pistol, killed three of its occupants that were attempting to surround him. Ordering his men forward he charged the third emplacement, catching several grenades in midair and hurling them back at the enemy. Inspired by his display of valor, his men stormed forward, overran the hostile position, but were again pinned down by increased fire. Securing a light machine gun and three boxes of ammunition, 1st Lieutenant Burke dashed through the impact area to an open knoll, set up his gun and poured a crippling fire into the ranks of the enemy, killing approximately seventy-five. Although wounded, he ordered more ammunition, reloading and destroying two mortar emplacements and a machine gun position with his accurate fire. Cradling the weapon in his arms he then led his men forward, killing some twenty-five more of the retreating enemy and securing the objective. 1st Lieutenant Burke's heroic action and daring exploits inspired his small force of thirty-five troops. For his valiant action, 1st Lieutenant Lloyd L. Burke received the Medal of Honor.

By the end of October, Line "Jamestown" and its Outpost of Line Resistance (OPLR) was seemingly secure in friendly hands. On 01 November, the 5th Cavalry went into Division reserve and remained there for several days while the 7th and 8th Cavalry continued organization of the lines. Interrogation of prisoners revealed the CCF planned to send reserve battalions forward to overwhelm UN patrol bases and then withdraw. Such strong forays began to occur in the left column of the 1st Cavalry Division sector. In response to these actions, the 5th Cavalry returned to positions between the 7th and 8th Cavalry Regiments.

On 10 November 1951, the 70th Tank Battalion status of attachment changed and it was permanently assigned to the 1st Cavalry Division. On 19 November, the 3rd Infantry Division assumed responsibility for the sector. Beginning a gradual phase out of the 1st Cavalry Division from the front lines, the 5th Cavalry moved off the line and the 7th and 8th Cavalry, along with the 61st and 77th Field Artillery were placed under operational control of the 3rd Infantry Division. On 21 November, the 7th Cavalry moved into a reserve status, far from the front.

Meanwhile, the 8th Cavalry was rotating platoons through its patrol bases in order to give them a rest. The majority of its forces were situated north of the 38th parallel, near Sokkogae, North Korea, southwest of Chorwon. One particular platoon outpost on Hill 255, located about 1900 yards west of "F" Company, was under the command of 1st Lieutenant James L. Stone. At about midday of 21 November, the platoon began receiving artillery and mortar fire. At 2205 hours, the enemy unleashed a tremendous preparatory barrage. At 2215 hours, the position was attacked by a battalion size, Chinese force. Hostile artillery blasted the defensive position and hundreds of screaming, grenade throwing Chinese swarmed up the slopes, blowing holes in the barbed wire with bangalore torpedoes.

In the engagement at "Three Sisters", one of the last major battles of the war, Lieutenant Stone's platoon, "F" Company, 8th Cavalry, came under heavy attack by overwhelming Chinese forces, 1st Lieutenant Stone stood erect and exposed to the terrific enemy fire calmly directed his men in the defense. He further exposed himself by moving to a defensive flame-thrower that had failed to function and repaired the weapon. Throughout a second attack, 1st Lieutenant Stone, though painfully wounded, personally carried the only remaining light machine gun from place to place in the position in order to bring fire upon the Chinese advancing from two directions. Throughout the battle, he continued to encourage and direct his depleted platoon in its hopeless defense. Although again wounded, he continued the fight with his carbine, still exposing himself as an example to his men. When this final overwhelming assault swept over the platoon position, his voice could still be heard faintly urging his men to carry on, until he lost consciousness. Only because of his driving spirit and heroic action was the platoon emboldened to make its brave but hopeless last ditch stand. For his valiant action, 1st Lieutenant James L. Stone received the Medal of Honor.

Elements of the 1st Cavalry Division that had not been placed under the control of the 3rd Infantry Division, moved south to the old reserve area east of Uijong-bu. There, the units carried out an initial program of care and cleaning of equipment. After time out for a second Thanksgiving, intensive training was reinstituted. By December 1951, the Division, after 549 days of continuous fighting, began planning for rotation back to Hokkaido, Japan. The First Team had performed tough duties with honor, pride, and valor with distinction.

Hill 255 - "Three Sisters" - Each regiment established platoon or company sized Outpost of Line Resistance (OPL) patrol bases which was a new tactic designed especially for the terrain and battle conditions of Korea. The bases were situated on prominent terrain features and utilized this height advantage to inflict a terrible toll on the enemy who attacked in "human wave" formations.

Chapter 5: Return to Japan

1951-1957

Japanese Islands - While retaining its time-honored culture, Japan rapidly absorbed Western technology during the late 19th and early 20th centuries. After its devastating defeat in World War II, General MacArthur had allowed the Emperor Showa, Hirohito, to remain and retain his throne as a symbol of national unity. The actual power of government rested in a network of powerful politicians, bureaucrats, and business. As a result, Japan had became a staunch ally of the United States during the Korean War.

Soon after the 1st Cavalry Division had been pulled off Line "Jamestown" and placed in reserve, word came down that they were to assume a new mission of establishing a strong defensive military presence in the mountainous, northern island of Hokkaido, Japan. Hokkaido is the second largest of the four main islands of Japan. The climate of Hokkaido is rather cold and harsh in the winter but not as hot and humid in the summer as it is in the other regions of Japan. The reasons for uprooting and transferring the 1st Cavalry Division to the northern island of Hokkaido, Japan were founded upon a long standing strategy to deter the interests of the Union of Soviet Socialist Republics (USSR) from expanding their control and influence further eastward.

In negotiations among the World War II allies, Joseph Stalin declared his interest in occupying the nearby islands of Sakhalin, Hokkaido, and Kurile in exchange for entry of the USSR into the war against Japan. President Roosevelt disagreed, but by the end of the war, Soviet troops had taken Sakhalin and the Kuriles and were preparing to invade Hokkaido. To preempt any actions, the 77th Infantry Division, recently battered in bloody fighting on Okinawa, was rushed to Hokkaido to disarm the Japanese garrisons and establish a firm control under the United States. Regiments of the 77th Division established themselves at Asahigawa, Hokkaido's capital, Sapporo, its principal city, and Chitose, the site of a former Japanese naval air training base.

In early 1946, when the 77th Division was demobilized, the 511th Parachute Infantry Regiment, 11th Airborne Division assumed responsibility for the entire island, establishing its headquarters on the site of a Japanese experimental dairy farm at Sapporo. The garrison, named Camp Crawford for a major who had won the Distinguished Service Cross, gradually grew to become a permanent installation with brick barracks and stucco administration buildings, family housing, an all-grade school, clubs, and support facilities. In 1948 the 31st Infantry Regiment, 7th Infantry Division moved from Korea to Camp Crawford, absorbing most of the personnel of the 511th Parachute Infantry Regiment.

The 31st Infantry, the famous "Polar Bear Regiment" that had served in Siberia during World War I and the Philippines in World War II, was stationed on Hokkaido from 1948 to 1950. The rest of the 7th Infantry Division was stationed near Sendai on the main Japanese island, Honshu. When the Korean War broke out in 1950, the 7th Infantry Division furnished troops to the 24th Infantry, 25th Infantry, and 1st Cavalry Divisions as those units deployed to Korea in early July and August.

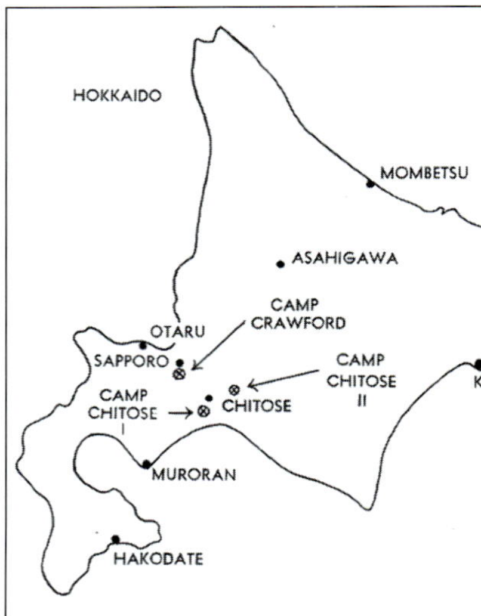

Early in December, the 1st Cavalry Division began moving by water to Hokkaido, replacing the 45th Inf Div, which moved to Korea. Division Headquarters and the 7th Cav Regt were located at Camp Crawford, just outside the island's largest city, Sapporo. The 5th Cav Regt was assigned to Camp Chitose I and the 8th Cav Regt to Camp Chitose II.

When the 7th Infantry Division began its rotation to Korea in the spring of 1951, the 45th Infantry Division, a National Guard unit from Oklahoma, arrived on Hokkaido to continue the show of force against the possibility of a Soviet invasion. The Division established its headquarters, trains, engineers, special troops, and one infantry regiment at Camp Crawford while the artillery, a tank battalion, and two infantry regiments were housed in Quonset huts at Camp Chitose. The Division trained hard throughout 1951, anticipating its commitment to combat which came in 1952 after the 45th relieved the 1st Cavalry Division in central Korea.

To carry out the rotation, the 45th Infantry Division had to be moved from Hokkaido, Japan to assume the assignments of the 1st Cavalry Division. Advance parties of both divisions, down to company level were flown to their new locations to facilitate the move. The transfer of the main body of the two divisions was completed by water in four principal steps.

The first of the regiments to leave Korea, the 5th Cavalry Regiment was relieved in reserve by the 180th Infantry Regiment. On 07 December, the 5th Cavalry Regiment left Inchon. Four days later, its convoy entered the harbor of Muroran, on the southeastern coast of Hokkaido, and movement to Camp Chitose was completed by train. On 18 December, the 7th Cavalry Regiment departed Korea. On 30 December, the 8th Cavalry Regiment, the first of the regiments to engage the enemy in Korea, turned over its equipment to the 279th Infantry Regiment and moved out in skirmish lines in a march over the hills and to the beach where they were loaded into LSTs. The last unit to embark was the 77th Field Artillery, which had moved off the line as the year of 1951 was ending. On 12 January 1952, the 77th left Korea and arrived at the port of Muroran on 16 January closing out the move of the 1st Cavalry Division from Korea to Hokkaido just 18 months after the July 1950 landing at Pohangdong.

The 1st Cavalry Division was assigned the mission to defend the Island of Hokkaido while maintaining maximum combat readiness in the event that it was necessary for their re-intervention in the Korea theater. Hokkaido, the most northern island of Japan, is certainly its coldest. By its unique strategic proximity to the Asian mainland, it was a key center of electronic surveillance for gathering intelligence on Russian activities. The northern tip of the island is only a few miles from Sakalin, a large Russian island, and from the Kuril Islands, also Russian. The majority of the electronic listening posts and radar centers on Hokkaido were under the control and managed by the US Air Force which, in addition to the task of monitoring Russian broadcasts, tracked movement of all Russian aircraft.

Under the command of Major General Thomas L. Harrold, the division controlled a huge encampment and training area of 155,000 acres. In their new locations, troopers found facilities comparable to those of state-side camps. 1st Cavalry Division Headquarters and the 7th Cavalry Regiment were stationed outside Sapporo at Camp Crawford with its steam heated, red brick barracks. Camp Crawford had been constructed in 1946 on the location of the Makomanai National Dairy, which was the largest dairy farm in Japan operated by the Japanese Agricultural Ministry. The 5th Cavalry Regiment was stationed 25 miles southeast of Sapporo, at Camp Chitose, Area I, adjoining the Chitose Air Force Base and adjacent to the thriving Japanese town of Chitose, for which it was named. About four miles east of Chitose, the 8th Cavalry Regiment and the 70th Tank Battalion occupied the new Quonsets of Camp Chitose II.

As soon as the main command group arrived, plans for the defense of the island of Hokkaido were formulated. Elements of the plan included Passive Air Defense, Civil Disturbance Alert, Dependant Evacuation Plan and a Chemical, Biological and Radiological Defense training. In addition, work began on a Master Training Program encompassing physical training, map reading, mine and field fortifications, signal communication, camouflage, first aid and tactical organization and deployment. The climate and

Aerial view of the 1st Cavalry Division Headquarters, Camp Crawford.(1952)

Red brick barracks occupied by the 7th Cavalry Regiment at Camp Crawford, Hokkaido

As soon as the main command group arrived, a winter training program was in place enabling the troopers to become accustomed with the harsh weather. (February 1952)

and landing exercise in connection with the naval operation. This combined Navy - Army exercise might also provide an opportunity to alarm the Communists that a new land invasion was underway.

However, one of the main reasons for the troop movement, lack of adequate garrison facilities in Japan, encouraged the Armys decision to rotate the three Regimental Combat Teams (RCT's) of the 1st Cavalry Division back to Korea, one at a time.

The Navy was heartily in favor of some action and suggested that an amphibious demonstration be mounted. This could conceivably lure enemy reinforcements out on the roads and expose them to attack by air and surface craft. Under Vice Admiral Robert P. Brisco, the Seventh Fleet commander, joint AMPHIBIOUS TASK FORCE SEVEN was set up with 15 October established as the target date. The demonstration was scheduled for the area near Kojo, North Korea and planning for the land, sea, and air phases proceeded at a swift pace. Two alternative assault plans were worked up, one for a landing by two divisions in column and one for an attack by a single RCT. For purposes of deception, only the highest echelon of command knew that the maneuver was to be only a demonstration.

terrain of Hokkaido were ideally suited for instruction in Arctic survival techniques, skiing and snow-shoeing. In parallel with these activities, the Division undertook an exhaustive aerial and ground reconnaissance and mapping of the island, By early 1952 the 1st Cavalry Division was well on its way to becoming accustomed to the frigid temperatures and heavy snows of Hokkaido.

On 07 March, the Island of Hokkaido gave the troops of the 1st Cavalry Division an unexpected welcome. In addition to the standard manifestation of being knocked down to the ground, an earthquake cracked the concrete floors of the Quonset Huts and cracked the overhead water and steam pipes. The engineers had a major task on their hands in restoring normal operations.

By the end April, the initial training program was completed by all of the Division units and a new dimension to the training was added. On 21 April plans took shape for Air Transportability training. The training was conducted by the a team of officers and enlisted men from the 187th Airborne Regiment, who concentrated on the principles of troop air movement. On 01 June, having learned how to go into battle by air, troopers began an extensive training in assaults by water with the initiation of an Amphibious training program conducted by members of the Mobile (Marine) Training Team "Able" (MTTC "A"). By the end of June, a new program of Light Aviation Training was introduced which enabled the training of personnel in the use of light aircraft in field exercises

Naval surface operations during the summer of 1952 had consisted mainly of routine patrol and blockade of the Korean coast, mine sweeping operations, and the shelling of targets along the coast to harass and interdict the enemy's lines of communication. But the biggest naval operation, *AMPHIBIOUS TASK FORCE SEVEN*, was a planned exercise/ demonstration at Kojo on the east coast of North Korea.

In July, it was decided that it might be wise in the interest of economy for the Army to hold a major troop movement

Supporting Divisional plans, prepared for re-deployment of Regimental Combat Teams to Korea for a sixty day tour of duty, were designated Operation DECOY. On 05 October 1952 the 8th Cavalry, along with the 99th Field Artillery Battalion, "C" Battery, 29th AAA AW Battalion and elements of the 2nd Engineer Amphibious Special Brigade departed Camp Chitose II for movement to Otaru in preparation for a practice amphibious landing on beaches near Kangnung, South Korea which was to be carried out on 12 October. History would repeat itself for the 8th Cavalry, for it had participated in the first amphibious landing of the Korean war two years and three months earlier.

The amphibious ships carrying the 8th Cavalry Regiment sortied from Hokkaido. On 12 October, D minus 3, the planned rehearsal was carried out at Kangnung, hampered by winds of 25 knots. With the loss of four Landing Craft, Vehicle, Personnel (LCVPs) after broaching on the beach, the rehearsal operations had to be broken off. The next day all troops re-embarked and sailed farther north. For the next three days, handicapped by the same weather conditions, the Far Eastern Air Force (FEAF) and the Fifth Air Force stepped up their operations around Kojo. Joining in, the planes from the supporting Navy carriers targeted enemy positions around Kojo and North Korean Naval surface craft.

On 15 October the 8th Cavalry RCT aboard the *USS Bayfield*, arriving off the east coast of North Korea near the

Amphibious Landing Exercise at Chigasaki Beach. (October 1953)

Quonset huts occupied by the 8th Cavalry Regiment at Camp Chitose II, Hokkaido. Japan from January to October 1952.

Kojo beach area, was joined by a larger force of more than a hundred ships, including the Carriers *USS Sicily* and *USS Badoeng Strait*, led by the Battleship *USS Iowa*. At 1400 hours landing craft were launched for preparation for the landing. As part of the feint, troops went over the side of the ship and as the boats circled around the ship the troops were picked up on the seaward side of the ship and sent below. Operating in this manner gave the North Koreans the illusion that the boats were still loaded as they headed for shore.

Seven waves of landing craft were sent in from the transport area to pass the line of departure and then retire, seaward. During the assault landing feint, owing to the skill of the coxswains, no boats were lost or seriously damaged. but two mine-sweepers were hit by shore fire and five carrier planes lost to enemy anti-aircraft fire.

The enemy response to the elaborate scheme was disappointing. Little evidence of significant troop transfers came to light and the Communist shore batteries threw only a few answering shells at the assault force. Whether this denoted a lack of ability to respond quickly or perhaps a preference to wait until the United Nations Command (UNC) troops had landed and then to launch a counterattack was impossible to surmise. Evidently the discovery that the operation was only a feint added to the frustration of all the UNC personnel who had not been in on the secret. The realism of the planning and mounting of the operation had built up UNC expectations and although the training was adjudged valuable, the damage to morale served to balance this off.

Following completion of the exercise the *USS Bayfield* moved south, across the 38th Parallel, to Pohangdong (the same beach that the 1st Cavalry Division had landed on in July 1950) and the 8th RCT conducted a beach landing as originally planned. After the landing exercises were completed the 8th RCT moved south to Camp Tongnae, Korea, a location outside of Pusan.

In November, elements of the 8th Cavalry Regimental Combat Team were attached to the 772nd Military Police Battalion. This attachment was a first of a series of security missions performed by units of the 1st Cavalry Division. During this period, cavalrymen wore Military Police brassards and rode trains and guarded marshalling yards. In the marshalling yards, the troopers rode "shotgun" along with the Korean police to provide security for rail cars being transferred or loaded. Those who drew UN Express passenger car duty turned out in Class "A" uniforms and performed "stateside" duty enroute.

Another variation of the Military Police duties was the guarding of hospital trains that carried either wounded UN or enemy soldiers. For the next two months the regiment performed security missions around the familiar cities of Pusan and Taegu, away from the main fighting.

On 08 December 1952, the 7th Cavalry Regiment (less the Tank Company and Tank Maintenance Section), the 77th Field Artillery Battalion and "B" Battery, 29th AAA AW Battalion departed Camp Crawford and moved to Otrau for loading and on 12 December, sailed for Pusan to relieve the 8th Cavalry Regiment. On 15 December, the 7th Cavalry Regiment Combat Team arrived at Pusan and effected a relief of the 8th Cavalry Regiment Combat Team who were guarding Prisoners of War (POW) at Camp Tongnae, Korea. On 18 December, the 8th Cavalry Regiment Combat Team arrived in Otaru and by 20 December, the 8th Cavalry Troopers, the 99th Field Artillery and "C" Battery, 29th AAA AW Battalion were all back in Hokkaido in time to celebrate Christmas and join in the winter training program.

Camp Tongnae, near Pusan, Korea. Home of the 3rd Battalion, 8th Cavalry Regiment. (October to December 1952)

The mission of the 7th Cavalry Regiment Combat Team was to furnish tactical support to Korean Communications Zone (KCOMZ). A secondary mission was to furnish security elements for critical areas, in conjunction with specialized training. The 1st Battalion was quartered at the Daisan School and assigned port security in the Pusan area. The 2nd Battalion was assigned security for the KCOMZ Headquarters area and providing train security guards for the Pusan-Taegu railways. The Heavy Mortar Company provided security for the United Nations Civil Assistance Command, Korea. (UNCACK) Headquarters.

On 19 December, the 3rd Battalion and "B" Battery, 29th AAA AW were attached to UN POW Command, Koje-do as part of POW Command Reserve and "E" Company was assigned security duty at Taejon. The 77th Field Artillery, encamped at Ichon-ni, continued training for all its batteries. The crews of the Air-Section, consisting of two L-19 aircraft, flew emergency flights in addition to observation flights for battery service sections. Each battalion was augmented by attachment of Medical Service personnel.

On 10 February 1953, the 5th Cavalry Regiment, 61st Field Artillery Battalion and Battery "A", 29th AAA AW Battalion, departed from Otaru, Japan for Pusan and Koje-do, Korea to relieve the 7th Cavalry Regimental Combat Team. On 15 February, the convoy carrying the 7th Cavalry Regimental Combat Team sailed from Pusan and docked at Otaru on 18 February. Final troop movement to Camp Chitose was made by rail. By 20 February the 7th Cavalry fresh from their Korean assignment, were assigned to the winter training program. In Korea, the 5th Cavalry Regimental Headquarters was stationed outside of Pusan. The 1st Battalion was assigned security missions in the area from Pusan to Teagu, the 2nd Battalion was assigned the from area from Teagu to Taejon, and the 3rd Battalion covered the area from Taejon to Seoul. On 27 April, all elements of the 5th Cavalry Regimental Combat Team, less the 3rd Battalion and Heavy Mortar Company, who remained on their security mission, returned to Camp Chitose I, Hokkaido.

In June the 82nd Field Artillery Battalion executed Operation FORTNIGHT, during which the battalion traveled more than 160 miles in a two-week exercise. In July, amphibious training was resumed and on 16 July, the 5th Cavalry Regiment completed a ten day motor and foot march reconnaissance to the Muroran Peninsula. The units remaining in Korea continued security missions under control of KCOMZ.

The Korean War wound down to a negotiated halt when the long awaited armistice was signed at 10:00 on 27 July 1953. A Demilitarized Zone (DMZ), a corridor - four kilometers wide and 249 kilometers long, was established dividing North and South Korea. The nominal line of the buffer zone was along the 38th parallel; however, the final negotiations of the adjacent geographical areas, gave the North Korean government some 850 square miles south of the 38th parallel and the South Korean government some 2,350 square miles north of it.

On 09 September 1953, the 3rd Battalion and Heavy Mortar Company of the 5th Cavalry Regiment returned to Hokkaido after seven months of duty in Korea. Planning

Prisoner of War compound near Camp Tongnae near Pusan, Korea.

A member of the Air Crew Section prepares an L-19 for an observation mission. (January 1953)

A total of 1,076 meetings had been held during the 2 years and 19 days before the Armistice Agreement was signed on July 27, 1953. The essential element of the agreement was the establishment of a 4km wide demarcation line that meandered along the 38th parallel bisecting the Korean Peninsula separating North Korea from South Korea.

The 1st Cavalry Division Headquarters Staff and the 5th Cavalry Regimental Commander presents the 5th Cavalry Drum and Bugle Corps with their 1st Place Award won in the Divisional Bands Competition. (November 1953)

Vehicles and personnel of the 7th Cavalry Regiment were moved by ship in the transfer from Camp Chitose to Camp Haugen, Japan. (24 August 1954)

was initiated for the 7th and 8th Regimental Combat Teams to conduct landing exercises at Chigasaki Beach on Honshu during October and November.

In October 1953, the Table of Organization for the 1st Cavalry Division Band was downsized from the authorized ninety-six to forty-eight troopers. The surplus band members were transferred to the 5th, 7th and 8th Cavalry Regiments and were allowed to form a Drum and Bugle Corps which represented the regiments at various functions. In July 1954, the Band, in line with their alternate security of the division mission, was assigned a secondary mission as a Smoke Generator unit.

In January 1954, commencing the third year on Hokkaido, the division began winter training for all units. On 15 February a new dimension was added to the intensive troop training program of the 1st Cavalry Division when "F" Company, 5th Cavalry Regiment took to the sky in a tactical airlift training operation. Using a system of shuttle flights, H-19 helicopters from the 6th Helicopter Company, transported the soldiers from their home base of Camp Schimmelpfennig to the training area of Ojoji-hara, some thirty miles away. Immediately upon arrival, the troopers, clad in overwhites, moved out with complete field gear, and set up operations. On 21 February, a new unit, the 41st Infantry Scout Dog Platoon joined the Division.

With the advent of warmer weather, the division resumed extensive Air Transport and Amphibious Training. Company, battalion, and regimental tests and full scale division maneuvers were scheduled for early fall.

In September 1954, the Japanese assumed full responsibility for defending the island of Hokkaido and the former home of the 1st Cavalry Division was turned over to the Japanese Ground Self Defense Forces. The entire First Team was relocated to the main island of Honshu. The 1st Cavalry Division Headquarters and the 5th Cavalry Regiment were located at Camp Schimmelpfennig outside Sendai. The 7th Cavalry Regiment and the 29th AAA AW Battalion occupied Camp Haugen, near Hachinohe. The 8th Cavalry Regiment was garrisoned at Camp Whittington, located near Koisumi, north of Tokyo.

On 26 September, during the relocation from Hokkaido to Aomori, the 99th FA Battalion experienced the loss of 35 men and one officer, along with two men from the 8th Cavalry Regiment, one man from the 15th Medical Battalion and approximately 1,500 Japanese civilians, in the sinking of the commercial ferry *Toya Maru* in the Tsugaru Straits during a typhoon. Eventually, all the artillery battalions, the 61st, 77th, 82nd and 99th, were garrisoned at Camp

Younghans, Yamagata Prefecture, near the town of Jinmachi approximately 35 miles west of Sendai.

Headquarters and Headquarters and Service Company and "D" Company, 8th Engineers along with a special ordnance detachment were stationed at Camp Loper Depot, a small camp one mile from the small sea port of Shigoma and approximately 20 miles from Camp Schimmelpfennig. The ordnance detachment took care of all ammunition stored in the underground magazines which, due to security issues, were "off limits".

In January 1955 the majority of the division participated in winter training. All of the regiments and the Division Artillery spent several weeks in the field at the Ojoji-Hara, Sekine, Fugi and Otakani Maneuver areas. Throughout the next three years the division was engaged in an almost endless series of field exercises along with the primary mission of guarding the northern sections of Honshu and as new mission requirements were established and needs changed, units were relocated to various regional camp facilities.

In 1956, while the 5th Cavalry Regiment remained at Camp Schimmelpfennig in Sendai and the 8th Cavalry Regiment remained at Camp Whittington located near Koisumi, North of Tokyo, the 7th Cavalry Regiment was relocated to Camp Otsu near Kyoto. The 1st Cavalry Division Artillery and Band were at Camp Drake outside Tokyo and the 70th Tank Battalion was stationed at Camp Fuji. The 13th Signal Company was garrisoned at Hardy Barracks in the Roppongi district of Tokyo.

In March 1956 Headquarters and Headquarters and Service Company and "D" Company of the 8th Engineers vacated Camp Loper with their heavy equipment being transported by LSTs. Personnel moved overland by truck convoy to Camp Drake, outside of Tokyo, where they remained until the division was reorganized in 1957.

In early May 1956, in preparation for a landing exercise, Marines of Mobile Training Team Bravo 2-56, Landing Force Training Unit, Pacific Fleet, set up headquarters at Camp Whittington and initiated pre-float training classes for all 8th Cavalry Regiment personnel. Following the training, from 28 May to 15 June, elements of the 1st Cavalry Division and the 3rd Marine Division held a joint amphibious landing exercise at Iwo Jima. Troopers from the 13th Signal Company and the 8th Cavalry Regiment along with Air Force personnel and Marine radio teams formed a strike team operating off the *USS Point Pleasant*, a Landing Ship, Dock (LSD). In perhaps the first use of helicopters in a landing exercise, the LSD landed on the beaches of Iwo Jima and the Intelligence and Reconnaissance (I&R) platoon from the 8th Cavalry Regiment performed a reconnaissance of Mt. Suribachi on choppers from the LSD.

It had been five years and eight months of occupational duty since December of 1951, when the 1st Cavalry Division had initially rotated back to Hokkaido, Japan from the Korean War. They had been assigned many occupational and training duties and their assigned subordinate units were dispersed and relocated from over a significant portion of Hokkaido to the main island of Honshu. Major station locations assigned to them slowly moved the Division commands back to Camp Drake, outside of Tokyo, where they had been initially headquartered for occupational duty at the end of World War II in 1945.

On 29 August 1957, compliance with a treaty, signed by the governments of Japan and the United States in 1957, which required the removal of all US ground forces from Japan's main islands, went into effect. The 1st Cavalry Division, headquartered at Camp Drake, Tokyo along with its subordinate units stationed throughout Honshu, Japan, were given orders to reduce strength to zero personnel and transfer to Korea (minus equipment). On 23 September 1957, General Order 89 announced the redesignation of the 24th Infantry Division as the 1st Cavalry Division and issued orders to reorganize the Division under the "Pentomic" concept, which utilized five "battle group" maneuvering units rather than the regimental units.

Closing ceremonies for Camp Crawford, Hokkaido were held as the 8th Cavalry Regiment moved to Camp Whittington located at Koisumi, Honshu, Japan. (September 1954)

1st CAVALRY DIVISION STATION LISTING

STATION LOCATION	01 Chitose, JP 02 Makonamai, JP 03 Taugu, KR 04 Pusan, KR			05 Koje do, KR 06 Sendai, JP 07 Musichiukawa, JP 08 Koisumi, JP		09 Gotambe, JP 10 Tokyo, JP 11 Otsu, JP	
DATE - Month/ Year	Feb 1952	Feb 1953	Feb 1954	Feb 1955	Feb 1956	Feb 1957	Aug 1957
Hq & Hq Company	01	02	02	06	06	10	10
Medical Det, Div HQ	01	02	02	06	06	10	10
5th Cavalry Regiment	01	01	01	06	06	06	06
7th Calvary Regiment	01	02	01	07	07	11	11
2nd Bn, 7th Cavalry		03					
3rd Bn, 7th Cavalry		05					
Tank Co, 7th Cavalry		04				09	09
8th Cavalry Regiment	01	02	02	08	08	08	08
Tank Co, 8th Cavalry		01				09	09
Division Artillery							
Hq & Hq Battery	01	01	01	06	06	10	10
Medical Det, Div Artillery	01	01	01	06	06	10	10
61st FA Bn (105mm How)	01	01	01	06	06	06	06
77th FA Bn (105mm How)	01	04	01	07	07	10	10
82nd FA Bn(155mm How)	01	01	01	06	06	11	11
99th FA Bn (155mm How)	01	02	02	06	06	08	08
29th AAA AW Bn (SP)	01	01	01	07	07	10	10
"B" Battery		05					
"C" Battery			02				
Support Command							
1st Cavalry Division Band	01	02	02	06	06	10	10
15th Medical Battalion	01	01	01	06	06	10	10
Ambulance Company				01		11	11
Medical Detachment				08	08	08	08
15th Quartermaster Company	01	02	02	06	08	10	10
15th Replacement Company	01	02	01	08	08	10	10
27th Ordnance Maint Battalion	01	02	02	06	06	10	10
1st Pltn, "A" Company			01	07	07	06	06
2nd Pltn, "A" Company						11	11
3rd Pltn, "A" Company				01	09	08	08
Seperate Battalions/Companies							
8th Eng'r Combat Battalion	01	01	01	06	06	10	10
"A" Company						06	06
"B" Company				07	07	11	11
"C" Company				08	08	08	08
13th Signal Company	01	02	02	06	06	10	10
16th Reconnaissance Company	01	01	01	01	09	09	09
70th Tank Battalion	01	01	01	01	09	09	09
545th MP Company	01	02	02	06	06	10	10
1st Traffic Pltn				07		06	06
2nd Traffic Pltn					07	11	11
3rd Traffic Pltn					06	06	08
If a station is not designated for a subordinate unit, it is co-stationed with its command unit							

Chapter 6: Demilitarized Zone

1957-1965

The concept of pentomic divisions was developed in 1956 to meet the needs of nuclear battlefields. The goal was to field highly mobile divisions with great fire power and supported by state-of-the-art communication systems and logistics. A division was composed of a maximum of 13,500 personnel assigned to five battle groups. Each battle group was a self-contained force trained to conduct independent operations when necessary. Specialized firepower support was provided by artillery and missile units armed with conventional ordinance and nuclear warheads.

On 15 October, in ceremonies held in Tonggu, Korea, the colors of the 24th Infantry Division were retired and the colors of the 1st Cavalry Division were passed to the Commanding General of the former 24th Infantry Division, Major General Ralph W. Zwicker. The First Team had returned, standing ready to defend Korea against Communist aggression.

In the reflagging and organization of the 24th Infantry Division to the 1st Cavalry Division under Pentomic TO&E 7T, the following subordinate units were activated, organized and assigned as organic elements of the Division. Open, activated, elements of the 1st Cavalry Division TO&E were filled by the transfer of equipment assets and specialized trained personnel resources from units of the 24th Infantry Division.

SEE FOLLOWING PAGES (67-68)

Staring Across The DMZ Center Line

A squad from "B" Company, 8th Cavalry Regiment training in an assault boat on the Imjin River.

Concurrent with the reorganization and reflagging of the 1st Cavalry Division to the pentomic concept, the 545th MP Company, the 61st, 77th, 82nd, and 99th Field Artillery Battalions, the 26th and 29th AAA Battalions and the 70th Tank Battalion, which had served so nobly in Korea and Japan, were inactivated and relieved from assignment to the 1st Cavalry Division.

The 1st Cavalry Division took over the facilities of a former 24th Infantry Division Headquarters Compound located in the western defense corridor at Bong il Chong in the Paju City area. Previous tenants, the 1st Marine Division, had relocated the summer villa and converted the rice patties at the entrance to the valley into an attractive lake. By the time the 1st Cavalry Division arrived, they were able to be billeted in permanent Quonset huts which had been constructed during a major program to improve the living conditions of the troops.

The redesignated and reorganized 1st Cavalry Division was assigned the mission of patrolling the "Freedom's Frontier" Demilitarized Zone (DMZ). The Order of Battle under the Pentomic System consisted of five Battle Groups per division and four rifle companies per group. Because of its extended area of operation, the 1st Cavalry Division was allowed the implementation of five Battle Groups, the 4th, 5th, 7th, 8th and 12th Cavalry, and a reconnaisance squadron.

Transfer of Equiptment Assets and Specialized Trained Personnel

1st CAVALRY DIVISION UNIT	RESOURCES UTILIZED
Hq & Hq Co, 1st Cavalry Division	Hq & Hq Co, 1st Cavalry Division
2nd Battle Group, 4th Cavalry Hq & Hq Co, 2nd Battle Group Mortar Btry, 2nd Battle Group "A", "B", "C" & "D" Cos, 4th Cavalry	Activated "B" Trp, 4th Cavalry Regiment Activated Activated
1st Battle Group, 5th Cavalry Hq & Hq Co, 1st Battle Group Mortar Btry, 1st Battle Group "A", "B", "C" & "D" Cos, 5th Cavalry	Activated "A" Co, 5th Cavalry Regiment Activated Activated
1st Battle Group, 7th Cavalry Hq & Hq Co, 1st Battle Group Mortar Btry, 1st Battle Group "A", "B", "C" & "D" Cos, 7th Cavalry	Activated "A" Co, 7th Cavalry Regiment Activated Activated
1st Battle Group, 8th Cavalry Hq & Hq Co, 1st Battle Group Mortar Btry, 1st Battle Group "A", "B", "C" & "D" Cos, 8th Cavalry	Activated "A" Co, 8th Cavalry Regiment Activated Activated
2nd Battle Group, 12th Cavalry Hq & Hq Co, 1st Battle Group Mortar Btry, 1st Battle Group "A", "B", "C" & "D" Cos, 12th Cavalry	Activated "B" Co, 12th Cavalry Regiment Activated Activated
Division Artillery Hq & Hq Btry, 1st Cavalry Division Artillery 2nd Bn, 19th Artillery (105mm) Hq & Hq Btry, 2nd Bn, 19th Artillery "A", "B", "C", "D", "E" & Svc Btrys, 2nd Bn, 19th Arty 2nd Bn, 20th Artillery (Rkt/How) Hq & Hq Btry, 2nd Bn, 20th Artillery "A", "B", "C", "D" & Svc Btrys, 2nd Bn, 20th Arty	 Hq & Hq Btry (Reorg) Activated "B" Btry, 19th FA Activated Activated "B" Btry, 20th FA Activated

Continued on next page

Transfer of Equipment Assets and Specialized Trained Personnel

1st CAVALRY DIVISION UNIT Cont.	RESOURCES UTILIZED Cont.
Division Trains	
Hq & Hq Detachment, Trains	Hq & Hq Co (Reorg)
Division Band	Band, 1st Cavalry Division (Reorg)
15th Administrative Service Co	15th Replacement Co
15th Aviation Company	Activated
15th Medical Battalion (Inf)	15th Medical Bn (Reorg)
Hq & Hq Det	Hq & Hq Co (Reorg)
"A" Co (Ambulance)	Ambulance Co (Reorg)
"B" Co (Clearing)	Clearing Co (Reorg)
15th Quartermaster Company (Inf)	15th Quartermaster Co (Reorg)
23rd Transportation Battalion (Inf)	Activated
Hq & Hq Co	Activated
"A" Co (Truck)	Activated
"B" & "C" Cos (Armed Carrier)	Activated
27th Ordnance Battalion (Inf)	27th Ordnance Bn
Hq & "B" Co (Main Spt)	Hq & Hq Det & "B" Co (Reorg)
"A" Co (Fwd Spt)	"A" Co (Reorg)
Separate Battalions & Companies	
Hq & Hq Co, 1st Cavalry Division	Hq & Hq Co, 1st Cavalry Division
1st Recon Sqdn, 9th Cavalry	Activated
Hq & Hq Trp	"A" Trp, 9th Cav & 16th Recon
"A", "B", & "C" Trps	Activated
3rd Medium Tank Bn, 40th Armor	Activated
Hq & Hq Co	"C" Co, 40th Armored Rgt
"A", "B", "C", "D", & "E" Cos, 3rd Med Tank Bn	Activated
8th Engineer Nattalion (Inf)	8th Engineer battalion (C)
Hq & Hq Co	Hq, Hq & Sv Co
"A", "B", "C", "D" & "E" Cos	"A", "B", "C", "D" Cos (Reorg) & "E"
13th Signal Battalion (Inf)	13th Signal Battalion (C)
Hq & Hq Co	13th Signal Co (Reorg)
"A" Co, Command Operations	Activated
"B" Co, Forward Communications	Activated
DMZ Police Company (Provisional)	Activated

Parade Ground Lake (1957)

Command Post DMZ Police Company

An L-19 "Bird Dog" from the 15th Aviation Company sets out on a mission of en route reconnaissance and convoy control.

Howitzer crew, of "B" Battery, 2nd Howitzer Battalion, 19th Field Artillery preparing to fire a 155mm Howitzer during Exercise "Trooper Turnout". (19 September 1960)

Entrance of Observation Post MAZIE, 1st Reconnaissance Squadron, 9th Cavalry Regiment (10 February 1960)

Each group operated as a self-contained independent force, the isolation of each much like the cavalry forts during the early Indian Wars.

At that time, the 1st Cavalry Division was the only US division in direct contact with the enemy. If attacked by North Korea, there was a last line of defense plan in place that specified the amount of time (4 hours) each element had to hold its position and fight in place before being relieved. The objective was that the 1st Cavalry Division had to hold its position for 24 hours while the 7th Infantry Division (located south of the DMZ) could be moved up for support. Additionally airborne and US Marine units from Okinawa, the 24th and 25th Infantry Divisions from Hawaii along with the 4th Infantry Division at Ft. Lewis, Washington would be deployed immediately should the situation warrant.

The 1st Cavalry Division spent the majority of its time in field operations patrolling along the southern border of the DMZ itself and adjacent areas in observation and listening posts which were manned 24 hours a day. During a typical thirteen month tour, troopers spent 193 days in the field. Two weeks of day time patrol was followed by two weeks of night time patrol. In addition, combat patrols searched the surrounding areas at night. Other assignments included the building and maintenance of defense positions (trenches and bunkers) and training exercises. Training remained a number one priority for the troopers and unit commanders.

In January 1958, the largest training exercise in Korea since the end of hostilities, Operation SNOWFLAKE, was conducted. US troops, combined with the Republic of Korea, Turkish and other UN units participated in maneuvers in deep snow and sub-zero weather. In May, this exercise was followed by Operation SABER which was designed to test the tactical, logistic and administrative condition of the Division. In August, Operation HORSEFLY utilized long range troop airlifts and a combined infantry tank force in an attack on an "enemy" airfield.

One of the most unique, and certainly most outstanding unit of the 1st Cavalry Division was the Demilitarized Zone Police Company (Provisional). The 150 officers and troopers of the company had the important mission of maintaining law and order in the United Nations Command section of the demilitarized zone forward of the Division sector. In addition, the unit had responsibility for civil control of Tae Song Dong, a small village located in the DMZ, which was under control of the United Nations.

An all-volunteer outfit, the company, composed of a headquarters platoon and three line platoons, performed its mission 24 hours a day, seven days a week, by manning observation posts and patrolling in and along the demilitarized zone. One platoon would cover the zone in the daytime and two were assigned to night operations. A patrol consisted of three men, carrying the equipment that they needed for their specific mission. They always reported in every thirty minutes. Diligent performance was achieved by screening applicants for ability, experience and intelligence. Duty was rugged, but high spirits were maintained by the keen sense of mission importance shared by members of the unit.

Unlike most of the other subordinate units of the Division, The DMZ Police Company had never been in combat nor did it have a long history. However, in its

South Check Point at Freedom Gate Bridge. One of the two bridges crossing the Imjin River into the area of the Demilitarized Zone patrolled by the 1st Squadron, 9th Cavalry. All traffic, from both directions on the one way bridge, must be checked for clearance to pass. (5 December 1962)

Historic Freedom Gate Bridge, built as a high level bridge across the Imjin River by the 84th Engineers, 2nd Engineer Gp, still stands as a memorial to all the prisoners who were returned at the end of hostilities.

DMZ Treaty Village, Today (NK Side)

short life span, it constantly performed one of the most important and toughest duties of the Free World defense. In 1959, the DMZ Police Company was inactivated and its security mission, along with assigned personnel, was transferred to the 1st Reconnaissance Squadron, 9th Cavalry.

In 1959, the 1st Cavalry Division made several modifications to its headquarters installation and filled in the lake at the entrance to create a parade ground. It was named "Brown Field" in honor of PFC Melvin L. Brown, "D" Company, 8th Engineers, 1st Cavalry Division, who was posthumously awarded the Medal of Honor for heroic actions on 04 September 1950. After being designated a headquarters compound for many years, the installation was redesignated "Camp Howze" on 25 March, 1960 in honor of Major General Robert L. Howze, Medal of Honor recipient and the 1st Commander of the 1st Cavalry Division from 1921 to 1925.

In 1962 the 1st Cavalry Division was the first division to be totally equipped with M14 rifles, M60 machine guns, M79 grenade launchers, Claymore mines, and Huey helicopters for "dust off" medical evacuations. In the spring of 1963, units received H-19, H-21 and H-37 helicopters. The Division conducted extensive air assault familiarization and training. In order to recruit potential helicopter pilots, personnel records were screened to determine individual results of initial induction testing. All troopers who had scored well were asked to volunteer for helicopter training. After passing a flight physical, all who volunteered (regardless of time remaining on their tour of duty) were transferred to Ft. Rucker, Alabama for pilot training.

Duty on the DMZ was hazardous and "alerts" were continuous. On 23 November 1962, "A' Troop, 9th Cavalry, manning Outpost Susan, was attacked by a barrage of grenades, that killed one trooper and wounded another. On several occasions in 1962 and 1963, troopers were attacked by marauding North Korean soldiers. Less than a year later, on 29 July 1963, a jeep patrol of the 9th Cavalry was ambushed, killing two of the unit. In the resulting pursuit of the raiders, another trooper was killed.

The 1st Cavalry Division remained headquartered at Camp Howze until 01 July 1965 when it was reorganized as the 1st Cavalry Division (Airmobile), officially activated at Ft. Benning, Georgia and prepared for a new mission. It was made up of personnel and equipment resources of the 11th Air Assault Division (Test) and brought to full strength by transfer of specialized elements of the 2nd Infantry Division. One month later, the 1st Cavalry was enroute to Vietnam to prove the effectiveness of airmobile operations. The personnel and equipment of the former 1st Cavalry Division, which remained in Korea, were reorganized and redesignated as the 2nd Infantry Division. They assumed the continuing mission of protecting the DMZ, where they remain on duty today.

NOTE - In July 1953 fighting was stopped by the armed truce. However, North Korea and the UN Command have remained officially in a state of war for forty-five years, signified by the fact that over 1,000 UN personnel have been killed while serving along the DMZ. As of today, because of communist obstructionist tactics, years have gone by and no peace treaty has ever been agreed to and signed. An ever present "alert" status is in effect, because of the presence of a North Korean military force of 1.1 million troops stationed near of the Demilitarized Zone. Across the DMZ, a South Korean force of 660,000 troops supported by 40,000 American soldiers stationed in the area confront the North Koreans.

On 15 June 2000, the first major breakthrough leading to more stable, peaceful conditions in the Korean Peninsula came about by the signing of a landmark agreement committing the Koreans to work toward reunification, to allow the reunions of families separated during the Korean War and to improve economic cooperation between the two nations. While the joint communique was short of specifics, it signaled that the leaders, South Korean President, Kim Dae-jung and North Korean leader, Kim Jong Il hoped to put relations on the Korean Peninsula on an even footing

H-37 Heavy Lift Helicopters

1st Battle Group, 8th Cavalry patrols protective barrier in Operation DORT. (April 1964)

A squad from "C" Troop, 1st Recon Squadron, 9th Cavalry receives a final briefing before moving out to stakeout positions.

and raised hopes for the eventual reunification of the nations. Later, South Korean officials outlined plans for implementing the historic agreement — including the creation of a military hotline and building a railway crossing at the nations' heavily patrolled border.

However, Kim Jong Il and Kim Dae-jung did not reach a consensus on two other major issues: 1) The status of nearly 40,000 American soldiers stationed in South Korea, and 2) the long range and nuclear missile programs of North Korea.

Chapter 7: Ft. Benning - Airmobile

1965

It was not one man's dream; the notion of using helicopters as the fast, new steed of modern cavalry. The efforts of many went into the formation of the first air assault division. More than a decade of flying experiments, equipment evaluations and development of new hardware and tactics preceeded the conversion of the 1st Cavalry Division to the Air Assault configuration. Tests were conducted with troop carrying helicopters and those equipped with a variety of machine guns and rockets.

At Ft. Benning, Georgia, in the mid-fifties, General John R. Tolson developed a tactical doctrine for the use of helicopters in combat and became one of the leading, early pioneers in modern airmobility. Several other military personnel played key roles in development of missions, configurations, and weapons for air cavalry operations. Lt. General James M. Gavin, Chief of Staff for Operations of the Department of the Army, campaigned for helicopter units that could take over, extend and add great speed to the traditional roles of the cavalry.

Soon thereafter, at the Aviation School at Ft. Rucker, Colonel Jay D. Vanderpool assembled a "Sky Cavalry" platoon using borrowed personnel and equipment. The "Sky Cavalry" began giving impressive demonstrations of the roles that helicopters could play in combat. Heralding the era of gunships, the "Sky Cavalry" carried and tested a variety of weapons and rockets.

Inpetus for exploring airmobility concepts was generated by the Rogers and Howze Boards. The main mission of the Rogers Board was to review aircraft requirements of the Army in three categories: observation; surveillance and transportation. The findings of the Rogers Board were fine as far as they went, but Robert S. McNamara believed there was more promise in air mobility than the Rogers Board revealed. In the spring of 1962, Secretary McNamara ordered a panel of distinguished military leaders and civilian experts to re-examine the needs of the Army. This panel became known as the Howze Board and it was granted sweeping powers to conduct tests, stage war games and study combat in various areas of the world.

The findings of the Howze Board outlined the requirements for an air assault division. Such a unit would have five times the number of aircraft of a regular division along with an extensive reduction in ground vehicles. After the report was issued, McNamara gave orders to field the air assault concept and run it through its paces in a tough series of field tests which would further reveal the strengths and limitations of an air assault force. The unit chosen for the exercise was the 11th Airborne Division, which was reflagged as the 11th Air Assault Division. It was commanded by Major General Harry William Osborne Kinnard.

The first division-wide test, AIR ASSAULT I, was conducted in September and October 1963 near Ft. Stewart, Georgia. Four major exercises tested the mettle of the new air assault division. Experiments and field tests continued. Finally in the summer of 1964, the division faced its toughest trial in AIR ASSAULT II", an extensive two-sided maneuver against the reinforced 82nd Airborne Division.

Troubling events in Vietnam accelerated the decision to convert the 11th Air Assault (Test) to a combat division.

A 2.75 inch rocket is loaded into an experimental weapon system mounted on a UH-1B, an early model of the soon to become famous "Huey", for a demonstration at Ft. Stewart GA.

Troopers board helicopters for the final phases of the airmobile field tests that were conducted at Ft. Benning, GA and Ft. Rucker, AL. prior to the deployment of the Division to Vietnam.

The birth of the 1st Cavalry Division (Airmobile) closed out the actions and accomplishments of 11th Air Assault Division (Test). On 03 July 1965 the changing of the "colors" took place at Doughboy Stadium, Ft. Benning, GA. As the colors of the 11th Air Assault Division were cased, inactivating the Division, the colors of the 1st Cavalry Division were brought onto the field and were passed to the commander of the former 11th Air Assault Division (Test), Major General Harry W. O. Kinnard. The "airmobile" concept had become a reality.

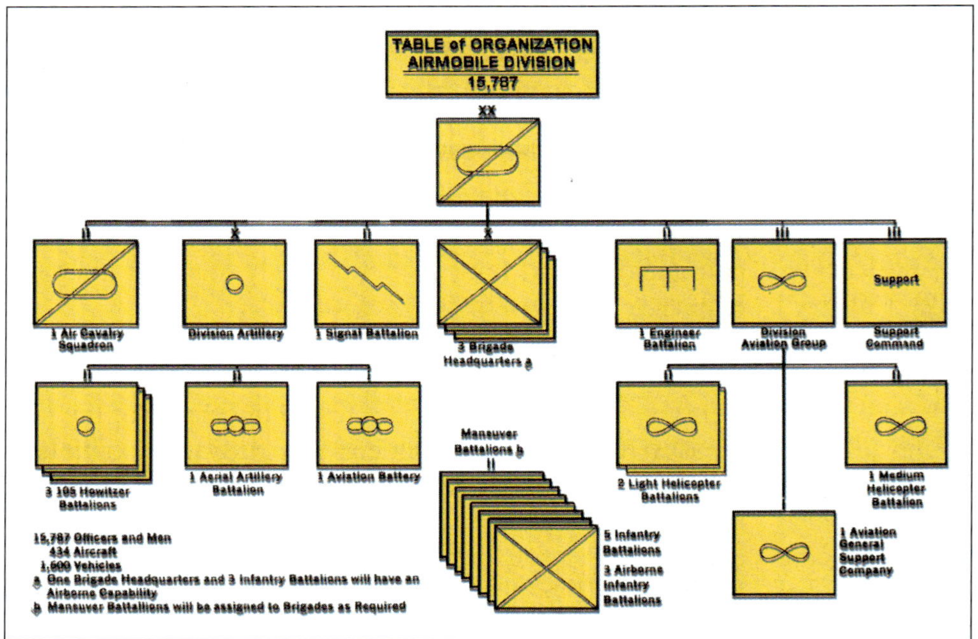

The Organizational Chart of the 1st Cavalry Division (Airmobile) as officially activated pursuant to General Order 185, Headquarters Third US Army. Assets and personnel were drawn from the resources of the 11th Air Assault Division (Test) and the 2nd Infantry Division.

The decision was initially made in March 1965. It was also decided that the new division would be formed around the colors of the 1st Cavalry Division. On 16 June, Secretary of Defense Robert S. McNamara made a public announcement that the Department of the Army had been given the green light to organize an air assault division. McNamara also stunned many observers, when he declared that the new division would be combat ready eight weeks after its organization.

The colors of the 1st Cavalry Division returned from Korea in 1965, but this new First Team was around only long enough to be reorganized and be re-equipped for a new mission. On 01 July 1965, the 1st Cavalry Division (Airmobile) was officially activated by General Order 185, Headquarters Third US Army, and was made up of the resources of the 11th Air Assault Division (Test), the 10th Air Transport Brigade and the 2nd Infantry Division.

The reorganized and redesignated subordinate units of the 1st Cavalry Division (Airmobile), in formation on the parade ground, had been restructured by the transfer of the equipment assets and specialized trained personnel of the following units:

SEE FOLLOWING PAGE

On 28 July, the First Team achieved REDCON-1, a readiness rating of highest combat priority and on that same day President Lyndon Johnson announced to the world: "I have today ordered to Vietnam the AirMobile Division." Work accelerated to a torrid pace as the First Team prepared to leave for its historic deployment. Troopers turned in their M-14 rifles and qualified on the new M-16E1. Specialized training was stepped up. The Airborne School at Ft. Benning graduated 659 new paratroopers for the division in just 10 days, which allowed the round out of the new airborne brigade which had just been approved.

The Aviation School at Ft. Rucker, Alabama held a special transition class to teach 89 aviators the skills that they would need to fly the UH-1 Huey helicopter. In addition, another 120 aviators were introduced to the aerial weapons firing course at Ft. Rucker.

In order to subdue the white name tags on faded fatigues, troopers used dark green dye and the legend has it that the Chattahoochie River ran dark green from the runoff of the dye used to camouflage the bright yellow, position revealing patch. General Kinnard, enroute to Vietnam, made a stopover in Japan to order a supply of black and dark green patches and insignia. This first use of a subdued rank and unit insignia set a precedent for all units of the Army to use.

In a matter of three and one-half weeks the newly formed units, organized into a 16,000 man division along the lines of the 11th Air Assault Division (Test), with a total of 434 helicopters, was prepared to enter combat, the ultimate test of its capabilities. Actual movement of personnel was called Operation PAT and was scheduled in three increments; an

1st CAVALRY DIVISION UNIT	RESOURCES UTILIZED
"HHC" 1st Cavalry Division, (Airmobile)	"HHC", 11th Air Assault Division (Test)
Division Reconnaissance 1st Squadron, 9th Cavalry	3rd Squardon, 17th Cavalry
"HHC", 1st Brigade, 1st Battalion (Airborne), 8th Cavalry 2nd Battalion (Airborne), 8th Cavalry 1st Battalion (Airborne), 12th Cavalry	"HHC", 1st Brigade, 11th Air Assault Division (Test) 1st Battalion, (Airborne), 188th Infantry 1st Battalion, (Airborne), 511th Infantry 1st Battalion, (Airborne), 187th Infantry
"HHC", 2nd Brigade, 1st Battalion, 5th Cavalry 2nd Battalion, 5th Cavalry 2nd Battalion, 12th Cavalry	"HHC", 2nd Brigade, 2nd Infantry Division 1st Battalion, 38th Infantry 2nd Battalion, 38th Infantry 1st Battalion, 23rd Infantry
"HHC", 3rd Brigade, 1st Battalion, 7th Cavalry 2nd Battalion, 7th Cavalry 5th Battalion, 7th Cavalry **(Note 1)**	"HHC", 3rd Brigade, 2nd Infantry Division 2nd Battalion, 23rd Infantry 2nd Battalion, 9th Infantry 1st Battalion, 11th Infantry **(Note 1)**
"HHB", Division Artillery 2nd Battalion, 19th Artillery (105mm) 2nd Battalion, 20th Artillery (AR) 1st Battalion, 21st Artillery (105mm) 1st Battalion, 77th Artillery (105mm) "E" Battery, 82nd Artillery (AVN)	"HHB", Division Artillery, 11th Air Assault 6th Battalion, 81st Artillery (105mm) 3rd Battalion, 377 Artillery (AR) 5th Battalion, 38th Artillery (105mm) 1st Battalion, 15th Artillery (105mm)
"HHC" & Band, Support Command 15th Medical Battalion 15th Supply & Service Battalion Aerial Eqpm't Sup't Co (Airborne) 15th Administrative Company 27th Maintenance Battalion 8th Engineer Battalion 13th Signal Battalion 15th Transportation Battalion 545th Military Police Company 191st Military Intelligence Detachment 371st Army Security Agency Company	"HHC" & Band, 11th Air Assault Division Support Command 11th Medical Battalion 408th Supply & Service Battalion 165th Aerial Eqpm't Supt Detachment 11th Administrative Company 711th Maintenance Battalion 127th Engineer Battalion 511 Signal Battalion 611th Aircraft Maintenance & Supply 11th Military Police Company 11th Military Intelligence Detachment "C" Co, 313th Army Security Agency Battalion
"HHC", 11th Aviation Group (Airborne) "HHC", 227th Aviation Battalion "A" Co, 227th Aviation Battalion "B" Co, 227th Aviation Battalion "C" Co, 227th Aviation Battalion "D" Co, 227th Aviation Battalion "HHC", 228th Aviation Battalion "A" Co, 228th Aviation Battalion "B" Co, 228th Aviation Battalion "C" Co, 228th Aviation Battalion "HHC", 229th Aviation Battalion "A" Co, 229th Aviation Battalion "B" Co, 229th Aviation Battalion "C" Co, 229th Aviation Battalion "D" Co, 229th Aviation Battalion 11th Aviation Company	"HHC", 11th Aviation Group (Test) "HHC", 227th Aviation Battalion "A" Co, 227th Aviation Battalion Aviation Co, 6th Special Forces Group Aviation Co, 7th Special Forces Group 110th Aviation Co, (Aerial Weapons) "HHC", 228th Aviation Battalion 132nd Aviation Company (ASH) 133rd Aviation Company (ASH) 202nd Aviation Company (ASH) "HHC, 229th Aviation Battalion "A" Co, 4th Aviation Battalion "A" Co, 5th Aviation Battalion 194th Aviation Company (AH) 131st Aviation Company (AW) 11th Aviation Company (General Supt)

Note 1:
On 01 April 1966, the 5th Battalion, 7th Cavalry was activated at Ft. Carson, Colorado, using the personnel and assets of the 1st Battalion, 11th Infantry. The advance party departed Peterson Field on 08 August and arrived Qui Nhon on 11 August. The main body departed Ft. Carson on 06 August and arrived Qui Nhon on 19 August 1966.

Advance Liaison Detachment, an Advance Party and the Main Body.

The first echelon to depart was the Advance Liaison Detachment composed of 32 officers and men. Led by Brigadier General John S. Wright, the detachment left on 09 August and arrived in Vietnam two days later. Beginning on 14 August, for six days, the Advance Party of 1,040 officers and men left Ft. Benning and, along with their equipment cargo of 132 tons, including nine UH1-B helicopters moved to Warner Robins Air Force Base at Macon, Georgia. From there, they were deployed by C-124 and C-130 aircraft of the Military Airlift command. Flying via Travis Air Force Base, California, Hickman Field, Hawaii and Clark Air Force Base, Philippines, they arrived in intervals at Nha Trang between the 19th and 27th of August 1965. Joining with the Advance Liaison Detachment, they established a temporary base camp near An Khe, 36 miles inland from the coastal city of Qui Nhon.

The third and final phase of troop and supply movement centered on the bulk of the 1st Cavalry Division departing Ft. Benning, Georgia and deploying by troop and cargo ships of the Military Sea Transport Service. Approximately 13,500 men and their cargo left Columbus, Georgia by train and bus to staging areas at port cities in the Southeastern United States. Six troop ships, seven cargo ships and four aircraft carriers were employed in the over water movement. The 1st Brigade loaded on the *Unites States Navy Ship (USNS) Geiger*, the 2nd Brigade and an Artillery Battalion loaded on the *USNS Buckner* and the 3rd Brigade on the *USNS Rose*. The remainder of the Division, including elements of the Support Command, Aviation Group and the various combat support units loaded on the *USNS Darby*, *Patch* and *Upshure*. On 16 August, the *USNS Buckner* and *USNS Darby* departed Charleston, South Carolina. The other four troop ships departed over the next four days from Charleston, South Carolina and Savannah. Georgia.

The 470 aircraft of the division were loaded on the carriers *USNS Kula Gulf*, *USNS Croaton*, *USNS Card* and the *USNS Boxer*. The *Boxer*, carried a total of 239 aircraft; fifty seven Chinooks (CH-47s), four Flying Cranes (CH-54s), six Mohawks (OV-1s), fifty Hueys (UH-1s), and one hundred twenty-two Sioux (OH-13s). On 11 August, the first of the carriers, the *USNS Croaton* sailed from Mobile, Alabama. The remaining three departed on subsequent days. The *USNS Card* left Mobile, Alabama and the *USNS Boxer* and *USNS Kula Gulf* sailed from the Mayport Naval Station at Jacksonville, Florida. The sea movement of the echelon traversed the world, moving both west and east. The *USNS Boxer* sailed east, moving through the Suez Canal. The other aircraft carriers and troop ships sailed west through the Panama Canal, stopping at Hawaii, Guam and the Philippines.

The USNS CROATON waiting at the dockside at Mobile, Alabama port for troops from the 1st Cavalry Division who are destined for South East Asia and Vietnam. (13 August 1965)

On 03 September, following the departure of the sea movement, a second echelon of the Main Body departed Ft. Benning, Georgia. The 18 Caribou aircraft of the 17th Aviation Company took off from Lawson Army Airfield enroute to Hamilton Air Force Base, California. The first, and longest leg of the trip took an average of 17 hours and was completed without an incident. On 06 September, the aircraft departed for Southeast Asia with scheduled refueling and maintenance stops at Hawaii, Wake Island, Guam and the Philippines, arriving at Vung Tau on 18 September. After a quick refueling stop, they moved on to An Khe.

Within a month of becoming the Army's first air mobile division, the First Team was on its way to combat as the first fully committed division of the Vietnam War.

The USNS BOXER sailing from Mayport Naval Station at Jacksonville, Florida port with elements of the 1st Cavalry Division who were destined for Vietnam.

Chapter 8: Vietnam War

1965-1972

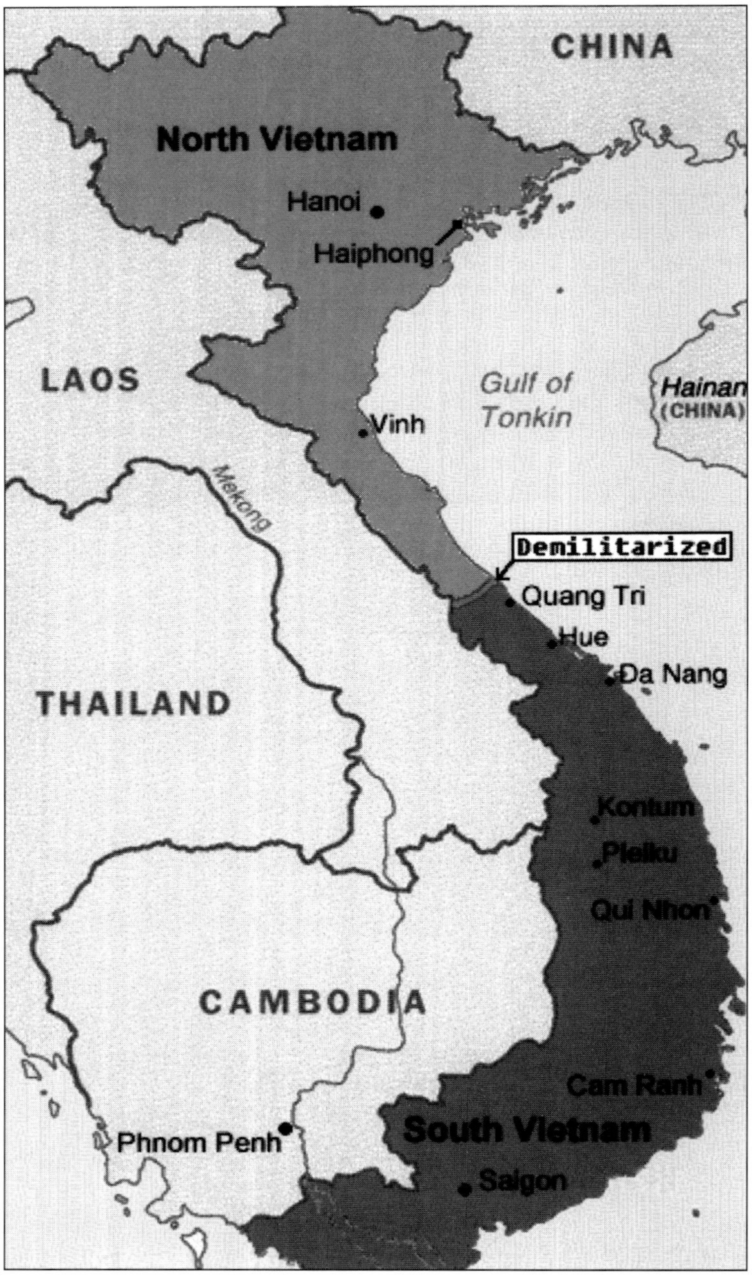

"What is happening in South East Asia today is not a local, temporary or isolated situation. It is part of the rivalry of power and the ideological contest which is taking place throughout the world. It is part of the stream of events continuing into the future. In both of those contests the most significant factor in Asia is China".

"In Vietnam, a Communist government has set out deliberately to conquer a sovereign people in a neighboring state by a carefully planned program of concealed aggression ... as real as that of an invading army. The war in Vietnam is not a spontaneous and local rebellion against the established government". (US State Department - 1965)

On 16 August, an advance party from "C" Company, 8th Engineer Battalion, on board C-130s, arrived at Nha Trang. By 21 August, the remainder of the advance party had arrived. They joined with advance liaison forces to establish a temporary base camp near An Khe, in the Central Highlands of southwestern Binh Dinh Province, thirty-six miles inland from the coastal city of Qui Nhon.

The focus of the facilities of the camp was on the helipad which needed to be built. It would become the biggest helipad in the world and would soon become the hub of activity for the 1st Cavalry Division. Surrounded by flat terrain, except for Hon Cong Mountain on the western perimeter, the area was covered by a dense undergrowth, bamboo trees, thorn thickets and ant hills twelve feet high. It would be cleared and contoured smooth as a "golf course" without the use of bulldozers or power equipment. The use of earth moving equipment in its construction would have stripped the land of its protective natural grasses, creating a vast dust bowl or a gigantic mud pie, depending on the season.

The major responsibility of constructing the support facilities required around the "golf course" and improving the airstrip at An Khe for C-130 operations was assigned to the 70th Combat Engineer Battalion. This unit was attached to the 1st Cavalry Division along with its subordinate units, "B" Company, 84th Construction Engineer Battalion, the 511th Engineer Company (Panel Bridge) and the 630th Engineer Company (Light Equipment). With an expansive area to protect, the sprawling base was designed with a surrounding perimeter defense area known as the Green Line.

Events surrounding base camp construction would lead to the camp being officially designated as Camp Radcliff on 21 February 1966, in honor of the first casualty of the 1st Cavalry Division in Vietnam. In August 1965 Major Donald G. Radcliff was the executive officer of the 1st Squadron, 9th Cavalry, 1st Cavalry Division, and a member of the site selection team that scoured the countryside around Binh Dinh Province to find the ideal location for the base camp of the 1st Cavalry Division.

When the site selection team was advised that the 7th Marines were planning a major strike against the enemy, Major Radcliff volunteered to fly a mission in support of Marine troop lifts. Intelligence indicated that the 1st Viet Cong (VC) Regiment was massing for an attack on the

Marine base at Chu Lai in Quang Tin Province. Rather than prepare defenses and brace for the attack, the Marines decided to meet the enemy on their own terms and launch a preemptive attack, code named Operation STARLITE. Operation STARLITE, the largest planned US military operation to that time, was to be a combined amphibious/ air assault operation against the VC Regiment twelve miles south of Chu Lai. The assault included two amphibious landing sites and three helicopter landing zones named Landing Zone (LZ) Red, LZ White and LZ Blue.

At dawn on 18 August 1965, the quiet shoreline of southern Quang Tin Province suddenly erupted in a volley of explosions from artillery and offshore guns, followed by massive aerial bombardment. At 0630 hours the Marines hit the beaches while an armada of helicopters swooped in from the west. The Marines encountered little resistance on the coast and started their march inland. The troops arriving at LZ Red met almost no resistance and disembarked without an incident. At LZ White the Marines drew fire from a nearby ridge line but managed to land and clear the area quite readily.

LZ Blue, however, was a different story. Major Radcliff was piloting a UH-1B helicopter gunship escorting the LZ Blue airmobile assault. Unknown to the trooplift, the landing zone was surrounded by the 60th VC Battalion, lying in wait. As the aircraft arrived at the landing zone, Radcliff realized that the lead troop-carrying helicopter was the target of heavy automatic weapons fire. He immediately pinpointed the VC position and placed accurate, devastating, suppressive fire on the opposing enemy forces. With his quick reaction, Major Radcliff saved countless lives and enabled the troop transport to land. As the troops deployed on the landing zone, Radcliff hovered nearby to insure their safety. The VC directed heavy fire at Major Radclift's helicopter, and as bullets tore through his aircraft, he was mortally wounded. The gallant, thirty-seven year old officer lost his life at the controls of his gunship during his baptism by fire in Vietnam.

General Harry W. O. Kinnard, the division commander, notified Lieutenant Colonel John B. Stockton of the loss of his executive officer. Stockton was on the *USNS Darby* with the main force of the aviation units of the 1st Cavalry. On 20 August 1965, although thousands of miles from the combat zone, the men and officers of the 1st Cavalry Division mourned the loss of their first comrade to fall in battle in memorial services on the deck of the *USNS Darby* as it passed through the Panama Canal.

For the Main Body of the division, deployed by sea, four weeks provided little idle or recreational time. The days were filled with additional training, preparation for jungle warfare and physical fitness drills. Weapons testing and familiarization continued from the aft decks of the *USNS Buckner* and *USNS Maurice Rose*, with homemade targets towed behind the ships. Two weeks before arrival in Vietnam, weekly doses of malaria suppressive tablets began — a routine that became a favorite method of counting off the passing months in the jungle.

On 05 September, the "Red Legs" of the Division Artillery sailed into Qui Nhon aboard the *USNS Upshur*. The Division Artillery consisted of the 2nd Battalion, 19th Artillery, the 2nd Battalion, 20th Artillery, the 1st Battalion, 21st Artillery and the 1st Battalion, 77th Artillery. Their first task was setting up unit headquarters at the division Base camp at An Khe.

On 09 September, the aircraft carrier *USNS Boxer* arrived with the giant CH-47 Chinooks of the 228th Helicopter Battalion which would be used in transporting much of the later arriving troops and cargo to An Khe. Landing preparation had begun several days before, at sea, with the partial removal of the cocooning materials. On 11 September, completely checked out, the first CH-47 took off from the deck. Sixty-nine hours later, the last CH-47 departed. Over the next week and a half, an organic air movement from Qui Nhon to the "golf course," involving twenty-one CH-47 Chinooks daily and over 1,100 flying hours, was completed on 22 September.

On 12 September the Division Support Command arrived in Vietnam, The Support Command units consisted of the 15th Medical Battalion, responsible for the health and life saving needs of the soldiers on and away from the battlefield; the 15th Transportation Corps Battalion, responsible for aircraft maintenance; the 15th Supply and Service Battalion, responsible for supplying everything from meals complete with toothpicks to gasoline; and the 27th Maintenance Battalion, responsible for performing timely maintenance of the ground vehicles and weapons of the Division. The functions of each unit were integrated into individual support organizations called Forward Service Support Elements (FSSE). Three of these elements were distributed throughout the operational areas of the 1st Cavalry Division.

Stepping foot on foreign soil created a mixture of emotions, relief and confidence. Those in the first landings knew one thing, the trek across the ocean was behind them and the "First Team" was ready to fight. (September 1965)

On 13 September, more elements of the 11th Aviation Group (Airmobile) assigned to the 1st Cavalry Division arrived in Vietnam. The 11th Aviation Group consisted of the 227th, 228th, and 229th Aviation Battalions and the 1st Squadron (Reconnaissance), 9th Cavalry Regiment. They were soon operating at full capacity in its missions of providing tactical mobility for combat troops and transporting equipment and supplies to units of the Division.

On 14 September, the 2nd Brigade disembarked from the troop ship *USNS Buckner* and marched ashore at Qui Nhon. The brigade consisted of three infantry battalions, the 1st and 2nd Battalions, 5th Cavalry and 2nd Battalion, 12th Cavalry. The Brigade loaded quickly into helicopters and moved inland to the main base camp of the Division at An Khe.

On 20 September, the 1st Brigade disembarked from the troop ship *USNS Geiger* and marched ashore at Qui Nhon. The brigade consisted of three airborne infantry battalions, the 1st and 2nd Battalions, 8th Cavalry and 1st Battalion, 12th Cavalry with the 2nd Battalion, 19th Artillery acting as the direct support unit. The Brigade loaded quickly into helicopters and moved inland to the main base camp of the Division at An Khe.

On 20 September, the 3rd Brigade disembarked from the troop ship *USNS Maurice Rose* and marched ashore at Qui Nhon. The brigade initially consisted of two infantry battalions, the 1st and 2nd Battalions, 7th Cavalry. (Note - The third element of the brigade, the 5th Battalion, 7th Cavalry, activated and assigned on 01 April 1966, would not arrive until 20 August 1966.) The Brigade loaded quickly into helicopters and moved inland to the main base camp of the Division at An Khe.

The Airmobile soldiers of the First Team had returned to war wearing the famous and feared patch of the 1st Cavalry Division. The First Team had prepared and was ready to enter its third war — and the longest tour of duty in combat history for any Army unit.

Now, the whole Division was on hand to establish the perimeter, to complete the clearing of the helipad and to erect buildings, tents, storage facilities and field fortifications. These efforts, undertaken in an oppressive climate, adverse jungle conditions and a hostile environment, were successful only through the intense and dedicated efforts of the troopers.

Troopers of the 8th Engineer Battalion received a priority project to level the summit of the adjacent Hon Cong Mountain that was designated as the location for the tactical communications hub of the Division. As soon as the dust settled from the blasting and clearing of the rock tip, the 13th Signal Battalion arrived with their equipment and began to set up operations.

One of the first assignments given to the 8th Engineer Battalion, in conjunction with the 13th Signal Battalion was construction of the signal complex. Access to the construction site was made possible by the use of CH-54 Flying Cranes to move the heavy earth moving equipment. All this was done while construction and upgrading of Camp Radcliff continued.

The date for the 1st Cavalry Division to officially assume complete control and responsibility for defense of An Khe and the surrounding Tactical Area of Responsibility (TAOR) was set for 28 September, however, the skytroopers wasted little time getting into action. The 1st Cavalry Division began its combat career on 18 September, when the 2nd Battalion, 12th Cavalry was put under the Operational Control (OPCON) of the 1st Brigade, 101st Airborne Division to assist in the initiation of Operation GIBRALTAR. This OPCON relieved the 1st Battalion, 327th Infantry, of the An Khe area defense responsibility to permit its commitment into the Vinh Thanh Valley, 10 miles northeast of An Khe. The 11th Aviation Group assembled all available aircraft to lift troops of the 101st Airborne Division into the operational area. "B" Battery, 1st Battalion, 77th Field Artillery was moved to support the operation while other batteries of that Battalion provided direct support for the 2nd Brigade in its assigned sector in the Base Camp defenses.

On 20 September the 2nd Battalion, 5th Cavalry replaced the 2nd Battalion, 12th Cavalry, assuming the OPCON responsibility to the 1st Brigade, 101st Airborne Division for its operational control for the An Khe base security. It remained until the 28th of September when the 1st Cavalry Division formally took over its TAOR.

On 10 October 1965, in Operation SHINY BAYONET, the First Team initiated their first brigade-size airmobile

Hon Cong Mountain Signal Complex In the background, a CH-47, Chinook, from "A" Company, 228th (ASHB) Aviation Battalion hovers overhead after delivering a load of equipment. (September 1965)

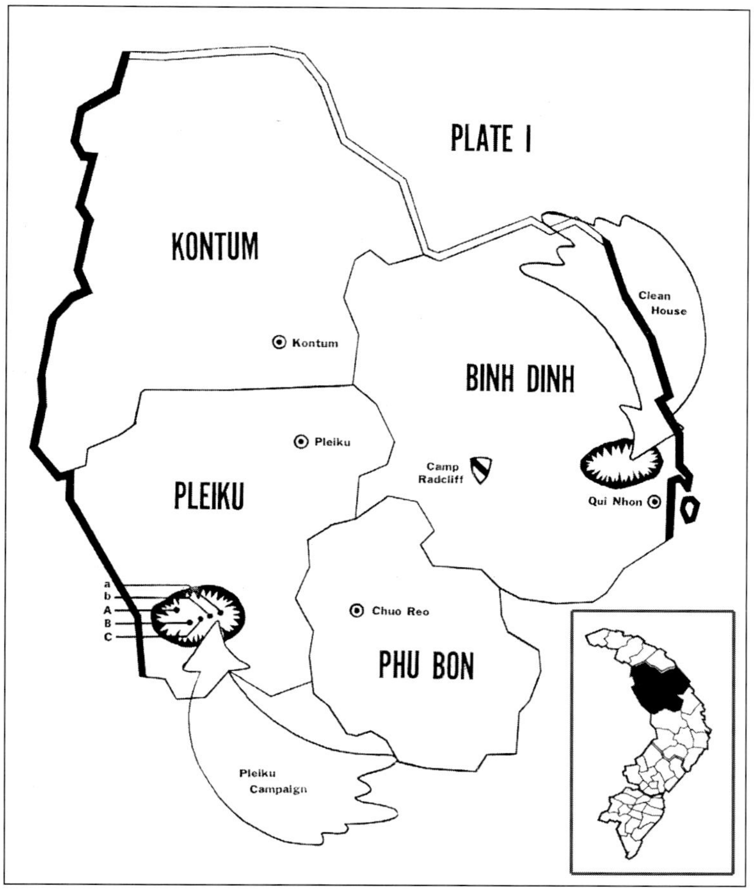

Pleiku Campaign Areas (19 October to 31 December 1965)

action against elements of the NVA 325th Infantry (Song Lo) Division located in the Soui Ca River Valley about 275 miles northeast of Saigon, near An Khe in Binh Dinh Province. Special agents and ARVN sources had located the base camp of two battalion of the Viet Cong 2nd Main Force Regiment. Operation SHINY BAYONET called for the ARVN 22nd Division to make initial contact and drive the VC toward the elements of the 3rd Brigade, 1st Cavalry Division who were deployed as a blocking force.

The air assault task force consisted of the 1st and 2nd Battalions, 7th Cavalry, 1st Squadron, 9th Cavalry, 1st Battalion, 12th Cavalry and the 1st Battalion, 21st Artillery. Rather than standing and fighting, the Viet Cong chose to disperse and slip away. Only light contact was achieved. The troopers had but a short wait before they faced a tougher test of their fighting skills; the 35-day Pleiku Campaign, Operation SILVER BAYONET.

In early October 1965 General Vo Nguyen Giap launched the full North Vietnamese Army Division concentrated in western Pleiku and nearby Cambodia, including the NVA 32nd, 33rd, and 66th Regiments, in a sweeping offensive against the thinly-guarded II Corp area west of Pleiku in the Central Highlands of South Vietnam. On 19 October, at Plei Me, in the western part of Pleiku Province, a strong force of NVA Regulars suddenly surrounded and besieged an outpost manned by US Special Forces and a South Vietnamese Civilian Irregular Defense Group (CIDG). This handful of soldiers and thirty miles of rugged, jungle terrain were all that stood between the onrushing NVA and their objective, the strategic town of Pleiku.

Putting the Plei Me camp under siege was a trap to lure a relief column into a massive ambush. The NVA 32nd Regiment was hidden all along the road leading from Pleiku to Plei Me, near the entrance to Ia Drang Valley. The enemy commanders thought that by smashing the relief column, the Americans and ARVN Forces remaining at Pleiku would not be strong enough to protect the other towns and villages of the Central Highlands. The ARVN intelligence had been able to intercept radio messages, decipher the scheme and establish an alternate counter attack. A column of South Vietnamese armor was dispatched to relieve the defenders at Plei Me. The ARVN armored column fought its way into Plei Me and two of the three enemy regiments broke contact and began to withdraw. Desperate to hold back any counterthrusts of Communist soldiers, the South Vietnamese II Corps commander requested American reinforcements.

On 23 October 1965, the 1st Cavalry Division, bivouacked at nearby An Khe, rushed to the aid of the beleaguered garrisons. Their first real combat test came at the historic order of General Westmoreland to send the First Team into an air assault mission, Operation SILVER BAYONET, to pursue and fight the enemy across 2,500 square miles of jungle. With its complement of over four hundred combat helicopters, the 1st Cavalry was completely air mobile. While the helicopter's mobility guaranteed quick movement and ready supply, it did not guarantee victory.

On 01 November, the first significant contact with the enemy was made when a rifle platoon of "B" Troop, 1st Squadron, 9th Cavalry overran a regimental aid station of the NVA 33rd Regiment six miles southwest of Plei Me, killing fifteen and capturing another fifteen. Two other rifle platoons from "B" Troop were landed to exploit the contact and began to sweep through the area. Soon after 1400 hours, scout helicopters reported a battalion-size force moving from the northeast. The fighting, at close quarters, was too close for rocket support from the Aerial Rocket Artillery, or for Tactical Air support. Reinforcements from the 1st

One of the many troop pick up zones utilized in the Pleiku Campaign. (25 October 1965)

An OH-13 observation helicopter, of the 1st Squadron, 9th Cavalry, lifts off for a reconnaissance mission near An Khe. By mid 1968, the OH-13 was being replaced by a more reliable, faster, maneuverable OH-6A (LOH), "Cayuse", better known as the "Loach". (November 1965)

and 2nd Battalions, 12th Cavalry and the 2nd Battalion, 8th Cavalry landed late in the afternoon.

Ground fire was intense on all reinforcement, resupply, and evacuation choppers, and seven ships were hit by hostile fire. By 1700 hours, "B" Company, 1st Battalion, 8th Cavalry joined in the battle. The enemy regiment was scattered and quickly smashed. The troopers inflicted many hundreds of casualties. This operation cost the NVA 33rd Regiment its aid station, many patients, and a significant amount of important medical supplies, as well as ninety nine killed and an additional 183 estimated wounded.

On 02 November, Division intelligence learned that a third regiment, the 66th had already arrived in South Vietnam and would move into assembly areas in the Chu Pong-Ia Drang area to join in the NVA Field Front. On the night of 3-4 November, rifle platoons of the 1st Squadron, 9th Cavalry again drew blood in a well executed ambush. A heavily laden NVA unit, estimated at company strength moving along an east-west trail, deciding to take a break just 100 meters short of an ambush site. The column loitered outside the killing zone for 90 minutes, while the troopers of the 9th Cavalry lay quietly in wait. At 2100 hours the NVA unit moved noisily along the trail westward. The first element was allowed to pass, and then the trap was sprung with eight Claymores along a 100 meter kill zone. It was perfectly executed and the weapons platoon of the enemy, with machine guns, mortars, and recoilless rifles was caught in a wall of lead as the 9th Cavalrymen fired continuously for two minutes. There was no return fire.

The troopers returned immediately to the patrol base and went to work to strengthen its perimeter. By 2230 hours the base perimeter was under heavy attack by an estimated two or three companies of NVA regulars. By midnight the perimeter was in grave danger of being overrun, but reinforcements of "A" Company, 2nd Battalion, 8th Cavalry, on stand-by alert at the Duc Co Special Forces (USSF) Camp, 12 miles of roadless jungle to the north, were alerted for commitment and were on their way. The operation was unique in that it was the first time a perimeter under heavy fire had been reinforced at night by heliborne troops air assaulted into an LZ that none of the assaulting forces had previously seen. It was also the first time that Aerial Rocket Artillery (ARA) had been employed at night within 50 meters of the friendly troops. By dawn the enemy attack had lost momentum, and contact diminished to occasional sniping from surrounding trees. The tangible results were 98 NVA killed by body count, ten were captured, along with two 82mm mortars, three 75mm recoilless rifles destroyed in place, as well as 100,000 rounds of 7.62 ammunition and large quantities of mortar and recoilless rifle ammunition evacuated. The implications of an ambush deep within what was expected to be secure territory must have stunned the NVA high command.

Within two days, on 6 November, Companies "B" and "C", 2nd Battalion, 8th Cavalry became heavily engaged with another NVA battalion located to the northeast of the Chu Pong Massif. Close-in fighting diminished as darkness fell and reinforcements were brought in. The enemy suffered another 77 dead, and many wounded were carried and dragged away from the battlefield.

Operation SILVER BAYONET was one the first battles of the 1st Cavalry Division that was closed out by heavy support from Operation ARC LIGHT, an Air Force strike force stationed at Anderson Air Force Base, Guam. Hundreds of escaping NVA troops died in the subsequent, blistering, precision bombing of the B-52s that moved across the ground like a giant carpet being unrolled. In the continuing three years, the B-52s dropped four times the bomb tonnage used in all of World War II on Vietnam.

On 09 November, after the completion of the search of the 1st Brigade throughout the operational area, the 3rd Brigade took control of Operation SILVER BAYONET. Five days later, on 14 November, the 1st Battalion, 7th Cavalry, reinforced by elements of the 2nd Battalion, air assaulted into the Ia Drang Valley near the Chu Pong Massif. During the insertion, with only two of the rifle companies on the ground, they were attacked by North Vietnamese regulars. Suddenly LZ X-Ray was "hot" from the start and only by heroic efforts of the lift ship pilots from the 229th Assault Helicopter Battalion were the remainder of the battalion and the first reinforcements from the 2nd Battalion, 7th Cavalry able to land.

At LZ X-Ray, the fighting was the most intensive combat in the history of the Division, from bayonets used in hand-to-hand combat, to artillery and tactical air support. During the engagement of the first day, the 1st Battalion, 7th Cavalry was reduced to approximately 340 officers and men; none missing. NVA casualties were much higher due to the awesome American fire support; six enemy were captured and evacuated.

On 14 November, the first day of battle at LZ X-Ray, Captain Edward W. Freeman, Alpha Company, 229th Assault Helicopter Battalion, 1st Cavalry Division (Airmobile) performed conspicuous acts of gallantry and intrepidity at the risk of his life and beyond the call of duty. As a flight leader and second in command of a 16-helicopter lift unit, he supported a heavily engaged American infantry battalion at Landing Zone X-Ray in the Ia Drang Valley, Republic of Vietnam. The unit was almost out of ammunition after taking some of the heaviest casualties of the war, fighting off a relentless attack from a highly motivated, heavily armed enemy force. When the infantry commander closed the helicopter landing zone due to intense direct enemy fire, Captain Freeman risked his own life by flying his unarmed helicopter through a gauntlet of enemy fire time after time, delivering critically needed ammunition, water and medical supplies to the besieged battalion. His flights had a direct impact on the battle's outcome by providing the engaged units with timely supplies of ammunition critical to their survival, without which they would almost surely have gone down, with much greater loss of life. After medical evacuation helicopters refused to fly into the area due to intense enemy fire, Captain Freeman flew 14 separate rescue missions, providing lifesaving evacuation-of an estimated 30 seriously wounded soldiers — some of whom would not have survived had he not acted. All flights were made into a small emergency landing zone within 100 to 200 meters of the defensive perimeter where heavily committed units were perilously holding off the attacking elements. Captain Freeman's selfless acts of great valor, extraordinary perseverance and intrepidity were far above and beyond the call of duty or mission and set a superb example of leadership and courage for all of his peers. Captain Freeman's extraordinary heroism and devotion to duty are in keeping with the highest traditions of military service and reflect great credit upon himself; his unit and the United States Army. For his valiant action, Captain Edward W. Freeman received the Medal of Honor.

It was during the battle at LZ X-Ray that "A" Company, 1st Battalion, 7th Cavalry was moving through the valley to relieve a friendly unit surrounded by an enemy force of estimated regimental size. First Lieutenant Walter J. Marm led his platoon through withering fire until they were finally forced to take cover. Realizing that his platoon could not hold very long, and seeing four enemy soldiers moving into his position, he moved under heavy fire and annihilated all four. Then, seeing that his platoon was receiving intense fire from a concealed machine gun, he deliberately exposed himself to draw its fire. Locating its position, he attempted to destroy it with an antitank weapon. Although he inflicted casualties, the weapon did not silence the enemy fire. Disregarding the intense fire directed on him and his platoon, he charged thirty meters across open ground and hurled grenades into the enemy position, killing some of the eight insurgents manning it. Although severely wounded, when his grenades were expended, and armed with only a rifle, he continued the momentum of his assault on the position and killed the remainder of the enemy. For his valiant action, First Lieutenant Walter J. Marm was awarded the Medal of Honor.

The next day, 15th November, the 2nd Battalion, 5th Cavalry marched into LZ X-Ray from LZ Victor, located two miles east. Joining in with the main body of "C" Company, 1st Battalion, 7th Cavalry, the 2nd Battalion, 5th Cavalry continued on unopposed to rescue the cut-off platoon. They brought the platoon back with all wounded and dead. Of the twenty-nine man platoon, nine were killed and thirteen were wounded. Survivors of "C" Company, 1st Battalion, 7th Cavalry was replaced on line by the fresh "B" Company, 2nd Battalion, 7th Cavalry. The battalions formed a strong perimeter and prepared for more action in the night.

On 16th November, 1st Battalion, 7th Cavalry, having seventy-nine men killed and one hundred-twenty one wounded was ordered back to the rear for reorganization. The battle continued for two more days. With the help of reinforcements and overwhelming firepower, the 1st Battalion, 5th Cavalry and 1st and 2nd Battalions, 7th Cavalry forced the North Vietnamese to abandon their attack and withdraw back into Cambodia.

By 1500 hours, the 1st Battalion, 7th Cavalry turned over LZ X-Ray to the 2nd Battalion, 5th Cavalry and 2nd Battalion, 7th Cavalry. After helping clear the area, the 1st Battalion, 7th Cavalry and the 1st Battalion, 5th Cavalry, less "A" Company were airlifted to the Camp Holloway airfield at Pleiku City.

On the morning of 17 November, with the close out of action at LZ X-Ray, the 2nd Battalion, 7th Cavalry, with "A" Company, 1st Battalion, 5th Cavalry attached, walked out of LZ X-Ray and headed for a location, identified as LZ Albany, to set up blocking positions to reinforce the exhausted skytroopers, prepare for extraction from the battle area and get out of the area targeted for an impending B-52 strike.

As the troopers moved safely out of the area, the new weapon, Operation ARC LIGHT, introduced to support the

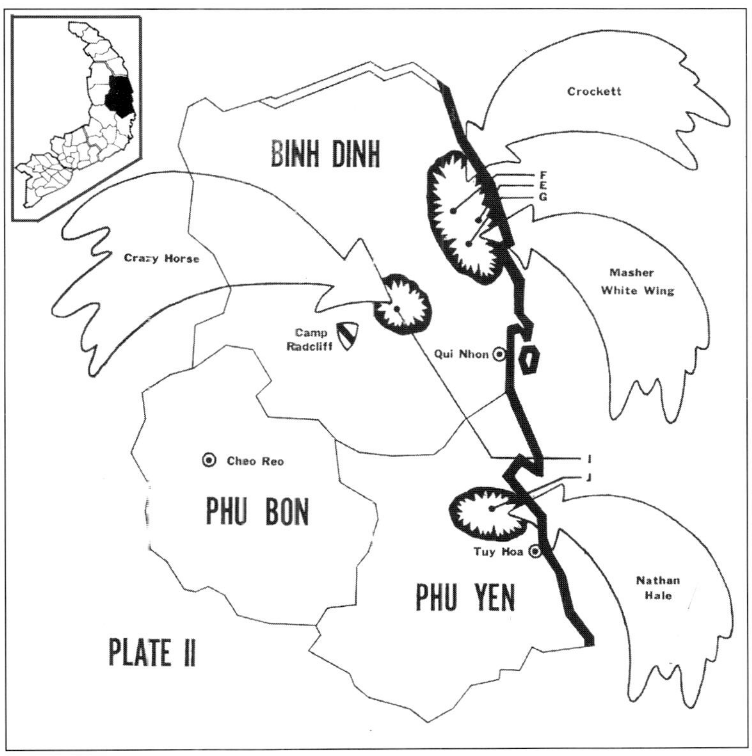

Bong Son Campaign Areas
(25 January to 01 July 1966)

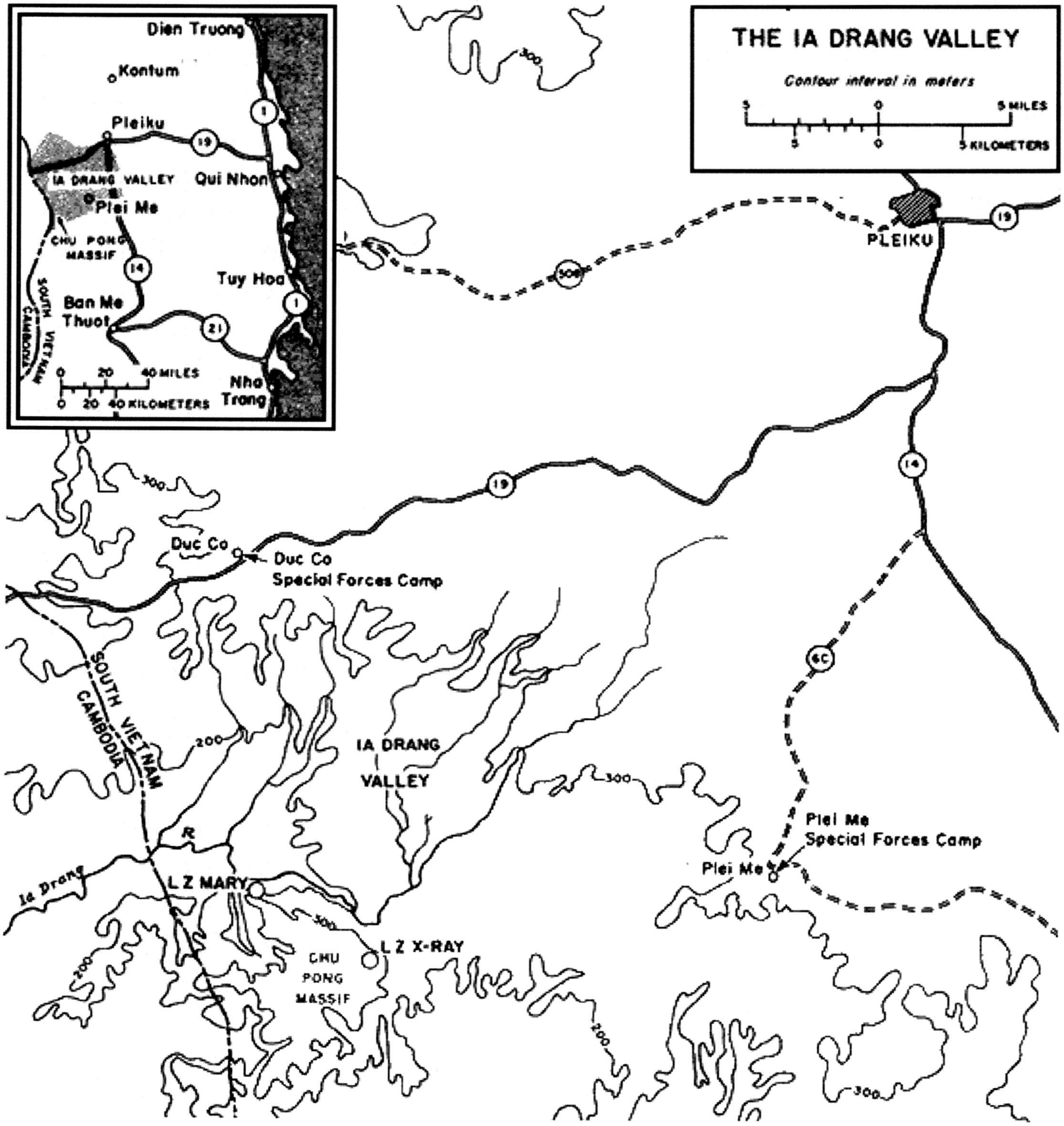

The Ia Drang Valley

The rigors of the three day long battle at LZ X-Ray show on the face of this 1st Battalion, 7th Cavalry Trooper, as he and other GarryOwen Soldiers inspect Communist weapons captured during the fierce fighting. (15 November 1965)

US Ground Forces, struck terror into the hearts of even the most hardened enemy soldier. Shortly after noon, a large area suddenly erupted with hundreds of thunderous explosions. The B-52 bombers had struck the area of the NVA withdrawal. The big bombers systematically worked over large areas of the Chu Pong Massif.

Early in the afternoon of 17 November, the 2nd Battalion, 7th Cavalry Command Group and "A" Company, 2nd Battalion, 7th Cavalry reached LZ Albany. The column was 550 yards long. "C" Company, 2nd Battalion, 7th Cavalry and "A" Company, 1st Battalion, 5th Cavalry put out flank security. Not known at the time, NAV soldiers of the fresh 8th Battalion, 66th Regiment (which had not seen action) deployed down the northeast side of the column. Survivors of the 33rd NAV Regiment deployed at the head of the 2nd Battalion column. As they drew near LZ Albany, the exhausted troopers were ambushed by the NVA units.

At 1320 hours mortar rounds exploded in the clearing and down the length of the column followed by a violent assault which fragmented the column into small groups. When the firing began, the troopers dropped into the tall elephant grass where it was impossible for the soldiers of either side to identify friend or foe except at extremely close range. Within minutes, the situation became a wild melee, a shoot-out, with the gunfighters killing not only the enemy but sometimes their friends just a few feet away.

For the next two hours, the battle roared. Douglas A-1E Skyraiders were brought in to dropped napalm and 250 pound bombs which slowed down enemy actions. Artillery fire was brought in. By dark, "B" Company, 2nd Battalion, 7th Cavalry had landed to reinforce LZ Albany. There was a small perimeter at LZ Albany and one at the tail of the column. In between, troopers were being hounded and killed throughout the night. Also in the night, a few isolated troopers escaped, trying to make it to the artillery position at LZ Columbus.

On 18 November daylight broke over a quiet and tense battlefield. Survivors began the grim task of recovering the dead from the intermingled bodies of both sides. By the 19th of November, evacuation of the wounded and dead was complete. On 20 November, after 3 days and nights on that bloody, hellish, haunted battleground, the survivors of the 2nd Battalion, 7th Cavalry were airlifted out. The casualties for the NVA was reported as 403 dead and 150 wounded. Total First Team casualties at LZ Albany were reported as 151 killed, 121 wounded and 4 missing in action. Nearly half of the 300 men killed in the Pleiku Campaign died at LZ Albany.

When the Pleiku Campaign, SILVER BAYONET, ended on 25 November troopers of the First Team had paid a heavy price for its success, having lost some three hundred troopers killed in action. However, they had killed 3,561 North Vietnamese soldiers and captured 157 more. The troopers destroyed two of three regiments of a North Vietnamese Division, earning the first Presidential Unit Citation awarded to a division in Vietnam. The enemy had been given their first major defeat and their carefully laid plans for conquest had been torn apart.

On 17 December, after a short rest, the 3rd Brigade went into action to conduct a four-day operation known as CLEAN HOUSE in the vicinity of Binh Khe, in Binh Dinh

A most important type of air support was the massive B-52 striking in a saturation bombing role known as "Arc Light Strikes", which were under the operational control of the Strategic Air Command (SAC) Air Force Strike Force stationed at Anderson Air Force Base, Guam. The detailed planning of where to place B-52 strikes was made through the Military Assistance Command, Vietnam (MACV).

Province's Soui Ca River Valley. From a position northeast of the valley, troopers moved down from high ground to sweep through suspected VC areas. By the Christmas Holidays, the 1st Cavalry Division returned to its original base of operations at An Khe on Highway 19.

Soon, the intelligence sections recommended a return to the Western Highlands early in 1966 in hopes of encountering the enemy reassembling in the unpopulated jungles. However a new threat emerged in Binh Dinh Province, a region of abrupt mountains and populated coastal plains. The Army of the Republic Vietnam (ARVN) 22nd Division, responsible for that area, was spread thin trying to keep Highway 19 open and secure. The intelligence staff of the 1st Cavalry Division had confirmed that the 2nd Regiment, Vietcong Main Force and North Vietnamese Army (NVA) 18th and 19th Regiments were operating in the area. These three regiments comprised the NVA Division known as the "Sao Vang" or "Yellow Star" Division.

On 04 January 1966, Operation MATADOR began in response to the intelligence reports. The 1st and 2nd Brigades were airlifted west of Pleiku and Kontum Provinces to begin a search and destroy mission. Conducted on the Vietnamese side of the Ton Le San River, this was the first time US troops actually went right to the Cambodian border. Previously they were under orders not to enter the three mile buffer zone along the border. During this operation, the 1st Cavalry saw the enemy flee across the border into Cambodia, confirming that the enemy had well-developed sanctuaries and base camps inside Cambodia. Operation MATADOR was closed out on 17 January.

The Chinese Lunar New Year begins on the phases of the moon and each of the years in the 12 year cycle is named after an animal. The Chinese year of 1966, the year of the "Horse", began on 21 January 1966. Just before midnight more than 400 helicopters of the 1st Cavalry Division flew over the darkness of the Central Highlands. One of the helicopters broadcast a message to the unseen people of the surrounding area. "Attention, members of the National Liberation Front. It is the Year of the Horse. It is also the year of the Horse Division. The 1st Cavalry Division will continue to strike from the sky as a constellation of death to all those who support Godless Communism. It will thunder across the heavens and strike everywhere, no matter where you hide. At midnight listen for the New Year and the thunder that heralds your destruction."

At midnight a 66 round mass barrage of high explosives was fired to bring in the New Year - followed by white phosphorus rounds for their pyrotechnic qualities. The year of the "Flying Horsemen" had began in Central Highlands.

On 25 January 1966, following the truce for the Tet holiday and Lunar New Year, MASHER/WHITE WING, which were the code names for the operations of the 3rd Brigade in Binh Dinh Province, began. The 3rd Brigade gathered its gear and weapons and began to move by highway and air to staging areas in eastern Binh Dinh Province. The opening phase of the mission included the 1st and 2nd Battalions, 12th Cavalry as well as the 1st Squadron, 9th Cavalry — a reconnaissance unit of scout, gun and infantry heliborne elements. The 1st Squadron, 9th Cavalry reconnoitered ahead of the convoy and along both sides of the road, searching for potential ambushes.

On 28 January, Operation MASHER, the first phase, began. The 3rd Brigade assaulted North of Bong Son and LZ Dog and soon encountered heavy resistance by the NVA. Contact by the enemy diminished in the first two days of February as the North Vietnamese continued their withdrawal to the north and west. In the first week of combat, the division had lost seventy-seven troopers and the enemy losses amounted to an estimated 1,350 killed in action. Two battalions of the NVA 22nd Regiment had been rendered ineffective.

On 07 February, Operation WHITE WING began the second phase of the "search and destroy" mission. On 16 February, following heavy enemy engagement, the battle weary 3rd Brigade, returned to the Division's home base of An Khe and was replaced in the field by the 1st Brigade. While the 1st Brigade patrolled in the valleys around LZ Bird, the 1st and 2nd Battalions, 5th Cavalry and 2nd Battalion, 12th Cavalry, of the 2nd Brigade encircled the "Iron Triangle", the regimental headquarters of the NVA. Aided by artillery and air support, the three battalions continued fighting for four days against a tenacious enemy defense.

On 17 February, "B" Company, 2nd Battalion, 5th Cavalry air assaulted into a LZ located in the "Crow's Foot" area of the Kim Son Valley and by 0915 hours came into contact with a VC company armed with heavy weapons and a large number of automatic weapons. Two additional companies of the battalion were quickly committed to exploit the contact. A third company assaulted to the southeast and immediately engaged another heavily armed unit. Intensive tube and aerial artillery fire were delivered on the area throughout the day. By 1800 hours, a sweep through the vacated defensive positions of the enemy revealed 127 KIA. A large number of mortars and recoilless rifles were left behind. It was concluded that the 2nd Battalion, 5th Cavalry had fought and decimated the anti-aircraft battalion of the Yellow Star Division as well as the signal company of the 2nd VC Main Force Regiment.

On 19 - 21 February, one of the main actions occurred in the area known as the "Iron Triangle", an elaborate, well fortified defensive position twelve miles south of Bong Son. During interrogation, a prisoner revealed the location of the NVA 22nd Regimental headquarters. Elements of the 2nd Brigade advanced into the area and were met by fierce resistance. Units of the NVA 22nd Regiment attempted to reinforce its headquarters, but were cut down in the crossfire of two companies of the 1st Battalion, 12th Cavalry. The next three days the area was saturated with artillery fire and B-52 strikes. All resistance collapsed after a final B-52 strike.

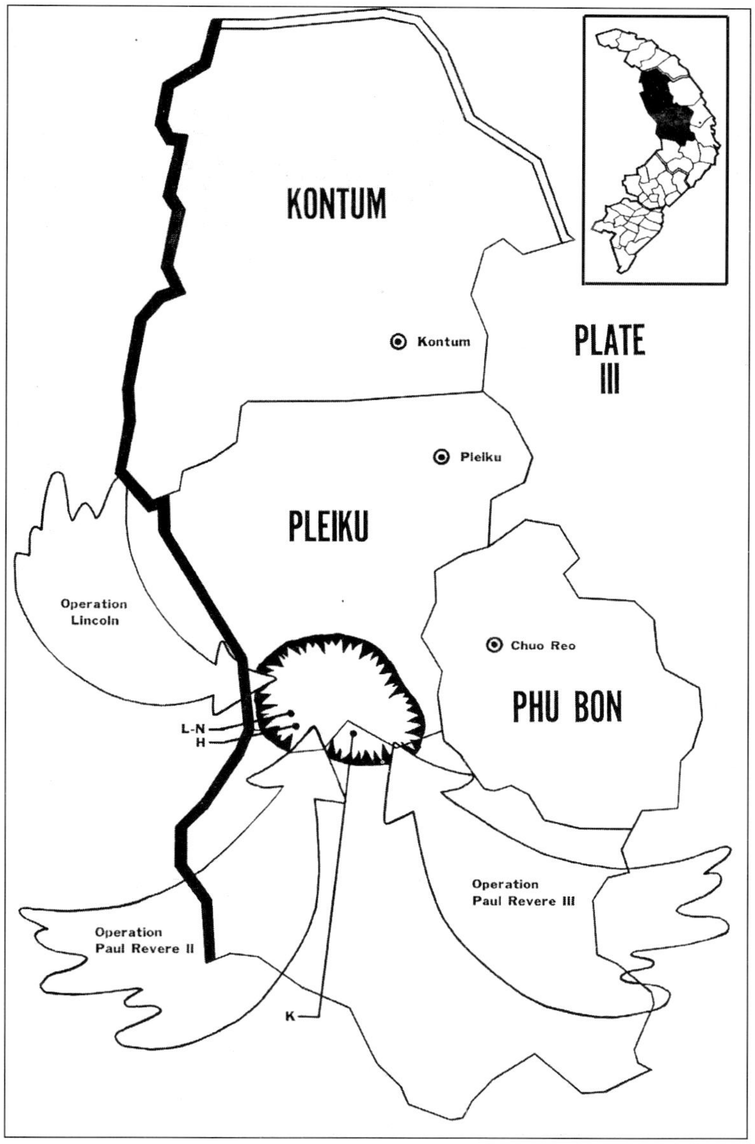

Pleiku Campaign Areas
(25 March to 27 December 1966)

On 01 March the final phase of WHITE WING commenced, moving into the jungle covered Cay Giep Mountains. B-52s blasted openings in the thick jungle canopy, permitting engineer teams to descend from helicopters to clear out landing zones for the 2nd Brigade. Sweeping down the slopes of the Cay Giep Mountains, the 2nd Brigade encountered little resistance as the main body of the NVA 6th and 18th Battalions had fled, departing two days earlier, following the first air assault.

On 06 March 1966, Operation MASHER/WHITE WING ended and was, by all tactical measures, pronounced a military success with the enemy losing its grip on the Binh Dinh Province, however, its name would be heard again and again during the next six years. The 1st Cavalry Division had once again made an effective use of mobility and firepower. Helicopters airlifted entire infantry battalions a total of seventy-eight times and moved artillery batteries fifty-five times. During MASHER/WHITE WING the 1st Cavalry Division had maintained constant contact with the enemy for the forty-one day operation, an unprecedented feat in the Vietnam War. In the actions, the Division clashed with all three regiments of the Sao Vang Division and rendered five of its nine battalions ineffective for combat.

On 06 May, in the midst of two more Operations, LEWIS and CLARK and DAVY CROCKETT, the Division experienced its first change of command when Major General Harry W. O. Kinnard turned over his post to his replacement, Major General John Norton. General Norton, a veteran paratrooper, was no stranger to the First Team or airmobility. He had commanded a battle group in Korea in 1959-60 and later, on the Howze Board had helped pioneer airmobility.

On 16 May, the next major Operation, CRAZY HORSE, commenced during the hot summer, with the temperature soaring to 110 degrees. The search and destroy mission extended into the heavy jungle covered hills between Soui Ca and the Vinh Thanh Valleys. The 1st Brigade went into action against the 2nd Viet Cong Regiment. Intelligence indicated that the Viet Cong were massing in a natural corridor known as the "Orgeon Trail", planning to attack the Special Forces Camp on 19 May, the birthday of Ho Chi Minh. Initial contact was made by Company "B", 2nd Battalion, 8th Cavalry at LZ Hereford. Company "A", 1st Battalion, 12th Cavalry was airlifted to a nearby point to join the battle. The two companies held off superior enemy forces throughout the night. The next morning more elements of the 12th Cavalry and the entire 1st Brigade became involved in CRAZY HORSE. The fighting now consisted of short but bitter engagements in tall elephant grass and heavily canopied jungle. The battleground covered approximately 20 kilometers with the Viet Cong holed up on three hills. Once they were surrounded, all available firepower was concentrated in their area. The Viet Cong regiment was hit with artillery, aerial rockets, tactical air strikes by F-4s and bombs from high flying B-52s. Many of the enemy soldiers, if not killed outright by the devastation, were cut down by heavy crossfire. Many important military documents, detailing the Viet Cong infrastructure in Binh Dinh Province, were discovered.

On 18 May, in the course of the fighting, a squad leader with "B" Company, 2nd Battalion, 12th Cavalry, Staff Sergeant Jimmy G. Stewart, demonstrated the leadership and courage necessary to engage and destroy the enemy. Early in the morning, a reinforced North Vietnamese company attacked "B" Company, which was manning a defensive perimeter. The surprise onslaught wounded five members of a six man squad caught in the direct path of the enemy's thrust. Sergeant Stewart became a lone defender of vital terrain — virtually one man against a hostile platoon. Refusing to take advantage of a lull in the firing which would have permitted him to withdraw, Sergeant Stewart elected to hold his ground to protect his fallen comrades and prevent an enemy penetration of the company perimeter. As the full force of the platoon-sized main attack struck his lone position, he fought like a man possessed, emptying magazine after magazine at the determined, on-charging enemy. The enemy drove almost to his position and hurled grenades, but Sergeant Stewart decimated them by retrieving

A CH-54 "Flying Crane" prepares to lift a 155mm Howitzer, out of Vinh Thanh Valley near An Khe. The Howitzer, manned by troopers of "C" Battery, 30th Artillery, supported the 1st Brigade in Operation CRAZY HORSE. (May 1966)

and throwing the grenades back. Exhausting his ammunition, he crawled under intense fire to his wounded team members and collected ammunition that they were unable to use. Far past the normal point of exhaustion, he held his position for four harrowing hours and through three assaults, annihilating the enemy as they approached and before they could get a foothold. As a result of his defense, the company position held until the arrival of a reinforcing platoon which counterattacked the enemy, now occupying foxholes to the left of Sergeant Stewart's position. After the counterattack, his body was found in a shallow enemy hole where he had advanced in order to add his fire to that of the counterattacking platoon. Eight enemy dead were found around his immediate position, with evidence that fifteen others had been dragged away. The wounded, whom he gave his life to protect, were recovered and evacuated. For his valiant actions, Staff Sergeant Jimmy G. Stewart received the Medal of Honor.

On 21 May, in a second major engagement of Operation CRAZY HORSE, the platoon of Sergeant David C. Dolby, a member of "B" Company, 1st Battalion, 8th Cavalry, suddenly came under intense fire from the enemy located on a ridge immediately to the front. Six members of the platoon were killed instantly and a number were wounded, including the platoon leader. Sergeant Dolby's every move brought fire from the enemy. However, aware that the platoon leader was critically wounded, and that the platoon was in a precarious situation, Sergeant Dolby moved the wounded men to safety and deployed the remainder of the platoon to engage the enemy. Subsequently, his dying platoon leader ordered Sergeant Dolby to withdraw the forward elements to rejoin the platoon. Despite the continuing intense enemy fire and with utter disregard for his own safety, Sergeant Dolby positioned able-bodied men to cover the withdrawal of the forward elements, assisted the wounded to the new position, and he, alone, attacked enemy positions until his ammunition was expended. Replenishing his ammunition, he returned to the area of most intense action, single-handedly killed three enemy machine gunners and neutralized the enemy fire, thus enabling friendly elements on the flank to advance on the enemy redoubt. He defied the enemy fire to personally carry a seriously wounded soldier to safety where he could be treated and, returning to the forward area, he crawled through withering fire to within 50 meters of the enemy bunkers and threw smoke grenades to mark them for air strikes. Although repeatedly under fire at close range from enemy snipers and automatic weapons, Sergeant Dolby directed artillery fire on the enemy and succeeded in silencing several enemy weapons. He remained in his exposed location until his comrades had displaced to more secure positions. His actions of unsurpassed valor during four hours of intense combat were a source of inspiration to his entire company, contributed significantly to the success of the overall assault on the enemy position, and were directly responsible for saving the lives of a number of his fellow soldiers. For his valiant actions, Sergeant David C. Dolby received the Medal of Honor. On 05 June 1966, Operation CRAZY HORSE was concluded.

Division Support Command (DISCOM) units, the 15th Medical Battalion, the 15th Transport Corps Battalion, the 15th Supply and Services Battalion and the 27th Maintenance Battalion supplied the troopers in the field with the food, clothing, ammunition and transportation they needed to do their jobs. The three DISCOM Forward Service Support Elements (FSSE) operated throughout the 1st Cavalry Division area of operations in support of Operations CRAZY HORSE, PAUL REVERE and THAYER in 1966. In response to the large commitments of Division forces in the central highlands, a fourth FSSE began operations in July and by November a fifth FSSE had been added. For their accomplishments from 01 june 1967 to 31 May 1968 DISCOM and its assigned units were awarded the Meritorious Unit Citation.

On 02 August, Operation PAUL REVERE II was launched to deny areas of the rich rice fields to the famished Viet Cong. Significant contact with the enemy did not occur until 08 August, at LZ Juliett. Company "A", 1st Battalion,

Chinook CH-47 unloads supplies at "C" Company, 1st Battalion, 7th Cavalry mountain top camp during Operation THAYER. (29 August 1966)

7th Cavalry came under heavy fire from a reinforced enemy battalion. In several hours of intense fighting, Alpha Company turned back repeated mass attacks. Timely artillery and air strikes eliminated the opportunity for the enemy to surround the Skytroopers. The roar of helicopters from two companies from the 1st Battalion, 12th Cavalry arriving at LZ Juliett frightened the enemy, causing them to flee.

On 02 August, the main body of the 5th Battalion, 7th Cavalry departed Ft. Carson, flew to the port of embarkment at Oakland, California and boarded the *USS Gaffey*. Following this movement, the advance party of the 5th Battalion, 7th Cavalry departed Peterson Field, Colorado on 08 August and arrived at Qui Nhon on 11 August. The main body made a brief stop at Okinawa on 15 August for a brief shore leave and after 20 days afloat, arrived at Qui Nhon on 20 August. They disembarked into LSTs and moved ashore. "HHC" and "D" Company joined in a wheeled vehicle convey to Camp Radcliff at An Khe and the remainder of the Battalion were flown in by C-130s. The 5th Battalion, rounding out the 3rd maneuvering element of the 7th Cavalry Regiment, began its initial training exercises at An Khe. On 05 September, following their first exposure to the jungle environment, the Battalion deployed for an additional training/operation mission of road security along Highway 19 from An Khe to Pleiku, with the Battalion Command Post (CP) located atop Mang Yang Pass. The first major deployment of the 5th Battalion was to LZ Hammond in preparation for Operation IRVING.

On 15 August Operation PAUL REVERE II ended with the battle of Hill 534, on the southern portion of Chu Pong Massif near the Cambodian Border. The operation had begun on 02 August, after Company "A" 1st Battalion, 5th Cavalry suddenly ran into a North Vietnamese battalion and Company "B", 2nd Battalion, 5th Cavalry began slugging it out with enemy troops in bunkers. A total of two battalions of Skytroopers were committed to the fight. When it ended the next morning, 138 NVA bodies were counted.

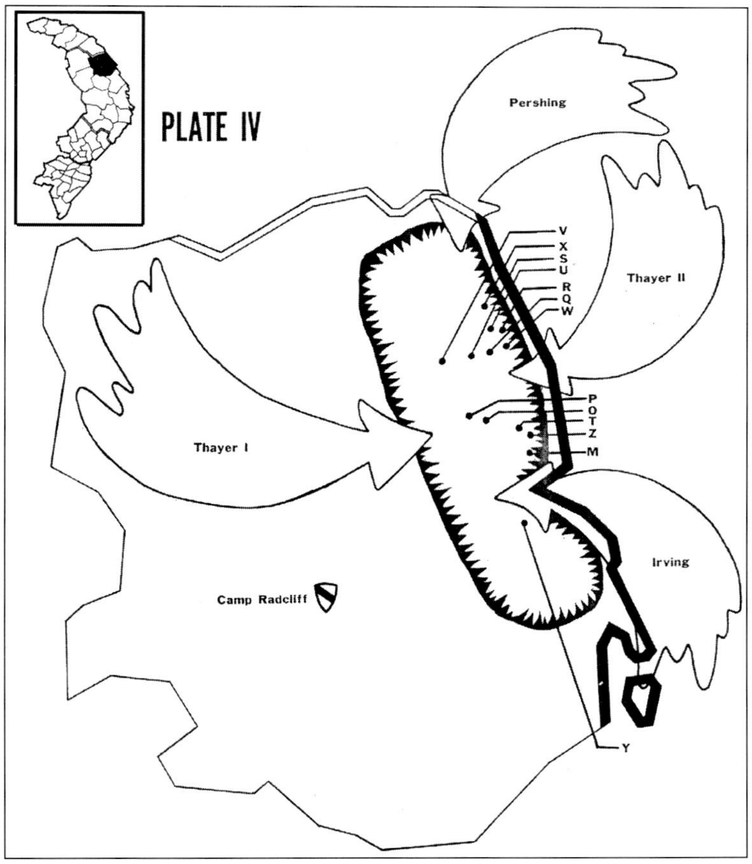

Bong Song Campaign Areas
(13 September 1966 to 21 January 1967)

At the end of PAUL REVERE II, which had killed a total of 861 of the enemy, a task force of the 2nd Battalion, 7th Cavalry was organized for Operation BYRD. The task force was dispatched to Binh Thaun Province, in the southern area of II Corps, to support the Revolutionary Development Program and bring the long months of Operation BYRD to a productive finish.

At that time the heavily populated province of Binh Thaun was almost totally under the power of two Viet Cong Battalions. The South Vietnamese government controlled little more than the provincial capital, Phan Thiet, a coastal town known for its fishermen and its fish sauce manufacturing industry. In sixteen months the 2nd Battalion, 7th Cavalry had fanned out from Phan Thiet and cleared the enemy from the populous "triangle" area that stretched north and west of Phan Thiet. They also cleared provincial roads that had been closed by the Viet Cong. Most significantly, the troopers reopened Highway 1, an action that restored commerce to life between Phan Thiet and Saigon.

On 13 September, Operation THAYER I began. It was one of the largest air assaults launched by the 1st Cavalry Division. Its mission was to rid Binh Dinh Province of NVA and VC soldiers and the political infrastructure of the Viet Cong. On 16 September, troopers of the 1st Brigade discovered an enemy regimental hospital, a factory for making grenades, antipersonnel mines and a variety of weapons. On 19 September, elements of the 2nd Battalion, 8th Cavalry traded fire with two NVA combat support companies.

On 21 September, near Bon Son in Binh Dinh Province, part of the squad of Private First Class Billy L. Lauffer, "C" Company, 2nd Battalion, 5th Cavalry, was suddenly struck at close range by an intense machine gun crossfire from two concealed bunkers astride the squad's route. Private First Class Lauffer, the second man in the column, saw the lead man fall and noted that the remainder of the squad was unable to move. Two comrades, previously wounded and being carried on litters, were lying helpless in the beaten zone of the enemy fire. Reacting instinctively, Private First Class Lauffer quickly engaged both bunkers with fire from his rifle, but when the other squad members attempted to maneuver under his covering fire, the enemy fusillade increased in volume and thwarted every attempt to move. Seeing this and his wounded comrades helpless in the open, Private First Class Lauffer rose to his feet and charged the enemy machine gun positions, firing his weapon and drawing the enemy's attention. Keeping the enemy confused and off balance, his one man assault provided the crucial moments for the wounded point man to crawl to a covered position, the squad to move the exposed litter patients to safety, and his comrades to gain more advantageous positions. Private First Class Lauffer was fatally wounded during his selfless act of courage and devotion to his fellow soldiers. His gallantry. at the cost of his life, served as an inspiration to his comrades and saved the lives of an untold number of his companions. For his valiant action, Private First Class Billy L. Lauffer received the Medal of Honor.

In the closing phases of Operation THAYER I and the prelude to Operation IRVING, enemy elements of the 7th and 8th Battalions, NVA 18th Regiment had been reported in the village of Hoa Hoi. On 02 October, at about 8:00 hours, a "Blue Team" platoon of "A" Troop, 1st Squadron, 9th Cavalry, flying a routine reconnaissance mission, was diverted and ordered to land and access the degree of enemy build up. Entering the village, they engaged a heavy concentration of the enemy forces and as the battle wore on, they called for backup at about 1000 hours. Quickly, the 1st Battalion, 12th Cavalry, deployed to encircle the village. At 1200 hours, "B" Company, 12th Cavalry, the first of the backup units to arrive, air assaulted into a landing area 300 meters east of the village in the face of heavy resistance. Immediately, the units came under intense small arms and mortar fire. "A" Company, 12th Cavalry landed to the southwest and began a movement northeast to the village. In the meantime, "C" Company, 12th Cavalry landed north of the village and began moving south.

The arrival of the 1st Battalion, 12th Cavalry took pressure off the "Blue Team" platoon of "A" Troop who, still heavily engaged with the entrenched enemy and taking heavy fire, were able to withdraw and make it back to the LZ, taking their dead with them. By late in the afternoon, "A" and "B" Companies, 1st Battalion had linked up and established blocking positions which prevented the enemy from slipping out of the village. During the course of the evening, "A" and "C" Companies, 1st Battalion, 5th Cavalry were airlifted into an area east of the village to assist in the containment of the enemy. Additional support of artillery forward observers from "A" Battery, 2nd Battalion, 19th Artillery helped as enemy locations were identified and called in during the night.

In the morning of 03 October, "C" Company, 1st Battalion, 12th Cavalry and "C" Company, 1st Battalion, 5th Cavalry attacked south to drive the remaining enemy forces into "A" and "B" Companies, 12th Cavalry which were braced in strong blocking positions to take the attack. This last action broke the strong resistance of the enemy and the mission was completed.

Lasting from 02 - 28 October, Operation IRVING was one of a trio of operations launched by Americans, South Vietnamese and South Korean Forces. The mission was established to entrap Viet Cong soldiers, who were fleeing Operation THAYER I, in a pocket between a group of hills and the coastline of Binh Dinh Province. The operations were complicated by a heavy concentration of civilians living in the operational area, but great care was taken to minimize civilian casualties.

Trapped in a tight cordon, the enemy lost 2,063 killed. The free world allies captured 2,071 NVA and Viet Cong troops - an unusually large number in Vietnam fighting. The 1st Battalion, 12th Cavalry was awarded the Presidential Unit Citation for valor and aggressive pursuit of the enemy on 02 October, after reinforcing the "Blue Team" platoon from "A" Troop, 1st Squadron, 9th Cavalry that was heavily engaged and outnumbered.

On October 25, Operation THAYER II continued the drive of pacification of the Binh Dinh Province. On 01 November troopers of "A" Troop, 1st Squadron, 9th Cavalry and elements of the 5th Battalion, 7th Cavalry became engaged in a sharp fight with the 93rd Battalion and the 2nd Viet Cong Regiment. The action took place in the vicinity of National Route 1 and Dam Tra-O Lake south of the Cay Giep mountains. In Thayer II the enemy suffered a punishing loss of 1,757 killed.

On 31 October, Operation PAUL REVERE IV was launched by the 2nd Brigade. Its units included; 1st Battalion, 5th Cavalry; 2nd Battalion, 12 Cavalry; "B" Troop, 1st Squadron, 9th Cavalry and the 1st Battalion, 77th Artillery. The operation called for extensive search and destroy in the areas of Chu Pong and the Ia Drang Valley as well as along the Cambodian Border. With only one exception only light contact with the enemy was achieved.

It was during this action that the platoon of Private First Class Lewis Albanese, "B" Company, 5th Battalion, 7th Cavalry was advancing through the densely covered terrain to establish a blocking position. The platoon received intense automatic weapons fire from close range. As other members of the platoon maneuvered to assault the enemy position, Private First Class Albanese was ordered to provide security for the left flank of the platoon. Suddenly, the left flank received fire from enemy soldiers located in a well-concealed ditch. Realizing the imminent danger to his comrades from this fire, Private First Class Albanese fixed his bayonet and moved aggressively into the ditch. His action silenced the sniper fire, enabling the platoon to resume movement toward the main enemy position. As the platoon continued to advance, the sound of heavy firing emanated from the left flank from a pitched battle that ensued in the ditch which Private First Class Albanese had entered. The ditch was actually a well-organized complex of enemy defenses designed to bring devastating flanking fire on the forces attacking the main position. Private First Class Albanese, disregarding the danger to himself, advanced 100 meters along the trench and killed six of the snipers, who were armed with automatic weapons. Having exhausted his ammunition, Private First Class Albanese was mortally wounded when he engaged and killed two more enemy soldiers in fierce hand-to-hand combat. His unparalleled actions saved the lives of many members of his platoon who otherwise would have fallen to the sniper fire from the ditch, and enabled his platoon to successfully advance against an enemy force of overwhelming numerical superiority. For his valiant action, Private First Class Lewis Albanese received the Medal of Honor.

In mid-morning of 21 November, "C" Company, 1st Battalion, 5th Cavalry was searching south of Duc Co along the border. Suddenly the 2nd Platoon began trading fire with a NVA force of significant size. The 3rd Platoon went to the aid of the 2nd Platoon. The two units, outnumbered by large numbers of North Vietnamese, fought desperately. The 3rd Platoon was overrun in fairly short order with only one man surviving - it happened before they were able to call in any effective artillery or air support. The 2nd Platoon took over 50% casualties but was not overrun - it had thirteen or fourteen killed in action and about as many wounded. As was typical in the early days of the Vietnam war, many of their M-16s had malfunctioned early in the fight. With the dense foliage, neither artillery nor helicopter gunships were very effective in providing support.

The remnants of the 2nd Platoon were saved by the arrival of a flight of Navy A-1E Skyraiders equipped with napalm which they dropped accurately during the attack. The 1st Platoon arrived a few minutes after the airstrike and linked up with 2nd Platoon. Except for a few stray rounds from the departing NVA, the battle was over. The foliage was too thick to cut an LZ and the wounded were lifted out one by one by Hueys equipped with winches. The killed in action were placed in a cargo net and were lifted out by a CH-47 Chinook helicopter. "A" Company located the ambush site of the 3rd Platoon and medevaced the one survivor. The "C" Regiment, 10th NVA Division paid a very high price for its victory, losing nearly 150 of their men.

On 09 December the Division launched Operation ROVER to evacuate all civilians from the Kim Son Valley. By 14 December, the valley was clear and all civilians had been evacuated from the area, bringing Operation ROVER to a close with the valley being declared a "free-fire" zone. Now the heavy work of the troopers had just began as they moved in to drive the elusive enemy from the area.

On 17 December, heavy contact was made in the Highway 506 Valley just east of the Kim Son Valley. "C" Company, 1st Battalion, 8th Cavalry observed an enemy squad moving into the valley and went after them. Air support was called in to help and drew ground fire from several positions. The Infantry Platoon of "A" Troop 1st Squadron, 9th Cavalry made an air assault into the valley and encountered heavy resistance. The 1st Battalion, 12th Cavalry was also brought in, along with four infantry companies and two platoons of the 1st Squadron, 9th Cavalry.

They attempted to encircle the enemy force of a "dug in" battalion. Night fell and an Air Force ship kept the area illuminated, but the encirclement was not complete and many of the enemy managed to escape. At least ninety-five did not as their bodies were found in a final sweep of the battle area on 19 December. On 27 December, Operation PAUL REVERE IV was closed out and 2nd Brigade troopers added their strength to Operation THAYER II.

The end of 1966 brought about an observance of a two day Christmas truce. On 27 December three NVA battalions from the 22nd Regiment used the truce to move into position for a surprise attack on LZ Bird in the Kim Son Valley which was well away from their usual haunts in the Hoai Nhon Delta area. The three enemy units threw a fierce infantry and mortar assault at LZ Bird. The LZ was only defended by "C" Company, 1st Battalion, 12th Cavalry and two 105mm artillery batteries, Battery "B", 2nd Battalion, 19th Artillery and Battery "B", 6th Battalion, 16th Artillery. All were under strength. It was within artillery fire support range of two other batteries. Initially the weather restricted air support operations. The NVA broke through the perimeter and occupied a few gun positions. The 12th Cavalry troopers fought back hand-to-hand and with everything they had. Finally, some of the 105s were cranked down to pointblank range and "beehive" rounds sliced through the attackers like scythes. For the next two days, troopers of the 1st Cavalry Division pursued the fleeing NVA and made contact several times. At least 266 NVA died in this battle.

It was also during this engagement on 27 December, Staff Sergeant Delbert Jennings, "C" Company, 1st Battalion, 12th Cavalry was defending an artillery position when attacked by a North Vietnamese Army regiment supported by mortar, recoilless-rifle, and machine gun fire. At the outset, Staff Sergeant Jennings sprang to his bunker, astride the main attack route, and slowed the on-coming enemy wave with highly effective machine gun fire. Despite a tenacious defense in which he killed at least twelve of the enemy, his squad was forced to the rear. After covering the withdrawal of the squad, he rejoined his men, destroyed an enemy demolition crew about to blow up a nearby howitzer, and killed three enemy soldiers at his initial bunker position. Ordering his men back into a secondary position, he again covered their withdrawal, killing one enemy with the butt of his weapon. Observing that some of the defenders were unaware of an enemy force in their rear, he raced through a fire-swept area to warn the men, turn their fire on the enemy, and lead them into the secondary perimeter. Assisting in the defense of the new position, he aided the air-landing of reinforcements by throwing white phosphorous grenades on the landing zone despite dangerously silhouetting himself with the light. After helping to repulse the final enemy assaults, he led a group of volunteers well beyond friendly lines to an area where eight seriously wounded men lay. Braving enemy sniper fire and ignoring the presence of booby traps in the area, they recovered the men who would have probably perished without early medical treatment. The extraordinary heroism and inspirational leadership of Staff Sergeant Jennings saved the lives of many of his comrades and contributed greatly to the defeat of a superior enemy force. For his valiant action, Staff Sergeant Delbert Jennings received the Medal of Honor and would later go

Members of the 8th Engineer Battalion, 1st Cavalry Division (Airmobile) have the job of clearing the road of any mines that might have been left by the Viet Cong. Here they are busy with the tedious task of probing with bayonets in an area of the road suspected to contain a mine. This action took place during Operation PERSHING, a search and destroy in the An Lao Valley and Bong Son Plain in South Vietnam. (16 May 1967)

on to become the Command Sergeant Major (CSM) of the 1st Cavalry Division.

For its heroic action, "C" Company, 1st Battalion, 12th Cavalry was awarded a Presidential Unit Citation for its action on 27 December. Many Aerial Rocket Artillery (ARA) and armed CH-47 sorties were flown in support of this battle. Not long after, the site at the bend in the river was deemed highly vulnerable and a new LZ Bird was built on a hogback several miles farther east.

As 1967 dawned, the 1st Brigade began making new contacts with the enemy units in central and southern Kim Son Valley. The 2nd Brigade began a sweep to the north, flushing the enemy from their position in the north end of the valley as well as the Crescent Area, the Nui Mieu and Cay Giep Mountains. On 27 January, as part of Operation THAYER II in the east central Hoai Nhon District of the Bin Dinh Province, the units of the 2nd Brigade, 1st Squadron, 9th Cavalry, 2nd Battalion, 12th Cavalry, 227th Assault Helicopter (AH) and 228th Assault Support Helicopter (ASH) Battalions and elements of the 40th ARVN Regiment and Vietnamese Marine Task Force Bravo conducted Operation BULLSEYE V.

Operation BULLSEYE V was initiated as a result of intelligence reports that there was a large concentration of enemy forces moving into position to attack the new logistical base at English Airfield and the artillery base at LZ Dog. Such an attack would cause general disruption to the area by destroying the Bong Song Bridge which would have impacted the success of THAYER III. In a three day sweep, supported by fire from US Naval gunships, the enemy 7th and 8th Regiments and a local VC Company, D-21, were routed out and scattered causing them to make a hasty and unconditional withdrawal from the area. In THAYER II, the enemy once again had suffered punishing losses of 1,757 men.

On 13 February 1967, Operation PERSHING began in a territory which was familiar to many skytroopers, the Bong Son Plain in northern Binh Dinh Province. For the first time, the 1st Cavalry Division committed all three of its brigades to the same battle area. The First Team was joined by the ROK Capital Division, the 3rd Brigade of the US 25th Infantry Division and the ARVN 22nd Division. A total of 22 Infantry Battalions were poised to drive the enemy from the Bong Son Plain.

On 14 February, the command post of the 1st Cavalry Division was moved to LZ Two Bits, its new home until the conclusion of Operation PERSHING, the fourth phase of the Binh Dinh Pacification Campaign. ARVN soldiers familiar with the methods of the Viet Cong operations in the Bong Son Plain helped the skytroopers locate and eliminate the numerous caves and tunnels infiltrated by the enemy. Skytroopers also met up with numerous North Vietnamese hiding in ditches and wells. The enemy units of the 3rd NVA Division had taken such heavy casualties that they were desperately trying to avoid contact until they could be reinforced.

On 16 February, in the Bong Son Plain Private First Class James H. Monroe, "C" Company. 1st Battalion, 8th Cavalry was deployed in a night ambush when the position was suddenly subjected to an intense and accurate grenade attack, and one foxhole was hit immediately. Responding without hesitation to the calls for help from the wounded men Private First Class Monroe moved forward through heavy small arms fire to the foxhole but found that all of the men had expired. He turned immediately and crawled back through the deadly hail of fire toward other calls for aid. He moved to the Platoon Sergeant's position where he found the radio operator bleeding profusely from fragmentation and bullet wounds. Ignoring the continuing enemy attack, Private First Class Monroe began treating the wounded man when he saw a live grenade fall directly in front of the position. He shouted a warning to all those nearby, pushed the wounded radio operator and the platoon sergeant to one side, and lunged forward to smother the blast of the grenade with his body. Through his valorous actions, performed in a flash of inspired selflessness, Private First Class Monroe saved the lives of two of his comrades and prevented the probable injury of several others. For his valiant action, Private First Class James H. Monroe received the Medal of Honor.

On 20 March, in the Binh Dinh Province, Specialist Fifth Class Charles C. Hagemeister, "A" Company, 1st Battalion, 5th Cavalry was engaged in combat operations against a hostile force, The platoon of Specialist Fifth Class Hagemeister suddenly came under heavy attack from three sides by an enemy force occupying well concealed, fortified positions and supported by machine guns and mortars. Seeing two of his comrades seriously wounded in the initial action, Specialist Fifth Class Hagemeister unhesitatingly and with total disregard for his safety, raced through the deadly hail of enemy fire to provide them medical aid. Upon

Escorted by a "Cobra" gunship, troops of the 5th Battalion, 7th Cavalry, move toward a landing zone where they will conduct a "search and destroy" mission. The infantrymen always know that the "snake" will be there, to cover them, if needed.

learning that the platoon leader and several other soldiers also had been wounded, Specialist Fifth Class Hagemeister continued to brave the withering enemy fire and crawled forward to render lifesaving treatment and to offer words of encouragement. Attempting to evacuate the seriously wounded soldiers, Specialist Fifth Class Hagemeister was taken under fire at close range by an enemy sniper. Realizing that the lives of his fellow soldiers depended on his actions, Specialist Fifth Class Hagemeister seized a rifle from a fallen comrade, killed the sniper, three other enemy soldiers who were attempting to encircle his position and silenced an enemy machine gun that covered the area with deadly fire. Unable to remove the wounded to a less exposed location and aware of the efforts of the enemy to isolate his unit, he dashed through the fusillade of fire to secure help from a nearby platoon. Returning with help, he placed men in positions to cover his advance as he moved to evacuate the wounded forward of his location. These efforts successfully completed, he then moved to the other flank and evacuated additional wounded men despite the fact that his every move drew fire from the enemy. Specialist Fifth Class Hagemeister's repeated heroic and selfless actions at the risk of his life saved the lives of many of his comrades and inspired their actions in repelling the enemy assault. For his valiant action, Specialist Fifth Class Charles C. Hagemeister received the Medal of Honor.

On 01 April, in the midst of Operation PERSHING, Major General John J. Tolson, a veteran paratrooper, aviator and career soldier of 31 years, became the new commanding officer of the 1st Cavalry Division (Airmobile). Major General Tolson took over command from General Norton.

In early April planning began for Operation LEJEUNE. The principal reason behind this operation was an urgent Marine requirement to free up some of their troops in Quang Ngai for further movement north. Operation LEJEUNE combined the efforts of four military services: The US Army, Marine Corps, Air Force, and the Navy. The tactical air force support was substantial. The Duc Pho area had been effectively controlled by the communists for more than ten years. Over the years the Viet Cong and its political arm of the South Vietnam Communists, the National Liberation Front (NLF), had increased their power by political indoctrination, torture, and terrorism; and until now had a well-developed infrastructure from the sea coast as far inland as Ba To.

On 04 April the first movement of troops in support of Operation LEJEUNE began by diversion of "B" Troop, 1st Squadron, 9th Cavalry from Operation PERSHING to LZ Montezuma at Duc Pho. Following on 07 April, the 2nd Battalion, 5th Cavalry deployed to a prepositioned location of LZ English. On 08 April the 11th Aviation Group, located at LZ Two Bits, airlifted the 2nd Brigade Task Force into the LEJEUNE area of operations to assume operational control of the area. The remainder of the Marine task force was placed under operational control of the 2nd Brigade until such time as they could be moved north.

It was apparent right from the beginning that the enemy in that area had never before been challenged by airmobile tactics. It was several days before they began to appreciate the versatility and flexibility of the operational capability of the 1st Cavalry Division. The enemy chose to disperse and hide. Contact was only sporadic with the heaviest combat action on 16 April.

On 16 April, near Duc Pho, Specialist Fourth Class George A. Ingalls, "A" Company, 2nd Battalion, 5th Cavalry accompanied his squad on a night ambush mission. Shortly after the ambush was established, an enemy soldier entered the killing zone and was shot when he tried to evade capture. Other enemy soldiers were expected to enter the area, and the ambush was maintained in the same location. Two quiet hours passed without incident, then suddenly a hand grenade was thrown from the nearby dense undergrowth into the center of the position of the squad. The grenade did not explode, but shortly thereafter a second grenade

Troopers from the Rifle Platoon, "B" Troop, 1st Squadron, 9th Air Cavalry, 1st Cavalry Division leap from their Huey to seize a hill top about three kilometers west of Duc Pho in Quang Ngai Province during Operation LEJEUNE.(24 April 1967)

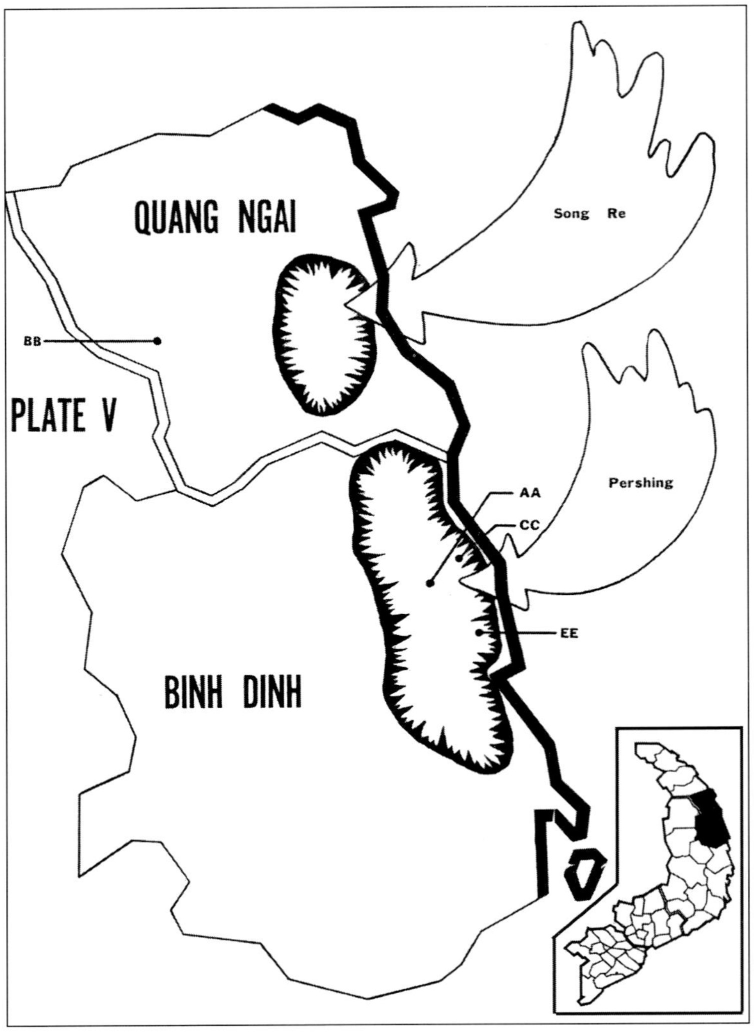

Bong Song Campaign Areas
(01 August 1967 to 18 February 1968)

landed directly between Specialist Fourth Class Ingalls and a nearby comrade. Although he could have jumped to a safe position, Specialist Fourth Class Ingalls, in a spontaneous act of great courage, threw himself on the grenade and absorbed its full blast. The explosion mortally wounded Specialist Fourth Class Ingalls, but his heroic action saved the lives of the remaining members of his squad. For his valiant action, Specialist Fourth Class George A. Ingalls received the Medal of Honor.

At noon on 22 April, Operation LEJEUNE was terminated. Although contacts were primarily light throughout the operation, 176 of the enemy had been killed and 127 captured. Operation LEJEUNE was unique in many ways. The deployment of the 2nd Brigade to the I Corps Tactical Zone was the first commitment of any large US Army unit in that area. More importantly, the engineering effort, including the lifting of 30 tons of equipment to build two tactical fixed-wing airstrips in a matter of a few days, was unparalleled in Army engineering history. Finally, the demonstrated "fire brigade" reaction capability of deploying a large task force in a day and a half to an entirely new area of operations proved again the flexibility of the airmobile concept. At Duc Pho the 1st Cavalry left behind two airstrips, an impressive line of communications, several critical connecting roads, and a damaged Viet Cong infrastructure. In light of the limited mission, Operation LEJEUNE was an unqualified success.

On 31 May a major engagement occurred when the 22nd NVA 9th Battalion came out of the mountains near An Qui to search for food. Nearly 100 died in the battle and those losses were soon followed by two other significant defeats near Dam Tra-O Lake and Soui Ca Valley. Captured NVA and VC soldiers revealed that morale was falling sharply within their units.

On 21 June, in the Binh Dinh province, Specialist Fifth Class Edgar L. McWethy Jr. "B" Company, 1st Battalion, 5th Cavalry, serving as a medical aidman, accompanied his platoon to the site of a downed helicopter. Shortly after the platoon established a defensive perimeter around the aircraft, a large enemy force attacked the position from three sides with a heavy volume of automatic weapons fire and grenades. The platoon leader and his radio operator were wounded almost immediately, and Specialist Fifth Class McWethy rushed across the fire-swept area to their assistance. Although he could not help the mortally wounded radio operator, Specialist Fifth Class McWethy's timely first aid enabled the platoon leader to retain command during this critical period. Hearing a call for aid, Specialist Fifth Class McWethy started across the open toward the injured men, but was wounded in the head and knocked to the ground. He regained his feet and continued on but was hit again, this time in the leg. Struggling onward despite his wounds, he gained the side of his comrades and treated their injuries. Observing another fallen rifleman lying in an exposed position raked by enemy fire, Specialist Fifth Class McWethy moved toward him without hesitation. Although the enemy fire wounded him a third time, Specialist Fifth

The "Chargers" of the 1st Battalion, 12th Cavalry quickly move across open ground after receiving sniper fire on an operation in the Bong Song Plain during Operation PERSHING. Armored tanks, in the background, provides a buffer zone for the advancing skytroopers. The tanks are from "D" Company, 1st Battalion (Mechanized), 50th Infantry. (August 1967)

Troopers of "C" Company, 2nd Battalion, 8th Cavalry move out from Landing Zone "Lou" to change positions with another company in the area. (24 August 1967)

Class McWethy reached his fallen companion. Though weakened and in extreme pain, Specialist Fifth Class McWethy gave the wounded man artificial respiration but suffered a fourth and fatal wound. Through his indomitable courage, complete disregard for his safety, and demonstrated concern for his fellow soldiers, Specialist Fifth Class McWethy inspired the members of his platoon and contributed in great measure to their successful defense of the position and the ultimate rout of the enemy force. For his valiant action, Specialist Fifth Class Edgal L. McWethy Jr. received the Medal of Honor.

On the same day, and same action of 21 June, near Binh Dinh Province, Specialist Fourth Class Carmel B. Harvey Jr., "B" Company, 1st Battalion, 5th Cavalry distinguished himself as a fire team leader during combat operations. Ordered to secure a downed helicopter, his platoon established a defensive perimeter around the aircraft, but shortly thereafter a large enemy force attacked the position from three sides. Specialist Fourth Class Harvey and two members of his squad were in a position directly in the path of the enemy onslaught, and their location received the brunt of the fire from an enemy machine gun. In short order, both of his companions were wounded, but Specialist Fourth Class Harvey covered this loss by increasing his deliberate rifle fire at the foe. The enemy machine gun seemed to concentrate on him and the bullets struck the ground all around his position. One round hit and armed a grenade attached to his belt. Quickly, he tried to remove the grenade but was unsuccessful. Realizing the danger to his comrades if he remained and despite the hail of enemy fire, he jumped to his feet, shouted a challenge at the enemy, and raced toward the deadly machine gun. He nearly reached the enemy position when the grenade on his belt exploded, mortally wounding Specialist Fourth Class Harvey, and stunning the enemy machine gun crew. His final act caused a pause in the enemy fire, and the wounded men were moved from the danger area. Specialist Fourth Class Harvey's dedication to duty, high sense of responsibility, and heroic actions inspired the others in his platoon to decisively beat back the enemy attack. For his valiant action, Specialist Fourth Class Carmel B. Harvey Jr. received the Medal of Honor.

In September, a fresh fighting unit linked up with the First Team. Now the bewildered enemy also faced the rumbling Armored Personnel Carriers (APC) of the 1st Battalion (Mechanized), 50th Infantry Division. The APCs figured prominently in the battle of Tam Quan in December. The lumbering vehicles eased the task of assaulting the bunkers and entrenchments of the NVA. Other units involved in the smashing victory Tam Quan included 1st Squadron, 9th Cavalry, "B" Company, 2nd Battalion, 8th Cavalry, 1st Battalion, 12th Cavalry and the ARVN 40th Regiment.

On 15 December, near My An, Binh Dinh province, Sergeant Allen James Lynch, "D" Company, 1st Battalion, 12th Cavalry distinguished himself while serving as a radio telephone operator. While serving in the forward element on an operation near the village of My An, his unit became heavily engaged with a numerically superior enemy force. Quickly and accurately assessing the situation, Sergeant Lynch provided his commander with information which subsequently proved essential to the unit's successful actions. Observing three wounded comrades lying exposed to enemy fire, Sergeant Lynch dashed across fifty meters of open ground through a withering hail of enemy fire to administer aid. Reconnoitering a nearby trench for a covered position to protect the wounded from intense hostile fire, he killed two enemy soldiers at point blank range. With the trench cleared, he unhesitatingly returned to the fire-swept area three times to carry the wounded men to safety. When his company was forced to withdraw by the superior firepower of the enemy, Sergeant Lynch remained to aid his comrades at the risk of his life rather than abandon them. Alone, he defended his isolated position for two hours against the advancing enemy. Using only his rifle and a grenade, he stopped them just short of his trench, killing five. Again, disregarding his safety in the face of withering hostile fire, he crossed seventy meters of exposed terrain five times to carry his wounded comrades to a more secure area. Once he had assured their comfort and safety, Sergeant Lynch located the counterattacking friendly company to assist in directing the attack and evacuating the three casualties. For his valiant action, Sergeant Allen James Lynch received the Medal of Honor.

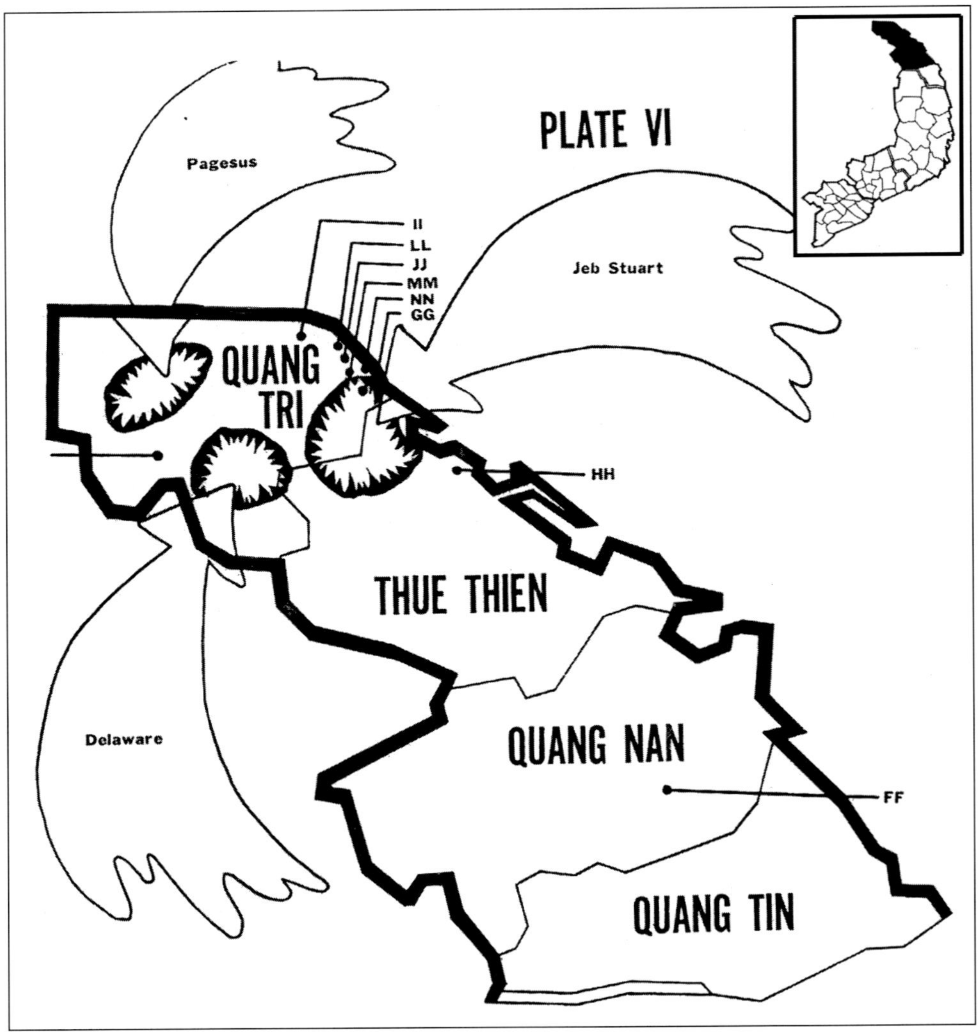

I Corps Campaigns
(01 April to 03 November 1968)

consider Operation PERSHING to have become a tedious, unglamorous mission that produced eighteen major engagements and numerous minor skirmishes in the eleven month campaign, it also revealed the bold, unselfish dedication to duty and country demonstrated by members of the First Team. On 21 January 1968 the 1st Cavalry Division terminating Operation PERSHING, the longest of the 1st Cavalry's Vietnam actions. When the operation ended, the enemy had lost 5,401 soldiers and 2,400 enemy soldiers had been captured. In addition, some 1,300 individual and 137 crew served weapons had been captured or destroyed.

On 22 January, Operation JEB STUART began in the Quang Tri area. It covered the initial move of major elements of the 1st Cavalry Division into northern I Corps following Operation PERSHING. Launched as a search and destroy mission aimed at enemy Base Areas 101 and 114, it had another goal of reinforcing the Marines in I Corps. The battles associated with the upcoming Tet Offensive would quickly overshadow the search and destroy nature of JEB STUART. As a result of Operation JEB STUART, the 1st Brigade was near Quang Tri City and was able to move quickly to support the Tet defenders.

On 12 January 1968, in the vicinity of Que Son Valley, Heip Duc Province, Sergeant William D. Port, "C" Company, 5th Battalion, 7th Cavalry was engaged in combat with the enemy. As the platoon of Sergeant Port was moving to cut off a reported movement of enemy soldiers, they came under heavy fire from an entrenched enemy force. The platoon was forced to withdraw due to the intensity and ferocity of the fire. Although wounded in the hand as the withdrawal began, Sergeant Port, with complete disregard for his safety, ran through the heavy fire to assist a wounded comrade back to the safety of the platoon perimeter. As the enemy forces assaulted in the perimeter, Sergeant Port and three comrades were in position behind an embankment when an enemy grenade landed in their midst. Sergeant Port, realizing the danger to his fellow soldiers, shouted the warning, "Grenade," and unhesitatingly hurled himself towards the grenade to shield his comrades from the explosion. Through his exemplary courage and devotion he saved the lives of his fellow soldiers and gave the members of his platoon the inspiration needed to hold their position. For his valiant action, Sergeant William D. Port received the Medal of Honor.

For nearly a year the Division scoured the Bong Son Plain, An Lo Valley and the hills of coastal II Corps, seeking out enemy units and their sanctuaries. Although many may

On 31 January, amid the celebration of the Vietnamese New Year of the "Monkey", the enemy launched the Tet Offensive, a major effort to overrun South Vietnam. Some 7,000 well equipped, crack NVA regulars blasted their way into the Imperial City of Hue, overpowering all but a few pockets of resistance held by ARVN troops and the US Marines. Within twenty-four hours, the invaders were joined by an equal number of NVA reinforcements. Almost simultaneously, to the North of Hue, five battalions of North Vietnamese and Viet Cong attacked Quang Tri City, the capital of Vietnam's northern province. The 1st Cavalry Division went on the move with four companies of skytroopers from the 1st Battalions of the 5th and 12th Cavalry who arrived at the village of Thon An Thai, just east of Quang Tri. Under heavy aerial rocket attack, the enemy quickly broke off the Quang Tri attack, dispersed into small groups and attempted to escape. Quang Tri was liberated within the next ten days.

On the day the Tet Offensive began, 31 January, Chief Warrant Officer Fredrick Ferguson, "C" Company, 227th Aviation Battalion overheard a call from wounded passengers and crewmen of a downed helicopter under heavy attack within the enemy controlled city of Hue. He unhesitatingly volunteered to attempt an evacuation. Despite warnings from all aircraft to stay clear of the area due to heavy antiaircraft fire, Chief Warrant Officer Ferguson began a low-level flight at maximum airspeed along the Perfume River toward the tiny, isolated South Vietnamese Army compound in which the crash survivors had taken refuge. Coolly and skillfully maintaining his course in the face of intense, short range fire from enemy occupied buildings and boats, he displayed superior flying skill and tenacity of purpose and landed his aircraft in an extremely confined area in a blinding dust cloud under heavy mortar and small-arms fire. Although the helicopter was severely damaged by mortar fragments during the loading of the wounded, Chief Warrant Officer Ferguson disregarded the damage and, taking off through the continuing hail of mortar fire, he flew his crippled aircraft on the return route through the rain of fire that he had experienced earlier and safely returned his wounded passengers to friendly control. For his valiant action, Chief Warrant Officer Fredrick Ferguson received the Medal of Honor.

On 02 February, the 2nd Battalion, 12th Cavalry made an initial assault outside of an ARVN outpost named PK-17, ten kilometers northwest of Hue. On 03 February, under dense fog conditions the battalion spotted the NVA troops at a rice farming hamlet named Thon La Chu. The strength of the enemy was estimated to be over one-thousand troops. The 2nd Battalion started moving across the field just before noon, every man a target. The advance, under small arms fire was slowed due to the lack of artillery support. The defenders had every advantage. The battalion was laying in open rice paddies, more or less pinned down. However, they were able to maintain their perimeter and hold off the enemy, even though they were short on ammunition, food and water. During the night the enemy surrounded their position.

In the morning, by the time they had fought to the tree line at the other side of the open field, nearly half of the 400 man battalion became casualties. Nine hours afterwards, the artillery units support fire began. The weather was lousy with monsoon rains, low clouds and fog, that there was great difficulty in trying to get casualties medevaced out. Finally, during the evening of the second day of the battle, two gunships of the 1st Squadron, 9th Cavalry were able to get in and airlifted some of the most seriously wounded to Camp Evans. That night, the battalion commander, Lieutenant Colonel Richard Sweet, made the decision to walk out of the encirclement under the cover of darkness.

That night, whatever force looks after infantrymen was with the remnants of the 2nd Battalion. The tattered, limping column made it to a mountain top where the NVA could not follow. At sunrise of 05 February, the battalion was positioned on a hill overlooking the North Vietnamese. Days later, the 2nd Battalion found their casualty rate was over 60 percent of its combat strength.

Following fierce fighting at Thon La Chu or "The Battle of Tee Tee Woods", the 3rd Brigade had been given the difficult mission of driving the Communists from Hue and the surrounding areas. The battle for Thon La Chu was far from being finished. They traded fire for two days and nights, then the skytroopers eluded the enemy with a risky night march. At sunrise of 05 February, the Brigade was positioned on a hill overlooking the North Vietnamese. By 11 February the enemy was blocked both on the north and south, but remained too strong and well entrenched for a frontal attack. On 21 - 22 February, the Brigade freed Thon La Chu and moved toward Hue where much of the fighting would be house-to-house. That they fought again for many more days is a tribute to inspired leadership and the fighting spirit of the young men. After the 1st Battalion,7th Cavalry overcame severe resistance, the southwest wall of the city was soon taken and they linked up with the 5th Battalion, 7th Cavalry.

The 1st Squadron, 9th Cavalry blocked the escape route to the west. Two more battalions of the 1st Cavalry Division also joined in the fight. The 5th Battalion, 7th Cavalry was stalled for a day by heavy enemy fire a kilometer outside Hue. The artillery blasted the NVA positions and the "GarryOwen" unit linked up with its Bravo Company which had made an air assault into Hue. By late February the NVA and Viet Cong invaders were driven from Hue ending the Tet Offensive. The NVA and Viet Cong had suffered a massive defeat, with 32,000 killed in action and 5,800 others captured.

Following the liberation of Hue, the 1st Cavalry Division began repositioning its forces to allow for the arrival of its other elements from the Central Highlands who had began their move north for a consolidation in I Corps in Thua Thien Province, the northeast province of South Vietnam. As the Tet offensive operations were being concluding at the end of February 1968, the last elements of the 1st Cavalry Division left the Central Highlands and completed their move north into the I Corps Area of Operation (AO).

The 1st Brigade continued security of Quang Tri and conducted operations south and west of the city. The 2nd Brigade, upon its arrival conducted operations to clear the enemy elements in the Hai Lang-My Chanh area. On 01 March, the 3rd Brigade, moved north of Hue City, along Route 1, and assumed the mission of base defense for Camp Evans, the center of operations the new headquarters of the division. The facility had been originally constructed by the Naval Mobile Construction Battalion One (NMCB1) for the 4th Marine Regiment. The new immediate mission of the 1st Cavalry Division was to keep pressure on the enemy units attempting to withdraw and escape from the Tet Offensive.

Throughout the weeks that the 1st Cavalry Division was engaged in the Tet Offensive, the 26th Marines at Khe Sanh were besieged by heavy artillery bombardment and ground attack. The North Vietnamese had ringed the lines of the Marines with a build up of entrenched infantry and a multitude of antiaircraft weapons. On 11 March, at the

"GarryOwen" Troopers of "D" Company, 2nd Battalion, 7th Cavalry dash toward a helicopter that will carry them to another location in their drive to relieve the Marines at Khe Sanh, the objective of Operation PEGASUS. (06 April 1968)

request of the Marine Command, the 1st Cavalry Division began a detailed plan, Operation PEGASUS, designed to relieve the Marines at Khe Sanh.

To accomplish the mission, the 1st Cavalry Division was augmented by the non-divisional units of the 1st Marine Regiment, 26th Marine Regiment, III Army of the Republic of Vietnam Airborne Task Force and the 37th Army of the Republic of Vietnam Ranger Battalion.

It became evident during the planning that the construction of an airstrip in the vicinity of Ca Lu would be a key factor for the entire operation. The airstrip, which became known as LZ Stud, had to be ready well before D-day (01 April 1968). Also, it was necessary to upgrade Highway Nine between the "Rock Pile" and Ca Lu to allow pre-stocking of supplies at LZ Stud. Construction of the airstrip and road improvements were assigned to a team of the 1st Cavalry Division engineers, the Seabees USN Mobile Construction Battalion #5, and the 11th Engineer Battalion,

Having established a forward base of operations, the second key element to the success of this plan was the closely integrated reconnaissance and fire support effort of the 1st Squadron, 9th Cavalry and air, artillery, and B-52 arc light strikes which were carried out during the 6 days preceding the launch of the main attack.

The 1st Squadron, 9th Cavalry operated from LZ Stud in gradually increasing concentric circles up to the Khe Sanh area, working with air cover from the 7th Air Force or the 1st Marine Air Wing. The Cavalry Squadron was almost the only means available to pinpoint enemy locations, antiaircraft positions, and strong points that the division would try to avoid in the initial assaults. The squadron was also responsible for the selection of critical landing zones. Their information proved to be timely and accurate.

At 0700 hours, on 01 April the attack phase of Operation PEGASUS commenced as the 3rd Brigade of the 1st Cavalry was airlifted by Chinooks and Hueys into LZ Stud in preparation for an air assault into two objective areas further west. Weather delayed the attack until 1300, when the 1st Battalion, 7th Cavalry air assaulted into LZ Mike, located on prominent ground south of Highway 9 and well forward of the Marine attack. The 2nd Battalion, 7th Cavalry followed into the same landing zone to expand the position. The 5th Battalion, 7th Cavalry air assaulted into an area north of Highway 9 approximately opposite LZ Mike.

The bad weather of D-day was to haunt the 1st Cavalry throughout Operation PEGASUS. "Good weather" was considered to be any condition when the ceiling was above 500 feet and slant range visibility was more than a mile and a half. The bad weather further proved the soundness of establishing LZ Stud as the springboard for the assaults. Troops, ammo and supplies could be assembled there ready to go whenever the weather to the west opened up. Marshalling areas further away would have drastically deteriorated response time.

On 02 April, the 1st Marine Regiment continued its ground attack along the axis of Highway 9. Two Marine companies made limited air assaults to support the Regiment's momentum. The 3rd Brigade air assaulted the 2nd Battalion, 7th Cavalry into a new position further to the west while the other two battalions improved their positions. The 2nd Brigade moved into marshalling areas in preparation for air assaults the next day.

The initial thrusts had met less enemy resistance than expected. As a consequence, the 2nd Brigade was thrown into the attack a day earlier than the original schedule with its three battalions, the 1st and 2nd Battalions, 5th Cavalry and 2nd Battalion, 12th Cavalry, moved into two new areas south and west of our earlier landing zones. Under enemy artillery during the assaults, their objectives were secured without serious difficulty.

On the 4th of April, the 2nd Brigade assaulted one battalion into an old French fort south of Khe Sanh. Initial contact resulted in four enemy killed. Continuing the attack the next day, heavy resistance was encountered.

On 06 April, units of the 1st Brigade entered the operation with the 1st Battalion air assaulting into LZ Snapper, due south of Khe Sanh and overlooking Highway 9. The circle began to close around the enemy. As units were airlifted into the various LZs along Highway 9, they did not have the knowledge of the final operational plans. It had been decided by the 1st Cavalry Divisional Tactical Operation Center (DTOC), at Camp Evans, to only provide for their general distribution once all of the units were at final attack positions. With all units in position, a courier was dispatched to carry the plans to the commands of all units and the Marines at Khe Sanh.

The heaviest contact on that date occurred in the 3rd Brigade's area of operation as the 2nd Battalion, 7th Cavalry continued its drive west on Highway 9. Enemy blocking along the highway offered stubborn resistance. In a day-long battle which ended when the enemy summarily abandoned his position and fled, the battalion had accounted for 83 enemy killed, captured one prisoner and 121 individual and ten crew-served weapons. The troops of the 1st Cavalry Division were airlifted to Hill

471 relieving the Marines at this position. This was the first relief of the defenders of Khe Sanh. Two companies of troopers remained on the hill while two other companies initiated an attack to the south toward the Khe Sanh Hamlet.

The 1st Cavalry Division forces on LZ Snapper were attacked by an enemy force utilizing mortars, hand grenades, and rocket launchers. The attack was a disaster for the enemy and twenty were killed. At 1320 hours the 84th Company of the Vietnamese 8th Airborne Battalion was airlifted by 1st Cavalry Division aircraft into the Khe Sanh Combat Base and linked up with elements of the 37th Ranger Battalion. The lift was conducted without incident and was the official link-up in forces at Khe Sanh.

On 07 April, the South Vietnamese III Airborne Task Force air assaulted three battalions into positions north of the road and east of Khe Sanh to block escape routes toward the Laotian border. Fighting throughout the area was sporadic as the enemy attempted to withdraw. American and South Vietnamese units began picking up significant quantities of abandoned weapons and equipment. The old French fort which was the last known enemy strong point around Khe Sanh was completely secured.

At 0800 hours on 08 April, the relief of Khe Sanh was effected and the 1st Cavalry Division became the new landlord. The 3rd Brigade airlifted its command post into Khe Sanh and assumed the mission of securing the area. This was accomplished after the 2nd Battalion, 7th Cavalry successfully cleared Highway 9 to the base and effected link-up with the 26th Marine Regiment. The 3rd Brigade elements occupied high ground to the east and northeast of the base with no enemy contact. At this time it became increasingly evident, through lack of contact and the large amounts of new equipment being found indiscriminately abandoned on the battlefield, that the enemy had fled the area rather than face certain defeat.

On 09 April, all 1st Marine Regiment objectives had been secured and Highway 9 was repaired and secured with only scattered incidents of enemy sniper fires. Enemy mortar, rocket and artillery fire into Khe Sanh became increasingly sporadic. Mop up operations continued.

On 10 April, pursuing the retreating North Vietnamese, the 1st Battalion, 12th Cavalry recaptured the Special Forces camp at Lang Vei, four miles west of Khe Sanh and seized large stockpiles of supplies and ammunition. This action became the last major encounter with in Operation PEGASUS as later in the day, orders were received to extract the 1st Cavalry Division as soon as possible to prepare for Operation DELAWARE/LAM SON 216, an assault into the A Shau Valley. Advance units started the pull out the next day and returned to the base areas at Quang Tri City and Camp Evans.

Limited operations continued until 15 April when Operation PEGASUS was officially terminated. The 1st Cavalry Division had scored a major airmobile victory by quickly reaching the besieged Marines Khe Sanh bastion without setbacks or heavy losses. The careful planning and preparation preceding the raid was backed up by aggressive and innovative tactics during its execution. The final statistics of Operation PEGASUS totaled 1,259 enemy killed and more than 750 weapons captured.

The next major encounter of the 1st Cavalry Division began with the strategic preparations for an assault into A Shau Valley. During the period 14 to 19 April, over 100 B-52 sorties, 200 Air Force and Marine fighter sorties and numerous aerial rocket artillery missions were flown against targets in the valley. In parallel, the 1st Squadron, 9th Cavalry of the 1st Cavalry Division began an extensive aerial reconnaissance of the A Shau Valley to select flight routes, locate antiaircraft and artillery weapons, and to develop targets for tactical air and B-52 strikes.

On 19 April, Operation DELAWARE/LAM SON 216 was launched into the cloud shrouded A Shau Valley. A Shau, located within 15 Kilometers of the Laotian border and 45 kilometers west of Hue, was situated between two high mountain ranges on the western edge of the Republic of Vietnam. On both sides of the valley, the mountains were over 1,000 meters in height, with the angle of slopes varying from 20 to 45 degrees. Three abandoned airfields were spread along the valley floor which ran northwest to southeast. The North Vietnamese forces had been in control of the valley since March 1966 when they overran the Special Forces camp located at the southern end. Since that time the valley had become a North Vietnamese Army base which was a major way-station and supply route known as the Ho Chi Minh trail which was used for the infiltration of personnel and supplies from North Vietnam through Laos along Route 547 into Thua Thien Province and the northern I Corps Tactical Zone.

A Shau Valley had been the jumping-off point for the Tet offensive against Hue. No free world forces had

A Chinook gunner, aboard the last ship to leave A Shau Valley, looks back on the scarred valley floor where so many fierce battles raged and so many troopers made the highest sacrifice possible for their comrades-in-arms. (17 May 1968)

penetrated A Shau Valley since a Special Forces Camp pulled out of the area in 1966 and the NVA began to think of it as an inviolable sanctuary. Operation DELAWARE/LAM SON 216 was a coordinated airmobile and ground attack using elements of three divisions - the 1st Cavalry, the 101st Airborne, and the 1st Army of the Republic of Vietnam Division. One prong was to be the attack along and astride Routes 547 and 547A, while the main attack was the assault into A Loui and Ta Bat on the valley floor.

For Operation DELAWARE/LAM SON 216, the 1st Cavalry Division utilized the 1st and 3rd Brigades and controlled coordination with the 101st Airborne and the 3rd Army of the Republic of Vietnam Regiment. The 2nd Brigade, 1st Cavalry Division remained in the Khe Sanh area to secure the base and continue limited operations in that area. This left the Camp Evans area almost completely void of combat units. Consequently, the 196th Light Infantry Brigade, Americal Division was given the mission of defense of the base at Camp Evans under operational control of the 1st Cavalry Division. As a further mission the 196th Light Infantry Brigade was also designated as a reserve unit for the I Corps.

The swift-moving skytroopers of the 1st Cavalry Division, fresh from their relief of Khe Sanh earlier in the month, leaped into the A Shau without ground support. Slinging artillery pieces below giant helicopters and pouring out murderous fire from rocket carrying gunships, they showed the 3rd NVA that there was no such thing as an enemy sanctuary in South Vietnam. While administering this lesson, the men of the 1st and 3rd Brigades seized tons of abandoned supplies and equipment.

In the third massive heliborne assault within a month, the 227th Assault Helicopter Battalion (AHB), 228th Assault Support Helicopter Battalion (ASHB) and 229th AHB prepared to commit their maximum available resources in support of the operation. The plan of operation was to assault into the A Shau Valley and simultaneously to insert a Long Range Reconnaissance Patrol to secure Signal Hill, a 4,879 foot peak 5 kilometers northeast of A Luoi Airfield to be used as a communications relay station. The weather, which played a harassing part throughout the twenty-nine day operation, was a heavy overcast on the plains and solid overcast above the valley floor. Cloud tops at about 4,500 feet surrounded the top of Signal Hill.

With the main assault aircraft on a temporary weather hold, "B" Company, 227th AHB, with four UH-1 helicopters, a command and control ship, and two escort gunships from "D" Company were suddenly in the position of the lead assault element. They departed Camp Evans, climbed through the thin overcast, and were vectored to the initial landing zone. The first view of the LZ showed it to be a bomb crater on a 40 degree slope surrounded by fifty foot trees. Touchdown was impossible and the Long Range Reconnaissance Patrol (LRRPs) would have to be rappelled from about thirty feet. The lead aircraft, unable to maintain a position over the LZ, made a "go around." The second aircraft attempting the drop, lost power and crashed into the LZ, sliding down the hill into the trees. The lead aircraft shifted its approach axis from west to east. On its second approach, the troopers successfully rappelled and extracted the miraculously uninjured crew of the downed aircraft. A total of fourteen sorties were lifted onto Signal Hill without further incident. From the initial assault it looked like a milk run, with no significant anti-aircraft fire being encountered.

Once on the valley floor, the intelligence efforts of the 1st Cavalry Division was supplemented by a small unit composed of electronic experts from the Department of Defense. Working quickly, they buried then-newly developed acoustic sensors in the A Shau fields. These sensors, through triangulation of sound waves propagated through the ground to listening devices - in a manner as in underwater sonar, provided the team with data on locations of enemy troop movements which supplemented other real time intelligence capabilities during the operation. Other sensors were permanently installed, remaining unattended, so that enemy movements could be monitored after US forces left the area.

On 20 April, the second day of the assault into the A Shau began with a hold due to a low overcast and fog on the valley floor. At approximately 0900 hours, "A" and "C" Companies, 227th AHB left to assault a ridge line four kilometers southwest of LZ Tiger to establish a third fire base, LZ Pepper. The helicopter approach was made from the north along the ridge into a slope studded with stumps. Again the aircraft experienced power loss at a hover and crashed, closing the LZ. Troopers of 5th Battalion, 7th Cavalry rappelled with chain saws to expand the LZ. Upon resumption of the lift, the new lead aircraft came too close to the trees on take off and also crashed. The crew got out unhurt and remained on Pepper for two nights due to bad

A member of the 11th Pathfinders Company, assigned to the 1st Brigade, directs the landing of a UH-1D Huey. The LZ, under the control of the 1st Battalion, 12th Cavalry was instrumental in the successful operation of Operation JEB STUART III. (August 1968)

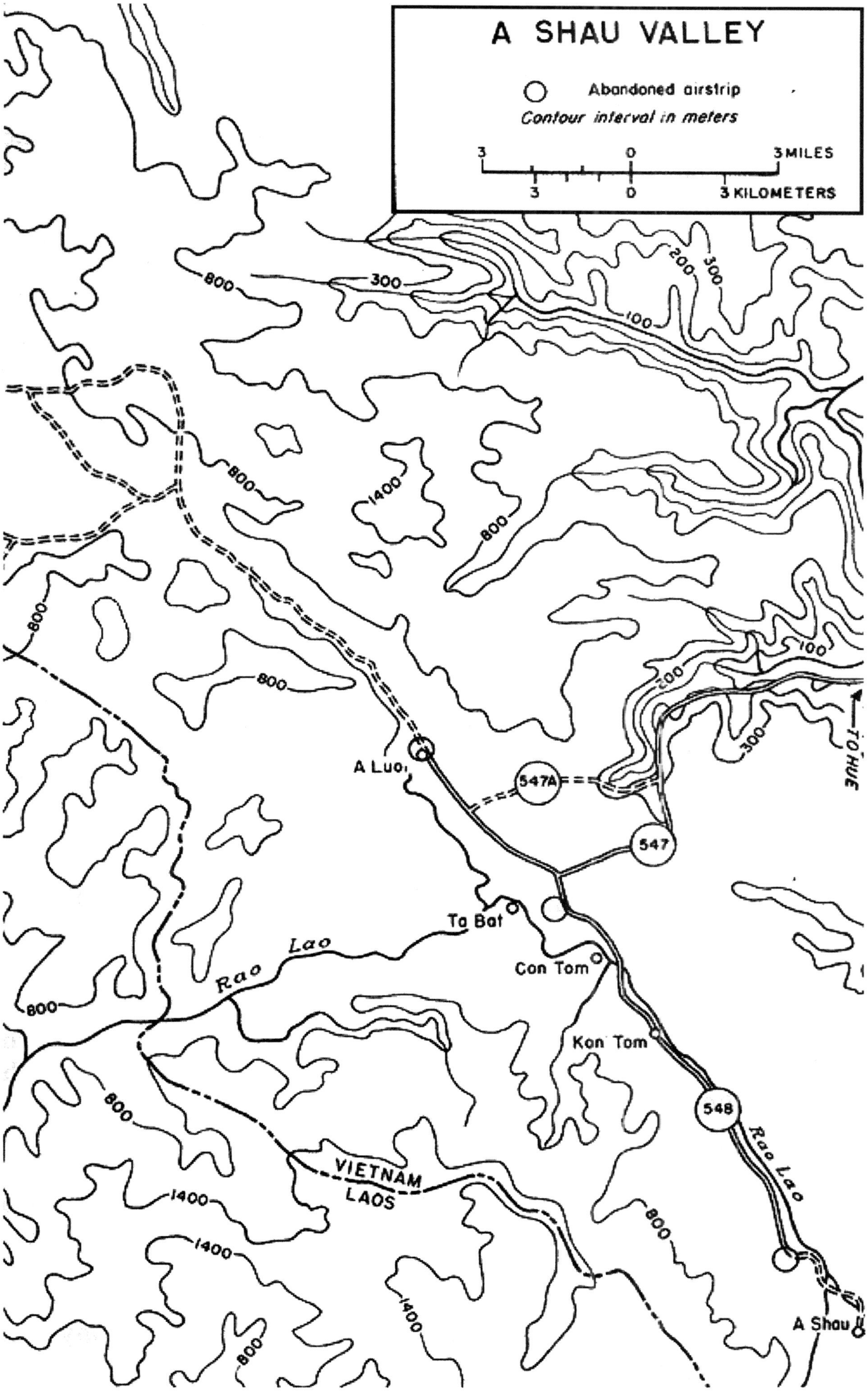

A Shau Valley

weather that restricted any further flying into the valley. Only one infantry company was airlifted into LZ Pepper before the deteriorating weather halted all flight operations.

Bad weather prevailed throughout the entire day of 21 April, preventing any air assaults from being conducted. "D" Company with 6 aircraft was on standby to haul a badly needed resupply to LZ Tiger. Shortly after noon the aircraft took off at 1 minute intervals with an eight hundred pound sling load under each aircraft. One by one, the aircraft were radar vectored to a point over LZ Tiger. Each aircraft had to feel its way through the holes in the overcast and create a separate approach axis. The gambit proved quite effective and a total of fourteen sling loads were taken into both LZ Tiger and LZ Pepper before the weather closed in again. For most of the crews, it was their first actual instrument flight. The next day, 22 April, brought much improved weather conditions and the airlift into LZ Pepper continued with all available aircraft being committed. At the end of the day, the 3rd Brigade was firmly entrenched in the A Shau Valley with three infantry battalions plus supporting artillery.

On 24 April, "A" and "B" Companies, 227th AHB, each with six aircraft, joined the 229th AHB to air assault elements of the 1st Brigade into LZ Cecile, two kilometers south of A Luoi Airfield. Again weather played a significant part forcing the lifts to climb as high as 11,000 feet to clear the cloud tops. Once over the valley and through the holes in the overcast the aircraft quickly maneuvered to the new LZ Cecile located at the southern end of a ridge about 2,200 feet high. Every aircraft was exposed to sniping fire on the approach and to automatic weapons positioned about 500 meters down the ridge as it departed no matter which way it broke. Very few aircraft sustained hits and none were lost as the 2nd Battalion, 8th Cavalry completed its liftoff to LZ Cecile prior to 1400 hours.

On 25 April, Captain James M. Sprayberry, "D" Company, 5th Battalion, 7th Cavalry, distinguished himself by exceptional bravery while serving as Executive Officer. His company commander and a great number of the men were wounded and separated from the main body of the company. A daylight attempt to rescue them was driven back by the well entrenched enemy's heavy fire. Captain Sprayberry then organized and led a volunteer night patrol to eliminate the intervening enemy bunkers and to relieve the surrounded element. The patrol soon began receiving enemy machine gun fire. Captain Sprayberry quickly moved the men to protective cover and without regard for his own safety, crawled within close range of the bunker from which the fire was coming. He silenced the machine gun with a hand grenade. Identifying several one man enemy positions nearby, Captain Sprayberry immediately attacked them with the rest of his grenades. He crawled back for more grenades and when two grenades were thrown at his men from a position to the front, Captain Sprayberry, without hesitation, again exposed himself and charged the enemy-held bunker killing its occupants with a grenade. Placing two men to cover his advance, he crawled forward and neutralized three more bunkers with grenades. Immediately thereafter, Captain Sprayberry was surprised by an enemy soldier who charged from a concealed position. He killed the soldier with his pistol and with continuing disregard for the danger, neutralized another enemy emplacement. Captain Sprayberry then established radio contact with the isolated men, directing them toward his position. When the two elements made contact, he organized his men into litter parties to evacuate the wounded. As the evacuation was nearing completion, he observed an enemy machine gun position which he silenced with a grenade. Captain Sprayberry returned to the rescue party, established security, and moved to friendly lines with the wounded. This rescue operation, which lasted approximately seven and one half

Artillerymen of the 2nd Battalion, 19th Artillery were located in the rolling and forested terrain of LZ Davis in the rugged foothills near Dak To. Their mission was to provide fire support for Operation COMANCHE FALLS. (October 1968)

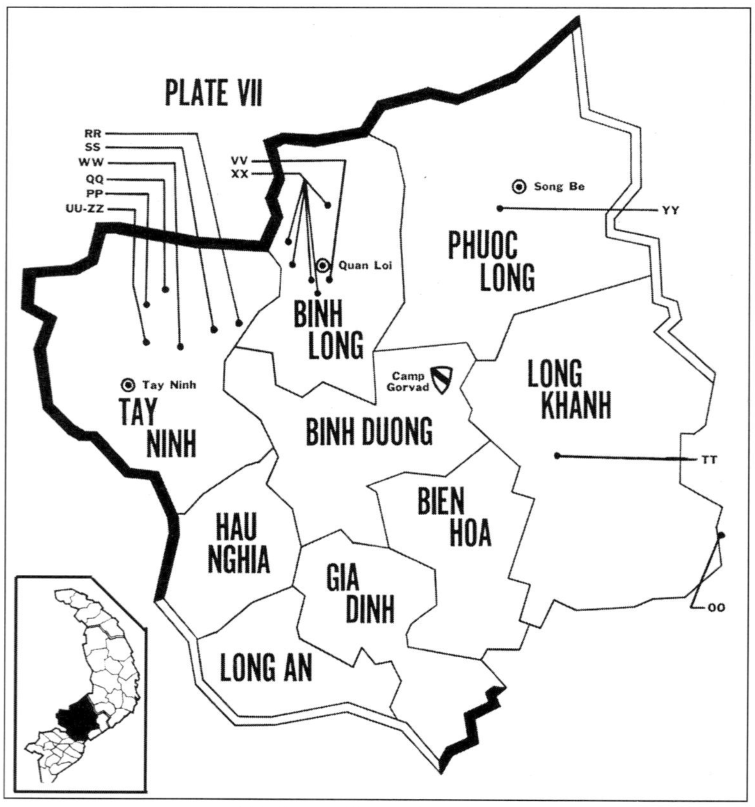

III Corps Campaign Areas (28 October 1968 to 26 March 1971)

A heavily armed Navy gunboat cruises the Vam Co Dong River, carrying the waterborne Jumping Mustangs conducting joint operations with the Navy. (December 1968)

A CH-54, "Flying Crane", helicopter from the 1st Cavalry Division (Airmobile) brings in a bulldozer to members of "A" Company, 8th Engineer Battalion at Landing Zone CAROLINE. (1969)

hours, saved the lives of many of his fellow soldiers. Captain Sprayberry personally killed twelve enemy soldiers, eliminated two machine guns, and destroyed numerous enemy bunkers. For his valiant action, Captain James M. Sprayberry received the Medal of Honor.

On the same day as the action of the 5th Battalion, the 1st Battalion, 7th Cavalry secured the abandoned airfield at A Luoi, enabling the 1st Brigade to land safely on the valley floor. Meanwhile, as the troopers moved out from the LZ, they found stocks on ammunition and abandoned anti-aircraft guns mounted on flatbed trucks. The retreating enemy soldiers had attempted to take as much of their stores as possible.

On the 29th of April, the 8th Engineer Battalion began working on the airstrip at A Luoi with heavy equipment that had been brought in by CH-54's. The field was actually ready to accept short take off and landing Caribou aircraft by noon on 01 May. However, it was not until the following day that the first cargo aircraft, a C-7A Caribou, landed at nearby LZ Stallion at 1120 hours. The 8th Engineers continued making runway improvements, working on the A Luoi airstrip so that it could handle the larger C-123 and C-130 cargo aircraft for resupply and extraction phases of the mission. Their task of upgrading the A Luoi airstrip to a C-130 capacity was completed on 03 May.

On 04 May, in A Shau Valley, First Lieutenant Douglas B. Fournet, "B" Company, 1st Battalion, 7th Cavalry distinguished himself in action while serving as rifle platoon leader of the 2d Platoon, "B" Company. While advancing uphill against fortified enemy positions in the A Shau Valley, the platoon encountered intense sniper fire, making movement very difficult. The right flank man suddenly discovered an enemy claymore mine covering the route of advance and shouted a warning to his comrades. Realizing that the enemy would also be alerted, First Lieutenant Fournet ordered his men to take cover and ran uphill toward the mine, drawing a sheath knife as he approached it. With complete disregard for his safety and realizing the imminent danger to members of his command, he used his body as a shield in front of the mine as he attempted to slash the control wires leading from the enemy positions to the mine. As he reached for the wire the mine was detonated, killing him instantly. Five men nearest the mine were slightly wounded, but First Lieutenant Fournet's heroic and unselfish act spared his men of serious injury or death. For his valiant action, First Lieutenant Douglas B. Fournet received the Medal of Honor.

For the next weeks, the brigades scoured the valley floor, clashing with enemy units and uncovering huge enemy caches of food, arms and ammunition. On 10 May the extraction phase of Operation DELAWARE began. The weather, which had been poor throughout the operation began to deteriorate. "A" Company, 227th AHB was assigned the mission to extract the 1st and 5th Battalions, 7th Cavalry, from the A Shau Valley. However, about noon, "A" Company, 2nd Battalion, 7th Cavalry made heavy contact about two kilometers northwest of LZ Pepper and the original mission was changed to extract the unit back to LZ Pepper. The extraction was accomplished smoothly with the assistance of two gunships, two fighters and eight ARA helicopters. Once this was completed the flight returned to LZ Stallion to wait for the 1st Battalion, 7th Cavalry to get into a pickup posture.

For the remainder of the extraction phase of Operation DELAWARE, the aircraft of the 227th AHB gave maximum air support to the ground commanders. With the rains increasing as they left, the 1st Brigade destroyed its firebases and left the way they came, by helicopter. The skytroopers had throughly disrupted the favorite Viet Cong sanctuary and denied its use for any preemptive strike preparations. On 17 May Operation DELAWARE/LAM SON was closed out.

Completion of DELAWARE/LAM SON created a lull in operations which did not last long. On the night of 19 May, North Vietnamese Army rockets slammed into the ammunition supply dump area of Camp Evans. The secondary explosions of ammunition, creating a fireball

Utilizing aircraft from two different units with two entirely different missions, a LOH from "E" Battery, 82nd Artillery, and an aerial rocket (ARA) Cobra of the 2nd Battalion, 20th Artillery, gave Echo Battery its own hunter-killer teams. Nicknamed "Blue Pecker" the combination flew daily patrols against enemy infiltration along the numerous trails of the III Corps jungle. (April 1969)

which topped out at nearly 15,000 feet, caused significant damage to the helicopters on the flight line which was located directly across from the ammunition supply dump. Immediately after the explosion, the artillery launched a helicopter in an effort to determine the direction of the attack. It remained up all night with only brief refueling stops. Damage to the remaining helicopters of the 227th AHB and 228th ASHB was so severe that there was only one flight out the next morning for emergency supplies.

Shock waves and flying debris from the blasts created a lethal environment, burning many of the tents, damaging bunkers of the pilots and caving in many of the roofs. Damage estimates that the big blasts (7 of them) were entire ammunition revetments being ignited internally after being exposed to the intense heat of the surrounding fires. Each blast was 200,000 pounds of explosives going off at once, with the second blast being the largest. In all, there was 10,800,000 pounds of ammunition destroyed that night. In addition to the ammunition explosions, there was a Petroleum, Oil and Lubricants (POL) supply dump for vehicles on the south by the main road going thru Camp Evans, and a POL supply dump to the north for aviation gas which, when ignited by the rocket attacks and surrounding ammunition blasts, added to the conflagration. The damage, although extensive to the facilities and equipment, resulted in only one causality. The entire 1st Cavalry Division operations was nearly brought to its knees by the direct hit of a few rockets.

Immediately after the initial explosions, security guards on the perimeter began to fire at the retreating intruders. For the remainder of the night, the camp was inundated with explosion after explosion. With the return of the early morning calm, security levels were increased and no indigenous people were allowed on Camp Evans at night. Utilizing the recent learning experiences of locating hazard materials in the proximity of personnel and critical equipment, the new ammunition and POL supply dumps were relocated to one of the most remote areas of Camp Evans.

On 27 June, as part of Operation JEB STUART III, the 3rd Squadron, 5th (Armored) Cavalry, 9th Infantry Division operating under the control of the 2nd Brigade, 1st Cavalry Division, had been assigned the mission of securing the Wunder Beach Complex and the access road to Highway 1, not far from Camp Evans. At 0900 hours "C" Troop, 3rd Squadron, 5th (Armored) Cavalry and "D" Troop, 1st Squadron, 9th Cavalry came under Rocket Propelled Grenade (RPG) fire as they were engaged in a detailed search of an area known as "The Street Without Joy".

As an indication of a battle to come, the residents of the nearby seacoast village of Binah An, Quan Tri Province, began to flee the area. In the attempt to detain and question the villagers, an NVA solder, hiding among the crowd, was captured and interrogated. He revealed that the entire 814th NVA Infantry Battalion was in the village. "A" and "B" Troops of the 3rd Squadron, 5th Cavalry, along with "D" Troop, 1st Squadron, 9th Cavalry, closed on the village, joining "C" Troop, 3rd Squadron 5th Cavalry. There was no good way of the enemy to escape during daylight hours due to the clear view and superior firepower of the surrounding forces.

In addition to the control fire directed at the enemy in the village, additional firepower of aerial rocket and Marine artillery, from Quang Tri, was made available along with Tactical Air Control (TAC) aircraft from Da Nang and a naval destroyer, with five inch guns, offshore. In the next seven hours, all of the firepower pounded the enemy to reduce the position of the enemy. During the afternoon, "D" Company, 1st and "C" Company, 2nd Battalions, 5th Cavalry, airlifted into an adjacent LZ and closed on the village. Due to the possibility of the enemy infiltrating the lines during the night, it was decided to overrun the position of the enemy and destroy their capability for effective operations during the night. The guided missile cruiser *USS BOSTON* arrived at dusk and in an all night bombardment her basic load of eight inch shells were exhausted. It was a nervous night for the enemy soldiers within the tight cordon. Unorganized, some of the survivors attempted individual escapes and were soon rounded up with tanks having turret mounted searchlights and two swift Navy patrol boats operating close to the shoreline. At 0930 hours, the next morning, a final assault was made on the enemy. In the after battle assessment, two hundred thirty-three of the 814th NVA Infantry Battalion were KIA and forty-four were taken as Prisoners of War (POW) with the 5th Cavalry units experiencing only three causalities. (Editor's Note: This was the first time that lineage elements of the original "A", "B" and "C" Troops, 5th Cavalry Regiment had fought as a consolidated unit since 1943 in World War II.)

On 28 June, in Quang Tri Province, Specialist Fourth Class Hector Santiago-Colon, "B" Company, 5th Battalion, 7th Cavalry distinguished himself at the cost of his life while serving as a gunner in the mortar platoon of "B" Company. While serving as a perimeter sentry, Specialist Fourth Class Santiago-Colon heard distinct movement in the heavily wooded area to his front and flanks. Immediately he alerted his fellow sentries in the area to move to their foxholes and remain alert for any enemy probing forces. From the wooded area around his position heavy enemy automatic weapons and small-arms fire suddenly broke out, but extreme darkness rendered difficult the precise location and identification of the hostile force. Only the muzzle flashes from enemy weapons indicated their position. Specialist Fourth Class Santiago-Colon and the other members of his position immediately began to repel the attackers utilizing hand grenades, antipersonnel mines and small-arms fire. Due to the heavy volume of enemy fire and exploding grenades around them, a North Vietnamese soldier was able to crawl, undetected, to their position. Suddenly, the enemy soldier lobbed a hand grenade into Specialist Fourth Class Santiago-Colon's foxhole. Realizing that there was no time to throw the grenade out of his position, Specialist Fourth Class Santiago-Colon retrieved the grenade, tucked it in to his stomach and, turning away from his comrades, absorbed the full impact of the blast. His heroic self-sacrifice saved the lives of those who occupied the foxhole with him, and provided them with the inspiration to continue fighting until they had forced the enemy to retreat from the perimeter. For his valiant action, Specialist Fourth Class Hector Santiago-Colon received the Medal of Honor.

On 14 July, command of the 1st Cavalry Division was transferred from General Tolson to Brigadier General Richard L. Irby, a veteran of World War II and Korea, who was serving as Assistant Division Commander. This change of command was soon followed by the passing of the reins, on 19 August, to Major General George I. Forsythe. General Forsythe had served both in the Pacific and Europe in World War II and then commanded the 502nd Airborne Infantry in 1956-57. In 1958 he had been assigned as senior advisor to the South Vietnamese Army.

On 25 October, the 1st Cavalry Division received orders to move to the III Corps Tactical Zone, a huge area around Saigon, which stretched west and north to the Cambodian border. From 28 October to 15 November, in Operation LIBERTY CANYON, the equivalent of a small sized town, packed, took wings and strategically repositioned itself at the other end of South Vietnam. The First Team established Division Headquarters at Phouc Vinh, assuming control and mission responsibility in four provinces; Phouc Long, Binh Long, Tay Ninh and Binh Duong. In a change of tactics, the brigades were deployed in a 150 mile arc north of Saigon. The 3rd Brigade established its base of operations at Quan Loi and the 2nd Brigade set up their base camp at Tay Ninh. The 1st Brigade set up its operations at a central location firmly between the 2nd and 3rd Brigades. The Division now confronted four enemy divisions; the 1st and 7th NVA and the 5th and 9th VC.

On 03 December, near Quan Loi, Sergeant John N. Holcomb, "D" Company, 2nd Battalion, 7th Cavalry distinguished himself while serving as a squad leader during a combat assault mission. Sergeant Holcomb's company assault had landed by helicopter and deployed into a hasty defensive position to organize for a reconnaissance-in-force mission when it was attacked from three sides by an estimated battalion-size enemy force. Sergeant Holcomb's squad was directly in the path of the main enemy attack. With complete disregard for the heavy fire, he moved among his men giving encouragement and directing fire on the assaulting enemy. When his machine gunner was knocked out, Sergeant Holcomb seized the weapon, ran to a forward edge of the position, and placed withering fire on the enemy. His gallant actions caused the enemy to withdraw. Sergeant Holcomb treated and carried his wounded to a position of safety and reorganized his defensive sector despite a raging grass fire ignited by the incoming enemy mortar and rocket rounds. When the enemy assaulted the position a second time, Sergeant Holcomb again manned the forward machine gun, devastating the attacking enemy soldiers and forcing them to again break contact and withdraw. During the enemy withdrawal an enemy rocket hit Sergeant Holcomb's position, destroying his machine gun and severely wounding him.

A Nighthawk Team of the 11th General Support Aviation Company takes off at dusk to begin its night mission. The Huey is armed with a minigun and a powerful 50,000 watt, night-probing searchlight and is backed up by the rocket artillery of the "Blue Max" Cobra from the 2nd Battalion, 20th Artillery. (1969)

Aerial View of Fire Support Base ILLINGSWORTH located near the Cambodian border and operated by members of the 2nd Battalion, 8th Cavalry Regiment. (21 March 1970)

Despite his painful wounds, Sergeant Holcomb crawled through the grass fire and exploding mortar and rocket rounds to move the members of his squad, everyone of whom had been wounded, to more secure positions. Although grievously wounded and sustained solely by his indomitable will and courage, Sergeant Holcomb, as the last surviving leader of his platoon, organized his men to repel the enemy, crawled to the platoon radio and reported the third enemy assault on his position. His report brought friendly supporting fires on the charging enemy and broke the enemy attack. For his valiant action, Sergeant John R. Holcomb received the Medal of Honor.

The beginning of 1969 found the 1st Cavalry Division and the ARVN forces engaged in Operation TOAN THANG II. In the first three weeks of operation the skytroopers netted one of the largest caches of munitions ever found in the Vietnam War. The spoils included more than 100,000 AK-47 rounds, 643 mortar rounds, 35,000 heavy machine guns rounds, and over a ton of explosives.

In February 1969, Operation CHEYENNE SABRE began in areas northeast of Bien Hoa with the mission to straddle and cut enemy infiltration routes. Most of the contact was light, but the 2nd Battalion, 7th Cavalry engaged the enemy, netting more than six hundred enemy killed. The spring and summer actions were punctuated by many North Vietnamese and Viet Cong attempts to overrun 1st Cavalry Division firebases.

On 21 March, in Tay Ninh Province, Specialist Fourth Class Donald R. Johnston, "D" Company. 1st Battalion, 8th Cavalry distinguished himself while serving as a mortarman, at a fire support base in Tay Ninh Province. His company was in defensive positions when it came under a devastating rocket and mortar attack. Under cover of the bombardment, enemy sappers broke through the defensive perimeter and began hurling explosive charges into the main defensive bunkers. Specialist Fourth Class Johnston and six of his comrades had moved from their exposed positions to one of the bunkers to continue their fight against the enemy attackers. As they were firing from the bunker, an enemy soldier threw three explosive charges into their position. Sensing the danger to his comrades, Specialist Fourth Class Johnston, with complete disregard for his safety, hurled himself onto the explosive charges, smothering the detonations with his body and shielding his fellow soldiers from the blast. His heroic action saved the lives of six of his comrades. For his valiant action, Specialist Fourth Class Donald R. Johnston received the Medal of Honor.

On 05 May, the 1st Cavalry Division welcomed its fifth Vietnam War commander, Major General E. B. Roberts. General Roberts was a newly minted aviator and former commander of the 1st Brigade. He had previously been assigned as Chief of Staff, 11th Air Assault Division (Test).

On 19 May, the 3rd Brigade moved to engage the 5th Viet Cong Division, which was trying to move toward the heavily populated cities of Long Binh and Bien Hoa. The 3rd Brigade killed a total of 132 VC and captured many tons of food and ammunition before returning to Quan Loi in June.

On 02 June, in Tay Ninh Province, First Lieutenant Robert L. Poxon, "B" Troop, 1st Squadron, 9th Cavalry distinguished himself while serving as a platoon leader on a reconnaissance mission. Landing by helicopter in an area suspected of being occupied by the enemy, the platoon came under intense fire from enemy soldiers in concealed positions and fortifications around the landing zone. A soldier fell, hit by the first burst of fire. First Lieutenant Poxon dashed to his aid, drawing the majority of the enemy fire as he crossed twenty meters of open ground. The fallen soldier was beyond help and First Lieutenant Poxon was seriously and painfully wounded. First Lieutenant Poxon, with indomitable courage, refused medical aid and evacuation and turned his attention to seizing the initiative from the enemy. With sure instinct he marked a central enemy bunker as the key to success. Quickly instructing his men to concentrate their fire on the bunker, and in spite of his wound, First Lieutenant Poxon crawled toward the bunker, readied a hand grenade and charged. He was hit again but continued his assault. After succeeding in silencing the enemy guns in the bunker he was struck once again by enemy fire and fell, mortally wounded. First Lieutenant Poxon's comrades followed their leader, pressed the attack and drove the enemy from their positions. For his valiant action, First Lieutenant Robert L. Poxon received the Medal of Honor.

On 18 July, Tay Ninh Province, Sergeant Rodney J. Evans, "D" Company, 1st Battalion, 12th Cavalry distinguished himself by extraordinary heroism while serving as a squad leader in a reconnaissance sweep through heavy vegetation to reconnoiter a strong enemy position. As the force approached a well-defined trail, the platoon scout warned that the trail was booby-trapped. Sergeant Evans led his squad on a route parallel to the trail. The force had started to move forward when a nearby squad was hit by the blast of a concealed mine. Looking to his right Sergeant Evans saw a second enemy device. With complete disregard for his safety he shouted a warning to his men, dived to the ground and crawled toward the mine. Just as he reached it an enemy soldier detonated the explosive and Sergeant Evans absorbed the full impact with his body. For his valiant action, Sergeant Rodney J. Evans received the Medal of Honor.

On the night of 12 August the VC threw simultaneous attacks against Quan Loi, LZ Becky, LZ Jon, LZ Kelly and LZ Caldwell. The VC were thrown back, experiencing heavy losses and more were gunned down by helicopter gunships as they fled in retreat. In all, the VC lost seven hundred men.

On 14 September, near Song Be, Sergeant Donald S. Skidgel, "D" Troop, 1st Squadron, 9th Cavalry distinguished himself while serving as a reconnaissance section leader. On a road near Song Be in Binh Long Province, Sergeant Skidgel and his section, with other elements of his troop, were acting as a convoy security and screening force when contact occurred with an estimated enemy battalion concealed in tall grass and in bunkers bordering the road. Sergeant Skidgel maneuvered off the road and began placing effective machine gun fire on the enemy automatic weapons and rocket-propelled grenade positions. After silencing at least one position, he ran with his machine gun across sixty meters of bullet-swept ground to another location from which he continued to rake the enemy positions. Running low on ammunition, he returned to his vehicle over the same terrain. Moments later he was alerted that the command element was receiving intense automatic weapons, rocket-propelled grenade and mortar fire. Although he knew the road was saturated with enemy fire, Sergeant Skidgel calmly mounted his vehicle and with his driver advanced toward the command group in an effort to draw the enemy fire onto himself. Despite the hostile fire concentrated on him, he succeeded in silencing several enemy positions with his machine gun. Moments later Sergeant Skidgel was knocked down onto the rear fender by the explosion of an enemy rocket-propelled grenade. Ignoring his extremely painful wounds, he staggered back to his feet and placed effective fire on several other enemy positions until he was mortally wounded by hostile small arms fire. His selfless actions enabled the command group to withdraw to a better position without casualties. For his valiant action, Sergeant Donald S. Skidgel received the Medal of Honor.

In the closing months of 1969, the 3rd Brigade successfully stymied enemy infiltration of the roads, trails, and narrow paths of the "Serges Jungle Highway" hidden beneath the triple-canopy jungle.

In November, an important discovery, that proved costly to the enemy, was made along the border between II Corps and III Corps. "B" Troop, 1st Squadron, 9th Cavalry found another canopy-concealed infiltration route and named it the "Jolley Trail" after the troop commander, Major Charles A. Jolley. By jungle standards the Jolley Trail was an elaborate, high speed roadway. It was paved with bamboo matting and lined every hundreds of yards with bunkers and bomb shelters. The trail was blasted by air strikes and patrolled repeatedly by 1st Cavalry helicopters and troopers on the ground.

On 02 December, in Phuoc Long Province, Second Lieutenant Robert R. Leisy, "B" Company, 1st Battalion, 8th Cavalry distinguished himself while serving as platoon leader during a reconnaissance mission. One of his patrols became heavily engaged by fire from a numerically superior enemy force located in a well-entrenched bunker complex. As Second Lieutenant Leisy deployed the remainder of his platoon to rescue the beleaguered patrol, the platoon also came under intense enemy fire from the front and both flanks. In complete disregard for his safety, Second Lieutenant Leisy moved from position to position deploying his men to effectively engage the enemy. Accompanied by his radio operator he moved to the front and spotted an enemy sniper in a tree in the act of firing a rocket-propelled grenade at them. Realizing there was neither time to escape the grenade nor shout a warning, Second Lieutenant Leisy unhesitatingly, and with full knowledge of the consequences, shielded the radio operator with his body and absorbed the full impact of the explosion. This valorous act saved the life of the radio operator and protected other men of his platoon who were nearby from serious injury. Despite his mortal wounds, Second Lieutenant Leisy calmly and confidently continued to direct the platoon's fire. When medical aid arrived, Second Lieutenant Leisy valiantly refused attention until the other seriously wounded were treated. His display of extraordinary courage and exemplary

One of the several huts containing ammunition for the VC and NVA was found in the CITY in a heavily camouflaged area six miles within the Cambodian border.

devotion to duty provided the inspiration and leadership that enabled his platoon to successfully withdraw without further casualties. For his valiant action, Second Lieutenant Robert R. Leisy received the Medal of Honor.

Although many of the First Team had made noteable, personal sacrifices in the heavy engagements of 1969, the year ended on a high note for the 1st Cavalry Division. The Division had throughly smashed the domination of the enemy over the northern areas of III Corps.

Success of the local interdictions brought about by the Operations of KENTUCKY COUGAR that were held in late 1969, led the North Vietnamese to put on a show of strength in neighboring Cambodia and Laos as well in South Vietnam. In the opening months of 1970, several pitched battles were fought with the North Vietnamese and Viet Cong who attempted to overrun bases. In every instance, the enemy was thrown back with heavy losses. There had been a slight increase in the fighting during the first week of each month. Many considered that the North Vietnamese were attempting to improve their bargaining position at any peace negotiations. These attacks shattered a general lull that had prevailed over the battlefield since the previous September. The shellings and ground assaults of the North Vietnamese proved that the enemy had retained the capacity to inflect heavy casualties. On 01 to 04 February, two hundred twenty-eight enemy troops died in a major attack on FSB Tina.

On 10 February, in Phuoc Long Province, Specialist Fourth Class John P. Baca, "D" Company, 1st Battalion, 12th Cavalry distinguished himself while serving on a recoilless rifle team during a night ambush mission. A platoon from his company was sent to investigate the detonation of an automatic ambush device forward of his unit's main position and soon came under intense enemy fire from concealed positions along the trail. Hearing the heavy firing from the platoon position and realizing that his recoilless rifle team could assist the members of the besieged patrol, Specialist Fourth Class Baca led his team through the hail of enemy fire to a firing position within the defensive perimeter of the patrol. As they prepared to engage the enemy, a fragmentation grenade was thrown into the midst of the patrol. Fully aware of the danger to his comrades, Specialist Fourth Class Baca unhesitatingly, and with complete disregard for his own safety, covered the grenade with his steel helmet and fell on it as the grenade exploded, thereby absorbing the lethal fragments and concussion with his body. His gallant action and total disregard for his personal well-being directly saved eight men from certain serious injury or death. For his valiant action, Specialist Fourth Class John P. Baca received the Medal of Honor.

In response to the attacks of increasing intensity by the North Vietnamese, the 2nd Battalion, 8th Cavalry opened and closed seven Fire Support Bases (FSB) in as many weeks; 11 to 25 February, FSB Jamie; 25 February to 02 March, FSB GarryOwen; 02 to 05 March FSB Heather; 05 to 08 March, FSB Victor; 08 to 13 March, FSB Flasher; 13 to 17 March, FSB Drum; 17 March to 02 April, FSB Illingsworth. Each firebase of the week was christened by choking, black dust as helicopters landed, each under hundred degree temperatures and each was soon littered by hundreds of empty artillery canisters.

Both FSB Illingsworth and FSB Jay, a similar stronghold to the south manned by the 2nd Battalion, 7th Cavalry, were only a hike away from Cambodia and were situated on a heavily trafficked infiltration routes of the "Dog Head" area. The disruption of these routes caused by patrols originating from FSB Illingsworth and FSB Jay was invaluable, but soon became a targeted objective of the North Vietnamese. It was anticipated that the enraged NVA would be tempted out of their sanctuary and openly attack the two positions.

At 0415 hours on 29 March, a tripflare suddenly ignited on the southern line of FSB Jay, and the firebase was suddenly deluged by rocket, mortar, recoilless rifle and rocket propelled grenade fire. The first rounds of the NVA barrage hit the antennae of the Command Post and artillery Fire Detection Center (FDC). All communications with FBS Jay ceased and the Brigade and the Division were not aware of the attack until FSB Illingsworth reported the glow of flares and flashes of explosives on the horizon.

A battalion of the 272nd NVA Regiment swarmed out of the trees and met the stiff resistance of "A" and "E" Company, 2nd Battalion, 7th Cavalry and "B" Battery, 2nd Battalion, 19th Field Artillery. The NVA attack fell apart before dawn, and at first light, the dust-caked, bedraggled survivors of FSB Jay began standing up again. Fifteen of the troopers were killed and fifty-three were wounded.

A survey of FSB Illingsworth, conducted the following morning, indicated that all equipment had been destroyed by the heavy fire of the enemy. In the foreground are two Cargo Carriers of "A" Battery, 2nd Battalion, 32nd Field Artillery. (02 April 1970)

The remains of seventy four of the enemy were scattered over the firebase.

FSB Illingsworth was situated on the boiler pan of a dried up lake bed and surrounded by jungle. During the day, bulldozers and crews from the 8th Engineer Battalion were airlifted to the base to construct ammunition bunkers. In addition to its own "C" and "E" Companies, 2nd Battalion, 8th Cavalry, the firebase was protected by "B" Battery, 2nd Battalion, 19th Field Artillery, one quad .50 caliber gun track from "B" Battery, 5th Battalion, 2nd Field Artillery (-), three 155mm SP Howitzers from the 1st Battalion, 30th Field Artillery, two 8 inch SP Howitzers, from "A" and "D" Batteries and two 175mm from "B" Battery, 32nd Field Artillery (-), six 105mm Howitzers from "B" Battery, 77th Field Artillery, a search light jeep from "I" Battery, 29th Field Artillery and support of "A" Troop, 11th Armored Cavalry Regiment.

Given the circumstances that FSB Jay was overrun only two days before, the firebase commander scheduled an alert exercise every two hours after midnight. Crews of the PP-55 Ground Surveillance Radar, upon detecting any movement beyond the perimeter would initiate calls to the 2nd Battalion, 19th Field Artillery Center at Tay Ninh who would initiate a prearranged exercise fire. Along the perimeter berm, the M-60 machine guns would display a beautiful show of tracers as they were being prepositioned. The woods, outlined by artillery flashes, became silent and soon the radar crew could detect no more movement. However at 0218 hours on 01 April, there was a different ending, the NVA was making a full frontal assault. A battalion of the 272nd NVA Regiment had deployed from behind the wall of the jungle that faced FSB Illingsworth on three sides. From their hidden locations, the NVA opened fire with automatic weapons and launched more than 200 rounds of mortar, rocket grenades, 107mm, 122mm and 240mm rockets that rained down on the base before they began to move in on the perimeter.

As the battle raged below, the artillery fires were adjusted by observers of "A" Troop, 1st Squadron, 9th Cavalry and 1st Brigade Scout helicopters. Resupply and medevac aircraft arrived and departed on prearranged flight corridors which allowed continuous, safe artillery firings. A predawn reinforcement of "B" Troop, 1st Battalion, 11th Armored Cavalry was given a safe corridor to the FSB by artillery fire on its north and south flanks. By 0500 hours, having been unable to get a single man over the perimeter berm without being killed within seconds, the NVA faded away into the jungle. The successful defense was greatly aided by early warnings, prompt responses of an effective pre-arranged artillery fire plan and the bravery of the troopers manning the FSB. During the attack, nineteen troopers were killed and forty-three were wounded. Enemy losses amounted to eighty-eight dead and the capture of numerous weapons and munitions.

During this engagement, on 01 April, in Tay Ninh Province, Sergeant Peter C. Lemon, "E" Company, 2nd Battalion, 8th Cavalry distinguished himself while serving as an assistant machine gunner during the defense of Fire Support Base Illingsworth. When the base came under heavy enemy attack, Sergeant Lemon engaged a numerically superior enemy with machine gun and rifle fire from his defensive position until both weapons malfunctioned. He then used hand grenades to fend off the intensified enemy attack launched in his direction. After eliminating all but one of the enemy soldiers in the immediate vicinity, he pursued and disposed of the remaining soldier in hand-to-hand combat. Despite fragment wounds from an exploding grenade, Sergeant Lemon regained his position, carried a more seriously wounded comrade to an aid station, and, as he returned, was wounded a second time by enemy fire. Disregarding his personal injuries, he moved to his position through a hail of small arms and grenade fire. Sergeant Lemon immediately realized that the defensive sector was in danger of being overrun by the enemy and unhesitatingly assaulted the enemy soldiers by throwing hand grenades and engaging in hand-to-hand combat. He was wounded yet a third time, but his determined efforts successfully drove the enemy from the position. Securing an operable machine gun, Sergeant Lemon stood atop an embankment fully exposed to enemy fire, and placed effective fire upon the enemy until he collapsed from his multiple wounds and exhaustion. After regaining consciousness at the aid station, he refused medical evacuation until his more seriously wounded comrades had been evacuated. For his valiant action, Sergeant Peter C. Lemon received the Medal of Honor.

On 01 May 1970, the day for symbolic parades of military might of Communists countries, the First Team and ARVN units hit what was previously a Communist sanctuary. President Nixon had given the go-ahead for the surprise mission. Pushing into the "Fish Hook" region of the Cambodian border and occupying the towns of Mimot and Snoul, troopers scattered the enemy forces, depriving them of much needed supplies and ammunition. On 08 May, the troopers of the 2nd Brigade found an enemy munitions base that they dubbed "Rock Island East". Ending on 30 June, the mission to Cambodia far exceeded all expectations and proved to be one of the most successful operations of the First Team. All aspects of ground and air combat had been utilized. The enemy had lost enough men to field three NVA divisions and enough weapons to equip two divisions. A year's supply of rice and corn had been seized. The troopers and the ARVN soldiers had found uncommonly large quantities of ammunition, including 1.5 millions rounds for small arms, 200,000 antiaircraft rounds and 143,000 rockets, mortar rounds and recoilless rifle rounds. The sweeps turned up 300 trucks, a Porsche sports car and a plush Mercedes-Benz sedan.

While in Cambodia, Major General George W. Casey, who had served as Assistant Division Commander of the

1st Cavalry Division during the final stages of the Cambodian interdiction, became Division Commander.

The campaign had severe political repercussions in the United States for the Nixon Administration. Pressure was mounting to remove America's fighting men from the Vietnam War. Although there would be further assault operations, the war was beginning to wind down for many troopers.

In July 1970 the 1st Cavalry Division continued the task of clearing the remaining NVA combat elements from the area of III Corps north, east and west of Saigon. Fire bases supported combat units with indirect fire protection against identified targets over a wide area of remote operations. In addition, most firebases were located close enough so that they could direct protective fire support on each other's perimeter if necessary. "B" Battery, 1st Battalion, 30th Artillery out of FSB Moe, a one battalion sized firebase located in Song Be Province northwest of Saigon approximately two kilometers from the Cambodian border, provided typical artillery support. In addition to other elements, the firebase was equipped with a medium artillery platoon consisting of three 155mm Howitzers and staffed by a platoon leader, an XO, a fire control officer, and fifty troopers. The 155mm Howitzer, although not state-of-the-art, was a useful heavy weapon in the Vietnam jungle because it could be easily airlifted by a CH-47 Chinook helicopter and dropped into a new clearing to set up a new firebase.

On 07 July, Major General George W. Casey was reported missing in action while enroute to visit wounded skytroopers in a field hospital at Cam Ranh Bay. His death was confirmed. Brigadier General J. R. Burton assumed temporary command until 20 July when Major General George W. Putman, Jr, a former First Team artillery commander and division chief of staff, took command and led the Division through its final days in Vietnam.

Sweeping through III Corps through the remainder of 1970 and early 1971, the troopers searched out and engaged enemy forces and captured caches of supplies, weapons and ammunition. The efforts of the 1st Cavalry Division were not limited to direct enemy engagements but also, using the experiences gained during the occupation of Japan and Korea, encompassed the essential rebuilding of the war torn country of South Vietnam. As a result of its performance, the Division was awarded two Presidential Unit Citations, the Valorous Unit Citation, the Vietnam Civic Action Medal and the Cross of Gallantry with Palm.

On 26 February, 1971 Captain Jon E. Swanson distinguished himself by acts of bravery while flying an OH-6A aircraft in support of Army of the Republic of Vietnam (ARVN) Task Force 333 in the Kingdom of Cambodia. With two well-equipped enemy regiments known to be in the area, Captain Swanson was tasked with pinpointing the enemy's precise positions. Captain Swanson flew at treetop level at a slow airspeed, making his aircraft a vulnerable target. The advancing ARVN unit came under heavy automatic weapons fire from enemy bunkers 100 meters to their front. Exposing his aircraft to enemy antiaircraft fire, Captain Swanson immediately engaged the enemy bunker with concussion grenades and machinegun fire. After destroying five bunkers and evading intense ground-to-air fire, he observed a .51-caliber machinegun position. With all his heavy ordnance expended on the bunkers, he did not have sufficient explosives to destroy the position. Consequently, he marked the position with a smoke grenade and directed a Cobra gunship attack. After completion of the attack, Captain Swanson found the weapon still intact and an enemy soldier crawling over to man it. He immediately engaged the individual and killed him. During this time, his aircraft sustained several hits from another .51-caliber machinegun. Captain Swanson engaged the position with his aircraft's weapons, marked the target, and directed a second Cobra gunship attack. He volunteered to continue the mission, despite the fact that he was now critically low on ammunition and his aircraft was crippled by enemy fire. As Captain Swanson attempted to fly toward another .51-caliber machinegun position, his aircraft exploded in the air and crashed to the ground, causing his death. Captain Swanson's extraordinary heroism and devotion to duty are in keeping with the highest traditions of military service and reflect great credit upon himself, his unit, and the United States Army. For this action Captain Swanson was posthumously awarded the Medal of Honor.

Although 26 March 1971 officially marked the end of duties in Vietnam for the 1st Cavalry Division, President Nixon's program of "Vietnamization" required the continued presence of a strong US fighting force to remain "in country". On 30 April, a 3rd Brigade (Separate) was organized and activated to carry out the continuing mission of the 1st Cavalry Division. The mission of the unit, composed of experienced, specialized fighting units, was to continue the interdiction of enemy infiltration and supply routes in War Zone D, known as "Cav" Country.

On 05 May 1971, the colors of the 1st Cavalry Division, minus those of the 3rd Brigade, were moved from Vietnam to Ft. Hood, Texas.

Meanwhile in Vietnam, the 3rd Brigade became well equipped with helicopters from the 229th Assault Helicopter Battalion and later, a battery of "Blue Max" aerial field artilley units and two air cavalry troops. A Quick Reaction Force (QRF) - known as "Blue Platoons", was maintained in support of any air assault action. The "Blues" traveled light, fought hard and had three primary missions; 1) to form a "field force" around any helicopter downed by enemy fire or mechanical failure, 2) to give quick backup to ranger patrols that made enemy contact, and 3) to search for enemy trails, caches and bunker complexes.

"Blue Max", "F" Battery, 79th Aerial Rocket Artillery, was another familiar aerial artillery unit. Greatly appreciated by troopers of the 1st Cavalry, its heavily armed Cobras flew a variety of fire missions in support of the operations of the 3rd Brigade. The pilots of "Blue Max" were among the most experienced combat fliers in the Vietnam War.

Many had volunteered for the extra duty to cover the extended stay of the 1st Cavalry Division.

Most of the initial combat for the new brigade involved small skirmishes, but the actions became bigger and more significant. Two engagements in May of 1971 were typical operations. On 12 May, the third platoon, Delta Company, 2nd Battalion, 5th Cavalry tangled with enemy forces holed up in bunker complexes. With help from the US Air Force and 3rd Brigade gunships, the troopers captured the complex. Fifteen days later, helicopters of "F" Troop, 1st Squadron, 9th Cavalry received ground fire while conducting a reconnaissance mission over a large bunker complex. Air strikes were called in and the troopers overran the complex.

Early in June, intelligence detected significant enemy movement toward the center of Long Khanh Province and its capital, Xuan Loc. On 14 June, Delta Company of the 2nd Battalion, 5th Cavalry ran into an ambush in heavy jungle and engaged a company-sized enemy unit. The troopers were pinned down in a well-sprung trap. Cavalry field artillery soon pounded the North Vietnamese positions and heavy Cobra fire from Blue Max, "F" Battery of the 79th Aerial Field Artillery, swept down on the enemy positions keeping pressure on the withdrawing North Vietnamese throughout the night. The timely movements of the Brigade had thwarted the enemy build up north of Xuan Loc.

3RD BRIGADE, 1ST CAVALRY DIVISION (SEPARATE)
Regimental Units 2nd Battalion, 5th Cavalry Regiment 1st Battalion, 7th Cavalry Regiment 2nd Battalion, 8th Cavalry Regiment 1st Battalion, 12th Cavalry Regiment
Brigade Aviation Units 1st Aviation Platoon "F" Troop, 9th Cavalry (Air) 229th Aviation Battalion 362nd Aviation Company
Brigade Artillery Units 1st Battalion, 21st Field Artillery "F" Battery, 26th Artillery (Prov) "F" Battery, 77th Artillery, Aviation "F" Battery, 79th Artillery (ARA) "Blue Max"
Brigade Support Units 14th Military History Detachment 26th Chemical Detachment 215th Support Battalion 501st Engineer Company 525th Signal Company
Temporary Duty Unit 2nd Squadron, 11th Armored Cavalry
Attached Units 34th Inf Platoon, (Scout Dog Patrol) 62nd Inf Platoon, (Combat Trackers) 75th Infantry Rangers, "H" Company 191st Military Intelligence Company 405th Radio Receiver Detachment 483nd Military Police Platoon Radio Research Detachment (Prov)

On 30 March 1972, General Giap of the North Vietnamese Army began an offensive across the DMZ in a final attempt to unify the North and South. By 03 April, these thrusts became a full scale attack. More than 48,000 NVA and VC troops hit Loc Ninh. Two days later, on 05 April, the North Vietnamese threw heavy assaults against An Loc and announced that by 20 April, An Loc would be the new capital of the South for the North Vietnamese.

In April and May, stepped up bombings by B-52's helped blunt the North Vietnamese invasion. Large groups of enemy soldiers were caught in the open fields and entire NVA units were destroyed. Helicopters and gunships from the 3rd Brigade saw heavy action at An Loc and Loc Ninh, engaging heavy armor as well as ground troops. The intensity of the fighting took a heavy toll on them. For example, on 12 May, five Cobra gunships were destroyed in less than 30 minutes by Chinese Surface to Air (SA-7) Missiles.

On 15 May, relief units, moving down Highway 13, broke through and helped lift the bitter siege of An Loc. The North Vietnamese were reeling from huge losses and began to withdraw to their sanctuaries in Cambodia and Laos. Their spring offensive aimed at cutting South Vietnam in half and capturing Saigon had been decisively smashed. The helicopter air effort of the 3rd Brigade had turned in a magnificent performance in support of the remaining advisors with the ARVN units. During the period of 05 April through 15 May 1972, more than one hundred T-54 tanks, armored personnel carriers, and anti-aircraft guns were knocked out in the area around An Loc.

Joseph Kline, Co "B", 104th Avn Bn, 101st Airborne Division captured the intensity and energy typical of subsequent air battles with his painting "Missile! Missile! Missile" which portrays the situation of Captains Mike Brown and Marco Cordon, "F" Btry, 79th ARA, whose Cobra took a direct hit by a SA-7 Missile. They were

"Missile! Missile! Missile"

flying at 4,000 feet while on a mission in support of an ARVN airborne brigade in the vicinity of Tan Khai on highway 13, approximately six miles south of An Loc. Only sound training and the instinct for survival led to their safe return.

By 31 March 1972, only 96,000 US troops were involved in the combat operations in Vietnam. In mid June 1972, the standdown ceremony for the 3rd Brigade was held in Bien Hoa and the colors were returned to the United States. The last trooper left from Tan Son Nhut on 26 June, completing the Division recall that had started on 05 May 1971. With the 3rd Brigade completing their withdrawal, the 1st Cavalry became the first US Army Division to go to Vietnam and the last to leave.

"Firsts" had become the trademark of the First Team.

Even as the war continued, peace talks in Paris progressed, with Henry Kissinger as US negotiator. A break in negotiations, followed by US saturation bombing of North Vietnam, did not derail the talks, and a peace agreement was reached and signed on 27 January, 1973 by the United States, North Vietnam, South Vietnam, and the Provisional Revolutionary Government of the NLF. The accord provided for the end of hostilities, the withdrawal of US and allied troops, the return of prisoners of war, and the formation of a four-nation international control commission to ensure peace.

Fighting between South Vietnamese and Communists continued despite the peace agreement until North Vietnam launched another offensive in early 1975. The requests of South Vietnam for aid were denied by the US Congress, and after Nguyen Van Thieu, the President of South Vietnam, abandoned the northern half of the country to the advancing Communists, a panic ensued. South Vietnamese resistance collapsed, and North Vietnamese troops marched into Saigon on 30 April, 1975. Vietnam was formally reunified in July, 1976, and Saigon was renamed Ho Chi Minh City.

After a 20-year hiatus of severed ties, the US initiated formal normalization of diplomatic relations with Vietnam on 11 July, 1995. The cornerstone of this relationship is based upon the continued cooperation of Vietnam on the issue of missing Americans soldiers. The US considers achieving the fullest possible accounting of Americans missing and unaccounted for to be its highest priority with Vietnam.

Chapter 9: Tri-Cap to Armor

1971-1990

"Pass in Review." During a TRICAP Ceremony, combat vehicles, flags and musical instruments stand tall to honor the arrival of the 1st Cavalry Division back to Ft. Hood, Texas and CONUS. (05 May 1971)

On 26 March 1971, a Stand Down Ceremony at Bien Hoa, marked the departure of the 1st Cavalry Division from Vietnam. With the simple but brief ceremony highlighted by the 1st Cavalry Division Band and the Color Gaurd, the division's tour of duty in Vietnam came to a close. After sixty-six months "in country" and continuously in combat, the First Team left the 3rd Brigade (Separate) to carry on.

On 05 May, the colors of the 1st Cavalry Division, minus those of the 3rd Brigade (Separate), returned from Vietnam to Ft. Hood, Texas and were passed to the commander of the former 1st Armored Division, Major General James C. Smith. After twenty-seven years of outstanding service, hardships and dedicated service overseas, the major part of the 1st Cavalry Division was back in Texas where it had been organized over fifty years before.

Following the ceremonies, the 1st Cavalry Division began a reorganization and redesignation of its subordinate units to become the 1st Cavalry Triple-Capability (TRICAP) Division. The basic restructuring was accomplished by the transfer of equipment assets and specialized trained personnel of the following units: *SEE FOLLOWING PAGE*

Ft. Hood, Texas had begun its own long history on 15 January 1942, when the War Department announced that a camp, to be a permanent station of the Tank Destroyer and Firing Center, would be built in the vicinity of Killeen, Texas. Orders were issued for the Real Estate Branch of the Engineer Corps to acquire 10,800 acres of land northwest of Killeen. On 17 February 1942, the Army announced that the camp would be named Camp Hood in honor of General John Bell Hood, the "Fighting General" of the famous "Texas Brigade" of the Confederate Army of Northern

General John Bell Hood, Confederate States Army (CSA) 1831-79, Confederate General in the American Civil War, Born at Owingsville, KY. Graduated from West Point in 1853 and served with the 2nd (later redesignated as the 5th in August 1861) Cavalry Regiment in Texas. In April 1861 he resigned his commission from the army and in 1862 entered the Confederate States Army service under his old commander that he served under in the 2nd Cavalry Regiment, Robert E. Lee. He was the youngest of the full Confederate Generals, unmatched for bravery and gallantry on the battlefield, and in his prime a hero of the Southern Confederacy. On the battlefields of 2nd Manasses, Antietam, 7 Days Battles Fredricksburg, Gettysburg, and Chickamauga, the outstanding leadership of Hood as a division and brigade commander brought him and his men high acclaim as a major factor in each engagement. Following the battle of Nashville, 15 to 16 December 1864, Hood resigned his command on January 1865 and surrendered at Natchez, MS. in May. Although surrounded by controversy and criticism of his performance in his last battle, he yet remains symbolic of a once young, strong, Confederate South in its valiant but hopeless fight for National Independence.

Transfer of Equipment Assets and Specialized Trained Personnel	
1st Cavalry Division Unit	**Resources Utilized**
Headquarters 1st Cavalry Division (TRICAP)	Headquarters 1st Armored Division
HHC, 1st Armored Brigade 1st Battalion, 13th Armor 2nd Battalion (Mech Inf), 12th Cavalry 1st Battalion, 81st Armor	HHC, 1st Armored Brigade 1st Battalion, (Armored) 13th Armor 5th Battalion, 6th Infantry 1st Battalion, (Armored) 81st Armor
HHC, 2nd Air Cavalry Combat Brigade 3rd Squadron, 1st Cavalry 2nd Battalion, 13th Armor 4th Squadron (Air), 9th Cavalry	HHC, 2nd Brigade 3rd Squadron, 1st Cavalry 2nd Battalion (Armored), 13th Armor
HHC, 3rd Brigade, (Special) 2nd Battalion, 5th Cavalry 1st Battalion, 7th Cavalry 2nd Battalion, 8th Cavalry 1st Battalion, 12th Cavalry	Note: The 3rd Brigade (Special) 1st Cavalry Division remained in Vietnam until 26 June 1972
HHC, 4th AirMobile Infantry Brigade 1st Battalion (AM/Inf) 5th Cavalry 2nd Battalion, (AM/Inf) 7th Cavalry 1st Battalion, (AM/Inf) 8th Cavalry	HHC, 3rd Brigade 2nd Battalion, 46th Infantry 2nd Battalion, 52nd Infantry
HHB, Division Artillery 1st Battalion, 6th Field Artillery 1st Battalion, 77th Field Artillery	HHB, Division Artillery 1st Battalion, 73rd Field Artillery 1st Battalion, 6th Field Artillery
HHC, Division Support Command 15th Medical Battalion 15th Supply & Transport Battalion 27th Maintenance Battalion 315th Composite Supt Battalion	HHC, Division Support Command 47th Medical Battalion 501st Supply & Transport Battalion 123rd Maintenance Battalion
Separate Battalions & Companies HHC, 1st Cavalry (TRICAP) Division 8th Engineer Battalion 13th Signal Battalion 15th Adjutant General Company 15th Data Processing Unit 15th Finance Company 545 Military Police Company 230th Aviation Battalion	Division Troops HHC, 1st Armored Division 16th Engineer Battalion 141st Signal Battalion 501st Adjutant General Company 10th Data Processing Detachment 501st Military Police Company
Note 1: The 230th Aviation Battalion was an accounting organization setup to receive assets of the Vietnam Aviation Organizations.	

1st Cavalry Division Headquarters Facilities. The 1st Cavalry Division took over the World War II type buildings to house its Headquarters operations. Due to the limited size and arrangements of the structures, a "distributed" staff structure was employed.

Virginia, who was later Commanding General of the Confederate Army of Tennessee.

Although not a native of Texas, General Hood was nevertheless considered a state hero for his connections with the Texas Brigade and his prior service and dramatic forays in Texas while serving in the 2nd Cavalry Regiment (redesignated 10 August 1861, as the 5th Cavalry Regiment) at Ft. Mason, Texas under Lieutenant Colonel Robert E. Lee. During the construction phase, the Army purchased an additional 16,000 acres of land in Bell County for training purposes and 34,943 acres in Coryell County for a cantonment to house the Tank Destroyer Basic Replacement Training Center and an area for its extensive field training.

The 1st Cavalry Division, following a reorganization using the assets and personnel of the 1st Armored Division, received an experimental designation of a TRICAP Division and was reassigned to III Corps. The Division consisted of the 1st Armored Brigade, the 2nd Air Cavalry Combat Brigade (ACCB) and the 4th Airmobile Infantry Brigade. Division Artillery provided the fire support and the Division Support Command provided normal troop support and service elements. Its new mission, under the direction of Modern Army Selected Systems Test, Evaluation and Review (MASSTER) was to carry on a close identification with and test forward looking combined armor, air cavalry, and airmobile concepts.

Not to be lost in the historical archives, in January 1972 the Horse Platoon was officially established and formed as a Unit-Of-Choice recruiting tool for the 1st Cavalry Division and was sponsored by the 1st Cavalry Division Association. The transformation and organization of the last horse mounted cavalry unit in the United States Army renewed and restored the spirit and proud heritage of the horse cavalry to the Division.

TRICAP was derived from combining the ground (mechanized infantry or armor) capability, airmobile infantry and air cavalry or attack helicopter forces. TRICAP I was held at Ft. Hood, Texas beginning in February 1972. On 28 February 1972, the 4th Squadron, 9th Cavalry was reactivated to join in TRICAP evaluations. The purpose of TRICAP I was to investigate the effectiveness and operational employment of the TRICAP concept at battalion and company levels when conducting tactical operations in a 1979 European International mid-intensity warfare environment. The exercise consisted of six phases; movement to contact, defense and delay, exploitation, elimination of penetration, rear area security and night elimination of a penetration in an adjacent area.

By 31 March 1972, only ninety-six thousand US troops were involved in the Vietnam combat operations. In mid June 1972, the standdown ceremony for the 3rd Brigade was held in Bien Hoa and the colors were returned to the United States. The last trooper left from Tan Son Nhut on 26 June, completing the division recall which had started on 05 May 1971. With the 3rd Brigade completing its withdrawal, the 1st Cavalry had become the first US Army division to go to Vietnam and the last to leave.

On 26 June 1972, the 1st Battalion, 7th Cavalry; 2nd Battalion, 8th Cavalry; and the 1st Battalion, 12th Cavalry along with the 3rd Brigade (Separate) was brought back to the United States, completing the last stage of the "Vietnam recall" for the 1st Cavalry Division. The return of the units brought about several changes to the organizational structure of the 1st Cavalry Division. On 28 June the 2nd Battalion, 8th Cavalry was inactivated at Ft. Hood and on 29 June the 1st Battalion, 12th Cavalry replaced the 2nd Battalion, 12th Cavalry Regiment which was inactivated at Ft. Hood. On 22 August, the 1st Battalion, 7th Cavalry was inactivated at Ft. Hood followed by the inactivation of the 4th Squadron, 9th Cavalry at Ft. Hood, Texas on 13 September 1972.

However, the period of inactivation for the 2nd Battalion, 8th Cavalry Regiment and the 1st Battalion, 7th Cavalry was short lived. On 20 April 1974, the 2nd Battalion, 8th Cavalry was reactivated, redesignated 2nd Battalion (Armor), 8th Cavalry and reassigned to the 1st Cavalry

Horse Platoon, 1st Cavalry Division

Division where it has been to the present. On 06 June 1974, the 1st Battalion, 7th Cavalry was reactivated as a M-60A1 tank battalion and assigned to the 2nd Brigade, 1st Cavalry Division.

Continuing the organizational changes, on 20 June 1974 the 2nd Battalion, 5th Cavalry, which was inactivated upon its return to Ft. Hood in July 1972, was reactivated, reorganized as an armored unit and reassigned to the 1st Cavalry Division as a replacement unit for the 1st Battalion (Armored), 81st Armor Regiment which was inactivated on 20 July.

The units of the aviation operations of the division underwent several reorganizations and eventually the aviation operations became centralized under the command of a "Provisional" Brigade in 1974.

On 21 February 1975, the 1st Cavalry Division was once again reorganized, becoming the Army's newest armored division. During the Division's most recent past, the unit successfully completed field testing of Tactical Fire (TACFIRE), a computerized detection system to increase the effectiveness of artillery.

On 21 February 1975, the TRICAP evaluations ended. It was concluded that the employment of the TRICAP concept at the battalion level appeared to have application in some tactical situations, but employment at company level appeared to be feasible only for short periods of combat and for special missions. The evaluation also indicated that air cavalry would normally be controlled above the company level. The battalion task force encountered no combat support problems directly attributable to the TRICAP concept. With the ending of TRICAP and acceptance of the summary evaluation, the 1st Cavalry Division was reorganized and redesignated to become the newest armored division in the Army.

Along with the reorganization on 21 February, the mission of airmobile anti-armor warfare was transferred to the 6th Cavalry Brigade (Air Combat) co-located at Ft. Hood, Texas. On 21 July, the remainder of the air-mobile units, the 4th Squadron, 9th Cavalry and 7th Squadron, 17th Cavalry along with the 55th Signal Company and 34th Support Battalion were transferred the 6th Cavalry Brigade.

On 01 July 1977, the 1st Cavalry Division entered Phase 1 of the Division Restructuring Study (DRS) to test, evaluate and determine the most effective use of manpower and weapons systems of future "heavy divisions" on the battlefields of the future. Four of the maneuvering regiments were reorganized to include one maintenance company per maneuver unit. The 3rd Battalion, 10th Cavalry was activated along with "E" Battery, 1st Battalion, 68th Air Defense Artillery, the 68th Chemical Defense Company, and one anti-armor company. On 08 August, the 227th Aviation Battalion was activated with one air cavalry troop, one attack helicopter company, one general support company, and a maintenance element.

In 1979, when the North Atlantic Treaty Organization (NATO) held its first winter exercise in five years, the troopers and armor of the 2nd (Blackjack) Brigade, 1st Cavalry Division, along with appropriate support units were selected to participate in the REturn of FORces to GERmany (REFORGER) exercises. This was the first time First Team troopers were deployed on foreign soil since Vietnam. They were sealifted to Germany for the massive exercises which demonstrated the US will to reinforce Europe, and practiced

Opening ceremonies for the new 1st Cavalry Division Headquarters Building were held in July 1983. Major General William C. Chase (Retired), who commanded the Division in the final days of World War II and through the occupation of Japan, participated in the ribbon cutting which was held during the 36th reunion of the Association.

deployment and how NATO would defend against a Soviet invasion of Western Europe.

Ft Irwin, California was opened in 1980 enabling the army to conduct open full scale, field exercise. In its first use of the facility, the 2nd Brigade, 1st Cavalry Division fielded some 2,500 troopers in Operation DESERT HORSE. The exercise, lasting six weeks, focused on combined arms training, live fire exercises and chemical warfare. It was the first time the Army conducted battalion on battalion exercises using the Multi Integrated Laser Engagement System (MILES) in the largest peacetime field operations to date

In May 1980 the 1st Cavalry Horse Platoon was recognized as a Special Ceremonial Unit (SCU) by the department of the Army. Its assigned mission was threefold; to represent the 1st Cavalry Division and the Army in public appearances throughout the country; assist in recruiting by demonstrating the professionalism of soldiers; and maintain the proud traditions forged by the cavalry in the colorful history of the United States.

In September 1980, as part of the continuous preparation for combat against unknown enemies of the future, the 2nd Battalion, 5th Cavalry Regiment was selected to field test the new XM-1 tank. At the same time the Division shed the battle weary M-55 Sheridan armored reconnaissance airborne assault vehicles for M-60 tanks. Meanwhile, the 1st Battalion, 8th Cavalry and supporting troops of the 1st Battalion, 7th Cavalry were deployed to Germany as part of the Emergency Deployment Readiness Exercise to put together a combat ready tank battalion using stored (prepositioned) equipment.

On 01 October 1981, more than four years after it had been organized under the 1st Cavalry Division as a provisional unit and equipped with the most modern intelligence gathering electronic equipment, the 312th Military Intelligence Battalion was redesignated and became an active Army element in a colorful ceremony at Ft. Hood.

On 04 September 1982, the 1st Battalion, 8th Cavalry Regiment received a sudden alert to proceed to Germany. Their mission was to reinforce the already selected REFORGER units and participate in the annual autumn exercises. On 06 September, reinforced by two platoons of "B" Company, 7th Cavalry, the advance party was on its way. Picking up prepositioned equipment at Germersheim, West Germany, they maneuvered and fought mock battles with the 2nd Armored Division and 11th Panzer Brigade (Germany) and various British units.

Also in September 1982, 3,000 troopers of the 1st (Ironhorse) Brigade took part in Operation DESERT CHARGER, the first National Training Center (NTC) rotation of the division at Ft. Irwin, located in the High Mojave Desert of California. This kicked off a long on-going series of tough, realistic desert battles which enabled the division to stay on the leading edge of the warfare technology of today. The first units to attend were the 2nd Battalion, 7th Cavalry and the 3rd Battalion, 10th Cavalry. The Division now conducts three NTC rotations per year.

As 1983 began, the First Team began a restructuring, taking on the "Division '86" configuration. The reorganization included deactivation of the 1st Battalion, 12th Cavalry Regiment on the 15th of June; activation of a new helicopter unit, "A" Company, 227th Aviation Regiment on the 16th of June which would be later reorganized and redesignated as the 228th Aviation Battalion on 01 Oct '83. Other changes included increasing the authorized sizes of the 8th Combat Engineer, 13th Signal and 15th Medical Battalions along with the DISCOM supply and transport elements and 1st Battalion, 68th Air Defense Artillery.

In September, the division deployed nearly 9,000 soldiers to Europe for the fifteenth annual REFORGER exercises. The Division, the first US combat team to return to the Netherlands since World War II, drew prepositioned equipment, moved to a staging area and conducted exercise, CERTAIN STRIKE on the plains of Northern Germany. This deployment, a demanding twenty-seven day exercise, offered many challenges and unforgettable experiences. The success of the exercise proved that the Division was fully capable of performing its wartime mission.

The First Team became the "First unit to train as a division-size element in Northern Europe." REFORGER '83 was designed to test and evaluate plans and support

The replacement of the multiple World War II structures, that housed the Headquarters Staff, with a new 124,000 square foot, three story facility that overlooked Cooper Field, enabled the centralization of the 1st Cavalry Division Command Staff.

In synchronized step, tankers, carrying only their personnel armament, move out to their chartered aircraft in preparation for departure for the REFORGER exercises in Europe. (03 September, 1983)

agreements between Americans and European military and civilian agencies. REFORGER and a second strategic exercise CRESTED CAP, a major deployment of US Force fighters, together served to demonstrate that the United States remained willing and capable of reinforcing Europe with NATO committed ground forces and air units should a Soviet threat arise.

For many troopers taking part in this exercise, a key logistic feature was the use of Prepositioned Material Configured into Unit Sets (POMCUS). Troopers flew from Ft. Hood to Germany carrying only their uniforms, individual weapons and any specialized gear they would need, such as sighting equipment for the TOW anti tank missiles. Once they reached Herongen, West Germany, members of the 2nd Battalion received 164 tanks, armored personnel carriers, trucks, jeeps and mess vehicles in less than five hours from a POMCUS site. This early training in the use of prepositioned materials and equipment was extremely useful in preparing to deploy a fully armed division to the Persian Gulf in 1990.

The Cavalry Division Band accompanied the body of troops deployed to Europe for the annual REFORGER '83 and REFORGER '87 exercises. During these two large-scale exercises, the band served the division by promoting esprit de corps throughout the theater of operations in Germany and the Netherlands.

On 01 September 1984, the units of the Aviation Operations of the "Provisional Brigade" formed in 1974, were reorganized and formed into the Combat Aviation Brigade. The Brigade, composed of the 1st Squadron, 9th Cavalry Regiment, the 227th Aviation and the 228th Attack Helicopter Battalions was constituted in the Regular Army and activated at Ft. Hood under the command of Colonel Robert A. Goodbary. At that time The two aviation battalions combined with the reconnaissance squadron of the 9th Cavalry, gave the Combat Aviation Brigade its scout, observation, attack, and support capability.

The mission of the newly formed Aviation Brigade was to find, fix and destroy enemy forces using reconnaissance, fire, and maneuver capability to concentrate and sustain combat power at the critical time and place. The Brigade provided timely reconnaissance and intelligence throughout the battlefield, mass attack helicopter fires and rapidly repositioned combat power anywhere within the division area. It is capable of quickly inserting troops, supplies, and equipment to sustain the battle and at the same time providing the command, control and liaison assets necessary to manage and coordinate the battle.

In October 1984, the 1st and 2nd Forward Support Battalions were activated using elements of the three functional battalions. In the following year, on 20 June 1985, the Army of Excellence (AOE) Reorganization transformed the elements of the remaining functional battalions; Maintenance, Medical and Supply/Transport into the 4th Main Support Battalion. The same reorganization added the 493rd Transportation Aircraft Maintenance Company to the Division Support Command.

On 20 June 1985, the 1st Cavalry Division Band was withdrawn from the Division Support Command and designated as one of the Separate Battalions and Companies of the Division.

On 16 October 1986, the 1st Battalion, 7th Cavalry furled its colors and assumed a new role as the 1st Battalion, 32nd Armor and remained assigned to the 2nd Brigade, 1st Cavalry Division. In the same divisional organization change, the colors of the 1st Squadron (Reconnaissance), 9th Cavalry were removed from service and the unit was redesignated 1st Squadron, 7th Cavalry, which for the first

Moving from the staging area to the exercise battle area, the 1st Battalion, 8th Cavalry conducts an orderly road march of M-1s through the outskirts of Neinburg, Germany. (October 1983)

There is always a special bond between soldiers. In addition to being a military success, REFORGER, also proved to be a brilliant, and unusually highly visible, public relations act of spreading goodwill throughout Europe.

time since 1943 was conducting missions as a true cavalry unit. At that time, the unit consisted of one ground troop and two air troops, with a combat power of twenty M3A1 Bradleys, eight AH-1P Cobra Attack Helicopters and twelve OH-58C observation helicopters.

Additional organizational changes were implemented on 16 October as the 2nd Battalion, 33rd Armor Regiment, 3rd Armored Division, stationed in Germany, was assigned to the 1st Cavalry Division. Immediately upon assignment, it was redesignated as the 3rd Battalion, 32nd Armor Regiment. The transfer of this unit came about by the participation of the 1st Cavalry Division in the Cohesion, Operational Readiness Training (COHORT) program. Designed to foster esprit de corps and a sense of operational continuity, the main objective of COHORT was have soldiers exposed to the experiences of other operations by the alignment of units in the Continental United States (CONUS) with units stationed overseas.

In exchange for the move of the 2nd Battalion, 33rd Armor, the 2nd Battalion, 5th Cavalry was relieved from assignment to the 1st Cavalry Division and assigned to the 3rd Armored Division in Germany. This action provided the opportunity to conduct tests of deployment and field exercises in Europe. On 16 December, the 2nd Battalion, 5th Cavalry was inactivated and relieved from assignment to the 3rd Armored Division. Their period of inactivation was short lived as on 16 January 1987, the 2nd Battalion, 5th Cavalry was reactivated and assigned to Ft. Hood and the 1st Cavalry Division where it has been to the present, filling out the organizational structure.

On 16 January 1987, the 2nd Battalion, 7th Cavalry was relieved from assignment from the 1st Cavalry Division and assigned to the 4th Infantry Division at Ft. Carson, Colorado. It would not be until the end of the Gulf War and the subsequent reorganization of 11 August 1993, when the 4th Battalion, 6th Infantry, 2nd Armored Division would be reflagged as the 2nd Battalion, 7th Cavalry and assigned to the 3rd (Greywolf) Brigade, 1st Cavalry Division, filling out the present organizational structure.

In concert with the divisional changes made in the maneuvering units, the support units were reorganized to fulfill the overall mission objectives. On 20 May 1987, the 1st Forward Support Battalion, 2nd Forward Support Battalion, 4th Main Support Battalion and 493rd Transportation Aviation Maintenance Company were redesignated the 115th Support Battalion (FWD), 15th Support Battalion (FWD), 27th Support Battalion (MAIN) and 227th Transportation Aviation Maintenance Company respectively.

In the summer of 1987, the 1st Cavalry Division deployed on REFORGER '87 with the 2nd Armored Division. With the decline of the role of the Warsaw Pact, the sizes of subsequent REFORGER deployments were reduced, but command and control elements continued to evaluate the need for equipment types and repositioning of "war stocks" along with development of contingency plans to ensure the reliability and effectiveness of combat readiness, should deployment become necessary.

On 16 November 1988 the 4th Missile Battalion, 5th Air Defense Artillery was redesignated as the 4th Battalion, 5th Air Defense Artillery, assigned as a Separate Battalion to the 1st Cavalry Division and activated at Ft. Hood, Texas. The addition of the 4th Missile Battalion provided the added capability of short range air defense for the Division to protect itself against air strikes in the field.

At Ft. Hood, the division through deliberate planning, evolved into the combat unit which would eventually be assigned a major role in Operation DESERT STORM. Along with the constant training of personnel, equipment was updated. The XM-1 tank, renamed the "M1 Abrams", was accepted and issued, along with the M2 Bradley Infantry (BFV) and M3 Cavalry Fighting Vehicle (CFV) . New technology was fielded in the Multiple Launched Rocket Systems (MLRS) and the AH-64 Apache helicopter with

Nearing its destination, a convoy passes by Houston, TX on its way to the port of embarkation, where a ship will take it to the port of ad Dammam, on the eastern coast of Saudi, Arabia.

its "Hellfire" guided missile. The old reliable Jeep bowed to the High Mobility Multi-purpose Wheeled Vehicle (HMMWV) configured as troop/cargo carrier, light armored personnel carrier, communications equipment carrier and ambulance. The Heavy Expanded Multi-purpose Tactical Truck (HEMTT), capable of hauling fuel, ammunition and various cargos, also proved to be invaluable in the Gulf War.

Along with the hardware technology changes, communication innovations made possible quantum leaps in command and control operations by the fielding of Mobile Subscriber Equipment (MSE) which, essentially cellular telephones for both fixed sites and mobile vehicles, provides secure mobile voice/data and facsimile service. The MSE is augmented by Single Channel Ground to Air Communication System (SINGARS), which provides unprecedented security using frequency hopping technology. The inventor behind this amazing technology of "Frequency Spread" was the incredible and talented actress Heddy Lamar who applied for the basic patent in 1942 and gave the US government royalty free use.

All of this new equipment saw hard operational use at Ft. Hood and by the deployment of brigades to the National Training Center at Ft. Irwin, located in the high Mojave Desert of California. This facility encompasses one thousand square miles for maneuver training against the best trained opposing force in the world. The mission of Ft. Irwin is to provide tough, realistic combined arms training at battalion task force level using both live fire and opposing forces. To carry out this mission, the National Training Center has a computerized, live-fire complex with sophisticated targetry, a dedicated, full-time opposing force, a state-of-the-art range instrumentation system that monitors training battles and full-time combat trainers who observe and control units during exercises.

This effective training could not have come at a more opportune time in the history of the First Team. On 07 August 1990, a deployment order for the Southwest Asia operations was issued. Plans calling for the division to deploy by 15 September extended the work day to fourteen, sixteen and in some cases twenty-four hours. On schedule, by mid September more than eight hundred heavy loaded vehicles were loaded at the Ft. Hood railhead to make the trip to the seaports of Houston, Texas and Beaumont, Texas. An additional four thousand two hundred vehicles formed road convoys that left every two hours, around the clock.

On 16 September an Air Force C5A Galaxy, carrying the advance headquarters staff, departed Robert Gray Army Airfield at Ft. Hood. In the final drama, 1st Cavalry Division soldiers assembled for manifest roll call. The moment came. Buses were loaded and were driven to awaiting 747s. The planes were boarded and the time for memories had begun.

The USNS Capalla, a fast sealift ship capable of 33 knots, prepares to load. The roll-on roll-off (RORO) vessel can quickly be loaded with more than 700 vehicles and represents the most modern of sealift capability owned by the Navy.

Chapter 10: Persian Gulf War

1990-1991

In August 1990, the 1st Cavalry Division was alerted for deployment to Southwest Asia as part of the joint and combined forces participating in Operation DESERT SHIELD. The focus at that time was the defense of Saudi Arabia against potential Iraqi attack. With minimum delay, chartered aircraft flew the First Team soldiers from Robert Gray Army Airfield, Ft. Hood, Texas, to Dhahran International Airport via Paris, France and Cairo, Egypt. There, they settled into warehouses and tents to await the arrival of their equipment.

On 30 October the 1st Battalion, 227 Aviation Regiment began terrain familiarization and crew training to adjust to the Saudi Arabian desert environment. The unit placed special emphasis on night proficiency and continued flight training into November, changing emphasis to team and company drills. Plans and rehearsals for the defense of Saudi Arabia were developed and constantly refined.

Practicing techniques of low altitude engagement, the AH-64 helicopter teams ran fire drills without producing high dust signatures. They practiced and refined these techniques during company battle drills day and night. During the first week of January the AH-64 team practiced live fire night gunnery on the Pegasus Range,

1st Cavalry Division Troopers board a chartered 747, one of the 117 civil aircraft that is part of the Civil Reserve Air Fleet (CRAF) used to rapidly deploy the armed forces anywhere in the far regions of the world.

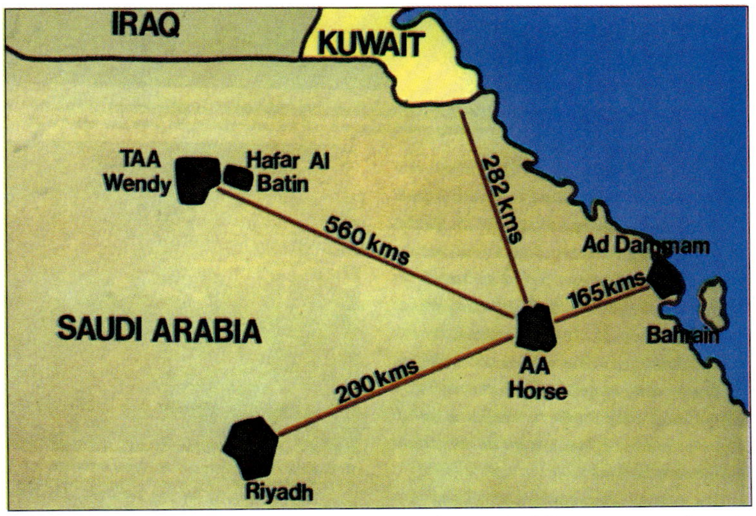

By 29 October, the last of the units moved out of ad Damman to close in on their equipment located at Assembly Area (AA) Horse, a patch of naked desert, broken by rolling hills and sand dunes that would become their temporary home.

As soon as the equipment arrived at the Port of Damman, it was offloaded, inspected and loaded on Heavy Equipment Transports (HET) for the 165km trip to the remote Assembly Area, (AA) Horse, in the Saudi desert 160 miles west of the airport.

Sand, whipped into a fog by the helicopter rotor down-wash, enveloped pilots in "brownouts" until they learned how to dispense with gradual approaches and land while moving forward.

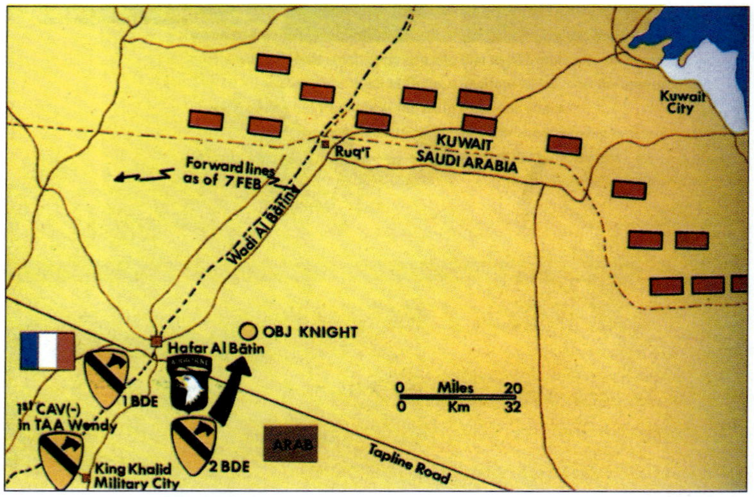

On 14 January, as defensive preparations along the Tapline road are completed, the 1st Brigade had joined in with the 6th (French) Light Brigade on the left and the 2nd Brigade, 101st Airborne Division (Air Assault), who held El Qaysumah Airfield and the Tapline on its right. The 2nd Brigade was scheduled to counterattack the enemy located at Objective Knight.

which had been built up out of the desert sands by the 8th Engineer Battalion.

After three months intensive training, the 1st Cavalry Division had been honed into one of the most modern and powerfully equipped divisions in the Army. The first glimpse of their capability came in December 1990 on the Pegasus Range. Every M-1 tank and Bradley crew test fired their new weapons as part of the new equipment transition training. Throughout this period, leaders of the Division were planning and rehearsing the First Team's initial role in the theater counterattack force - the force that would defeat any Iraqi attack into Saudi Arabia.

In January 1991 the Division was attached to VII(US) Corps and the focus of the First Team clearly began to shift toward offensive action. The Division moved five hundred kilometers to another assembly area near King Khalid Military City (KKMC) in northern Saudi Arabia. This repositioning put the Division in a key strategic location covering the historic Wadi al Batin approach into Saudi Arabia and threatening Iraq along the same avenue into western Kuwait. The time spent near KKMC was short and the Division once again picked up its seventeen thousand soldiers, which were now accustomed to "jumping". The Division moved north toward the juncture of the Saudi, Iraq and Kuwait borders through a series of defensive positions designed to thwart any preemptive attack along the Wadi.

On 11 January 1991, beginning to focus on offensive action, the 1st Battalion, 227th Aviation moved northwest to Tactical Assembly Area (TAA) Wendy, located in the vicinity of KKMC.

For the 1st Battalion, 227th Aviation the war began during the day, on 17 January, in an attack against fifty Iraqi tanks crossing the border into Saudi Arabia. Shortly after departure, the mission was scrubbed without any direct engagement. The Iraqi tanks had chosen to withdraw rather than to fight. With the Division positioned at TAA Wendy, plans for the defense of Wadi al Batin and areas north of Tapline Road were refined.

Meanwhile, the air war began and other Allied ground forces began to reposition its units for the offense. The First Team began a calculated war of deception along the Saudi border. The goal was to reinforce Saddam Hussein's belief that the main ground attack of the Allies would come up the Wadi al Batin, a natural invasion route, causing him to reposition additional forces there. The deception consisted of three major thrusts.

First, there were the firings of the Multiple Launched Rocket Systems (MLRS) of the 1st Cavalry Division that repeatedly lit the sky and battered command, control and troop targets deep in Iraq.

Second, the cannon batteries of the Division fired Copperhead rounds (computer controlled, rocket assisted projectiles) and thousands of high explosive rounds along

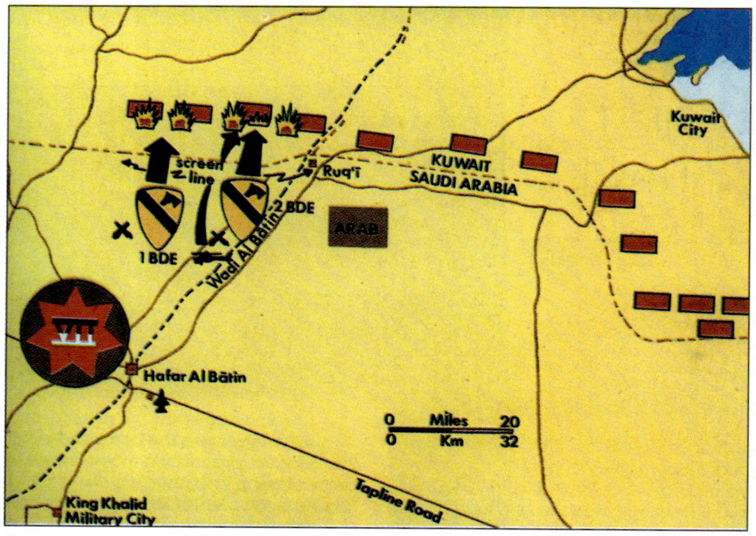

During the period 07 to 25 February, the lines of the 1st Cavalry Division had crept north, just below the Kuwait border, since the initial defense posture. Both the 1st and 2nd Brigade Task Force (TF) and supporting artillery conducted reconnaissance, artillery raids and "Berm Buster" obstacle reduction missions.

At 1530 hours on 24 February, the Tiger Team Battle Team, assigned to the 2nd Marine Division, rolled across the border and into Kuwait.

with improved conventional munitions deep onto Iraq positions.

And finally, the Aviation Brigade flew obstacle reduction and serial reconnaissance missions and identified and designated targets for destruction by the artillery units of the Division.

On 23 January the 1st Cavalry Division began its methodical "creep" forward toward the border with the 1st Squadron, 7th Cavalry moving its ground and aerial screens. These efforts were rewarded by the capture of the first enemy defector. The final week of January was characterized by intense vehicle maintenance and unit training in preparation for the ground phase. On 01 February, the 1st Squadron, 7th Cavalry began pushing a platoon out to observe the Ruqi Highway as the Saudi border guards began to withdraw from their posts. In further preparation, on 04 February, the Division conducted hellfire gunnery at night on the VII Corps Jayhawk Range.

On 05 February the 1st Squadron, 7th Cavalry reported nine vehicles moving north, in front of the screen. An AH-1 Cobra Helicopter, on the screen, received small arms fire from dismounted personnel near a desert observation tower. The Cobra returned fire with five rockets, scoring two direct hits, but not destroying the observation tower. On 07 February the 1st Battalion, 82nd Field Artillery fired a Copperhead round and made a direct hit on the observation tower and destroyed it.

On 10 February the 1st Battalion, 227th Aviation moved North of Hafar al Batin to AA Bart. Although plans had been made to conduct screening operations, none were flown as part of the continuing plans of deception that the 1st Cavalry Division would be the main ground attack of the Allies.

The first enemy encounter of Desert Storm's major ground offensive was on 19/20 February 1991, when the 2nd (Blackjack) Brigade of the division, supported by Apache helicopters of the Aviation Brigade, conducted Operation KNIGHT STRIKE I, which was named after the "Black Knights" of the 1st Battalion, 5th Cavalry. As Task Force 1-5, they moved ten kilometers into Iraq on a "reconnaissance in force", confirming and destroying enemy positions.

Task Force 1-5 moved north in a "diamond" with the cavalry fighting vehicles of the Scout Platoon on point, followed closely by Bradleys of Alpha Company leading and centered, tanks of Bravo Company on the left, tanks of Delta on the right and Bradleys of Charlie Company trailing. Tucked into the interior of the formation were two platoons of the 8th Engineers and Vulcan and Stinger Teams from the 4th Battalion, 5th Air Defense Artillery. Howitzers from the 3rd Battalion, 82nd Field Artillery remained positioned at the berm, where they would soon fire over five hundred rounds in support of Task Force 1-5.

Task Force 1-5 moved out, past the berm for a distance of 10 kilometers where Alpha Company made contact. The Bradleys instantly established a base of fire, while the tank companies raced forward. The Task Force struck savagely, destroying an enemy battalion in minutes. A-10s of the Air Force swept away over one hundred Iraqi artillery pieces entrenched and not observable from the ground positions. Task Force 1-5 started taking prisoners.

Then the strike turned ugly. Artillery rounds struck while the scouts and Alpha Company were collecting prisoners. Suddenly, rounds were falling on the Engineers and Alpha Company was taking direct hits. The scouts also took fire. Task Force 1-5 quickly regained firepower superiority while Charlie Company moved up to help with the prisoners. Shortly before 1400 hours, the artillery laid down a smoke screen to cover the orderly withdrawal south to join the Division to prepare for the subsequent series of final attacks. The enemy had drawn blood, killing three members of Task Force 1-5 during the battle.

After thirty-eight days of continuous air attacks on targets in Iraq and Kuwait, the commander of the Allied

As the Tiger Brigade moved into Kuwait, it approached the apocalyptic world of hellfire with more than five hundred oil wells set ablaze by the Iraqis before they capitulated. The smoke drifted north, darkening the sky over the position of the 1st Cavalry Division near Basrah and the Euphrates River valley.

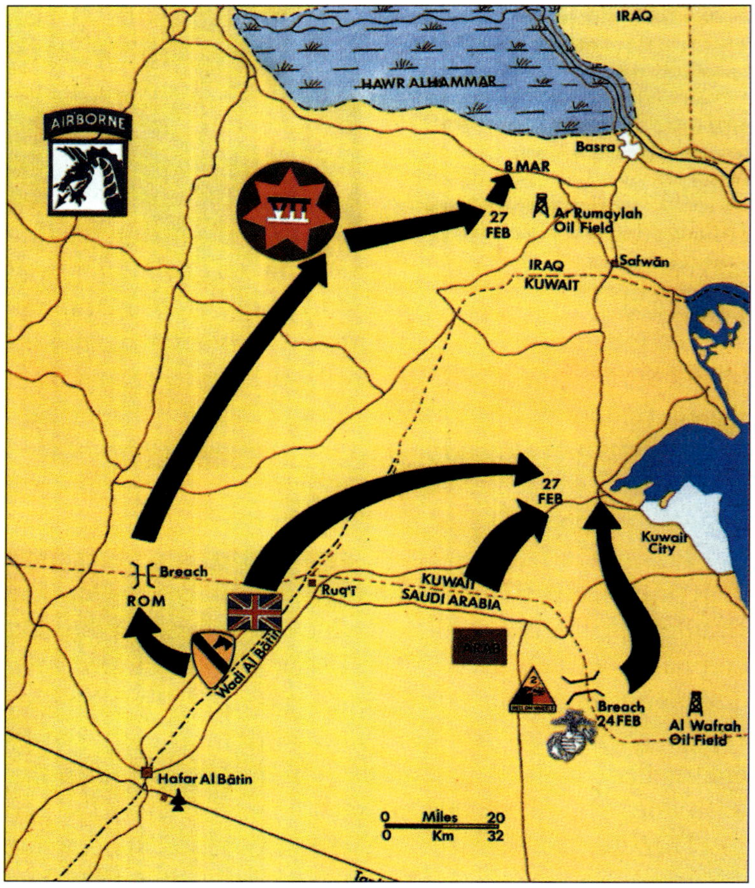

On 24 February, the Tiger Brigade attacked into Kuwait with the 2nd Marine Division and continued north meeting resistance in the vicinity of Mutla Ridge on 26 February and closed all escape routs from Kuwait City. Meanwhile, on 25 February, the 1st Cavalry Division withdrew from Iraq, began a swing to the left, at 1200 hours on 26 February conducted a refueling on the move (ROM) and late on 26 February crossed the breach sites of the 1st Infantry Division. By 1200 hours 27 February, the 1st cavalry Division had moved 300km deep in Kuwait territory.

Forces, General Norman Schwarzkopf unleashed all-out attacks against Iraqi forces very early on 24 February 1991. Aviation crew chiefs went to work by flashlight to ready the helicopters that would fly into killing range of the Iraqi 27th and 28th Infantry Divisions. On that day, the mission of the 1st Cavalry Division was to conduct a "feint" attack up the Wadi al Batin, again reinforcing the illusion that it was the Allies main ground attack. Meanwhile, far to the west, the VII Corps and the XVIII Airborne Corps had already begun a deep strike into Iraq.

In the early afternoon of 24 February 1991, the Division initiated Operation QUICK STRIKE. The 3rd Battalion, 82nd Field Artillery, reinforced by "A" Battery, 21st Field Artillery MLRS, laid down heavy fire in support of the 2nd "Blackjack" Brigade's "feint" attack up the Wadi al Batin. Blackjack moved out at approximately 1700 hours in a move north in a limited attack to fix the focus of the enemy on the Wadi.

After being postponed three times, the 1st Battalion, 227th Aviation launched early in the damp, cold dawn of 25 February without close air support. The mission was composed of sixteen AH-64s to concentrate fire power on the prime target areas.

Night fell as the rain and sand whipped the armor of the advancing wedge. Blackjack fought to the fire trenches of the enemy. The oil-filled trenches, hundreds of meters long and placed in two staggered rows. blocked progress up the Wadi. They overlapped so that the only way through the burning trenches was around the ends and into enemy prepared kill sacks.

The Iraquis lit the trenches as the 2nd (Blackjack) Brigade approached. They burned convulsively in the winds. Clouds of acid oil smoke and flying sand reduced visibility to the front. The Brigade came on and as they did, the enemy fought back. The Brigade advanced toward the fire trenches, belching pulsing strips of towering flame and smoke, vivid against the grey dawn.

Enemy resistance stiffened and they refused to yield. The enemy reacted as anticipated. Iraqi divisions focused on the movements of the First Team along the Wadi. It caused the Iraqi Forces to concentrate in the direction of the Wadi, tying down four Iraqi divisions, leaving their flanks thin and allowing the VII Corps to attack virtually unopposed and conduct a successful envelopment of Iraqi forces to the west.

The engagements of the Blackjack Brigade during the deception had destroyed elements of the Iraqi 12th Armored Division, 25th, 27th, 28th, 31st and 41st Infantry Divisions and a Corps artillery group. With the main attack in the west proceeding better than expected, the Brigade was ordered back at noon. In heavy rain, the brigade returned to Saudi Arabia to refuel its vehicles and prepare to launch into the final assault with the rest of the 1st Cavalry Division.

In parallel with the ground thrust, the Division carried out the air assault portion of Operation QUICK STRIKE as planned. Passing over the berm and over the 2nd Brigade, the pilots of the 1st Battalion, 227th Aviation could make out tankers below waving and cheering the "tank killers" on. Moments later, eight Iraqi soldiers held up tattered white flags. Bravo Company's "Grim Reaper" scouts landed their OH-58s to watch them until members of the 2nd Brigade could round them up and collect them. Meanwhile, the 1st Battalion, 227th Aviation was hitting

Playing in the last act of deception, the Apaches of the Aviation Brigade delivered dozens of Hellfire missiles into the bunkers of the enemy that were located along the attack route of the 2nd Brigade.

Tankers were pooled, forming Task Force COFER, that pushed forward with the Division as it moved, enabling the supply of one thousand tons of fuel that it consumed daily.

the enemy hard, destroying tanks, trucks, mortar and artillery pieces, fuel tankers and soldiers.

As the fire escalated, a radio call froze everyone for an instant. "We're hit, we're hit, we're going down". It was the Commander of Charlie Company, Captain Mike Klingele. Their wingman, 1st Lieutenant Robert Johnston, saw the crash and suppressed the enemy who had already began to move in on the wreck. Captain Klingele and his crewman Chief Warrant Officer Mike Butler were able to free themselves from the cockpit and began to run towards Johnston's ship. While other Apaches suppressed the enemy, Johnston took off with the two hanging by straps attached to the pylons.

The Iraqis, fearing the main effort was about to be launched, set fire trenches ablaze in front of the 2nd Brigade. The smoke, combined with the uncertain tactical situation, made a recovery of the downed AH-64 by a CH-47 Chinook too risky. The helicopter was destroyed in place with the launching of two Tube-launched Optically-tracked Wire-guided (TOW) missiles in order to prevent the possible salvage and analysis of secret, onboard weapon systems by any enemy country.

The 1st Battalion, 227th Aviation regrouped and made two more runs against the Iraqis before they were relieved by the 1st Battalion, 3rd Aviation Regiment. The tally for the day was thirty one bunkers, one tank, three howitzers, five trucks, a radar site and two grateful survivors.

Having fulfilled its assigned mission of deception, the 1st Cavalry Division was ready the following day when General Norman Schwarzkopf issued the command; "Send in the First Team. Destroy the Republican Guard. Let's go home". From the approximate center of the allied line, along the Wadi al Batin, the 1st Cavalry Division, led by Major General John H. Tilelli, Jr., attacked north into a concentration of Iraqi divisions, whose commanders remained convinced that the Allies would use the Wadi al Batin and several other wadies as avenues of attack.

Prior to the launch of the ground war, the Division Support Command (DISCOM) of the 1st Cavalry had established Refuel on the Move (ROM) sites south of the breach crossings. At the ROMs, the DISCOM and elements of the 43rd Corps Support Group waited out the afternoon of 26 February for the approach of the Division.

At each of four sites, twenty four parallel columns of vehicles pulled up, halting alongside a fueling point where a support soldier manned a fuel hose. As each vehicle came to a stop, a crewman vaulted out, grabbing the refueling hose and jammed it into the filler tube. Engines stayed running; the whine of fifteen hundred horsepower turbines rose above the shouting exchange of the soldiers and the desert wind. An entire company refueled every fifteen minutes. As endless lines of armor continued to file in; DISCOM continued to pump. By the end of the refueling operation, DISCOM had pumped 400,000 gallons of fuel into 6,100 vehicles headed into the desert of Iraq.

The 1st Cavalry Division crossed the line of departure and hit the Iraqi 27th Infantry Division. That was not their first meeting. the Division had actually been probing the Iraqi defenses for some time. As these limited thrusts continued in the area that became known as the "Ruqi Pocket". The 1st Cavalry found and destroyed elements of five Iraqi divisions, evidence that they had succeeded in their theater reserve mission of drawing and holding enemy units.

By the morning of 27 February, the combat units of the 1st Cavalry Division had made good progress going through the 1st Infantry Division breach and moving up the left side of VII Corps' sector. By mid afternoon, after a high-speed 190 mile (306 Km) move north and east, slicing into the rear of the enemy, General Tilelli's brigades joined in with the 24th Division across the VII Corps' boundary. The dust

Typical of the detailed, one-on-one operation is exemplified by the photograph of Corporal McCarty, Headquarters and Headquarters Company, 2nd "Blackjack" Brigade guarding two Iraqi Republican Guard prisoners who had just crawled out of their desert bunker and surrendered, while Captain Michael A. Eyre searches them for hidden weapons.

On 13 March, the remaining 17,000 troopers still in the war areas of the desert, in a stately column of armor, in single file, passed through the breach, to Saudi Arabia. The column of more than one thousand vehicles extended into the northern horizon.

storms had cleared early in the day, revealing the most awesome array of armored and mechanized power fielded since World War II. In a panorama extending to the horizon 1,500 tanks, another 1,500 Bradleys and armored personnel carriers, 650 artillery pieces, and supply columns of hundreds of vehicles, stretching into the dusty brown distance, rolled east through Iraqi positions, as inexorable as a lava flow.

On 27 February the 1st Battalion, 227th Aviation moved north across the Saudi Arabia - Iraq border as an element of the ground war. Aircraft stopped to refuel at Objective LEE, then continued northeast in Iraq to AA John. Upon arrival the aircraft remained on standby and were not committed to battle. On 28 February, the ground elements of the 1st Battalion, 227th Aviation closed with the aircraft as the forty eight hour cease fire went into effect. As the sun rose over the silent battlefield, the Aviation Brigade found itself squarely in the middle of the former sector of the Tawakalna Division. The desert sand was laced with unexploded cluster munitions from the intense air campaign against the Republican Guard.

On 28 February 1991, one hundred hours after General Norman Schwarzkopf had initiated the ground attack, President George Bush ordered a cease-fire. In the one hundred hours of battle before the cease-fire went into effect, the Iraqis had lost 3,847 of their 4,280 tanks, over half of their 2,880 armored personnel carriers, and nearly all of their 3,100 artillery pieces. Only five to seven of their forty-three combat divisions remained capable of offensive operations.

Immediately after the cease fire and the cessation of hostilities, the Division units began their assessment of battle damage and the round-up of scattered enemy soldiers. Much of the battlefield search and cleanup had to be done by a painstaking search and screening of prior Iraqi defensive positions. In the days following the ceasefire, the busiest soldiers were those engaged in the monumental task of tracking down, disarming, transporting, and caring for an estimated sixty thousand prisoners of war.

The units of the 1st Cavalry Division established defensive positions where the ceasefire had stopped their attack and then expanded north to "Highway 8", clearing bunkers and looking for enemy equipment and soldiers. The 1st (Ironhorse) Brigade stretched through the historic Euphrates River Valley.

On 04 March the Division conducted a thorough battle assessment. As part of the clean up operations, two OH-58 helicopters, equipped with loud speakers, along with AH-64 helicopters swept a large area of southern Iraq evaluating battle damage using gun cameras to record the wreckage. The mission also uncovered several Iraqi soldiers left behind in the retreat of the Republican Guard. On 07 March, as part of a symbolic early redeployment of troops, the Executive Officer (XO) and fourteen of the personnel from the 1st Battalion, 227th Aviation, departed with the advance party for Ft. Hood, Texas.

No one really thought they would send the main body of troops quickly home, but on 09 March, a day after the 1st Brigade moved north to secure highway 8, nine hundred 1st Cavalry Division Troopers left the battlefields for home. They stepped off their fleet of chartered aircraft to be met by a large crowd of family members and well wishers. Called a "symbolic redeployment", it made good the pledge of President Bush to return the troops as fast as possible. On 09 and 10 March the 1st Battalion, 227th Aviation supported demolition operations of abandoned equipment and undamaged Iraqi equipment.

In the following days, the 1st Cavalry Division moved south into Saudi Arabia and its new home, AA Killeen. named after the major town adjacent to Ft. Hood. AA Killeen was located on the plain of the Wadi al Batin, just west of Hafer al Batin. There on the plain of the Wadi al Batin - the Division began to prepare for their redeployment home.

Things began moving faster now. It was time to put the tracked vehicles aboard trucks for the trip south to the port for cleaning. All vehicles would be "scrubbed" in Dhahran. Soon as the vehicles left, soldiers were picked up and they left AA Killeen and the Wadi al Batin. Few, if any looked back. The destination of the convoy was a sprawling, white concrete highrise apartment complex at Kohbar in Dhahran. Built for someone who had never decided to move in or rent out, the deserted buildings with airconditioning, running water and toilets, were perfect for a redeploying army.

On Broadway, New York City, NY, the 1st Cavalry Division Band takes the Big Apple by storm, to the swaggering melody of the Division song, "GarryOwen".

At last the day had come, when duffle bags were locked for the last time and piled onto a stake truck, the last cool Saudi Pepsi went down and the last farewell was said to buddies. The trooper moved to the airport for their final staging for their flight home. All cheered as the wheels of the aircraft left the runway at King Fahd Airbase. By now, the passangers, perhaps a bit groggy knew they were on the way home. After 18 hours, a familiar sight of grey green, criss-crossed with trails and patches of caliche clay, swung into view, the plane touched down to a smooth landing on Robert Gray Airfield and everyone knew that they were home.

During Operation Desert Storm, the First Team accumulated several new "Firsts":

- "First" to defend along the Saudi-Iraq border.
- "First" to fire Copperhead artillery rounds in combat.
- "First" to conduct intensive MLRS artillery raids.
- "First" to conduct mobile armored warfare in Iraq.

In addition to the above "Firsts", the Division Support Command and the 13th Signal Battalion constantly met unprecedented logistical and communications requirements.

The timely and reliable distribution of coordinated battlefield command and control information was a factor in the successful execution of the air and ground attacks of the First Team. The mission of the 13th Signal Battalion was to setup, operate and maintain a reliable tactical communications network in the theater of operations of the 1st Cavalry Division. This activity was implemented by the use of an "antenna farm", erected deep in the Ruqi pocket near the Kuwait border and Mobile Subscriber Equipment (MSE), a Corps area communication system that provided secure mobile voice/data/facsimile service.

Reunion, and the best in America, seemed to be the theme for the Spring and Summer of 1991. To be a soldier and home was one of the finest of things. The First Team paraded down a dozen avenues. Atlanta, Houston, Dallas, Washington, DC and New York. The visit to New York City was a tremendous welcome from the press and police. At Battery Park, the parade formation turned into Broadway, a wide red carpet. and into the sunlit snowstorm of ticker tape.

Chapter 11: Force Restructuring

1991-1993

Honor Guard, 1st Cavalry Division

Upon return to the United States, The 1st Cavalry Division initiated the first of a series of reorganizations in the period May 1991 to August 1993. The Division, rebuilt as a new contingency force, was ready to deploy anywhere in the world on a moments notice.

On 21 May 1991, the 1st Cavalry Division became the largest division in the Army, with the reactivation of its 3rd (Greywolf) Brigade. Organic units included in this reactivation were formally assigned to the 1st (Tiger) Brigade, 2nd Armored Division that, at times, had been attached to the 1st Cavalry Division during the Gulf War. Units filling out the 3rd Brigade were 3rd Battalion, 41st Infantry Regiment, 1st and 3rd Battalions, 67th Armor Regiment, 1st Battalion, 3rd Field Artillery Regiment and the 502nd (later redesignated as the 215th) Forward Support Battalion.

In October of 1992 the new Engineer Brigade, 1st Cavalry Division was activated and organized. Though the "Engineering Restructuring Initiative" of the Army, the nucleus of the Brigade was formed around the division's historic 8th Engineer (Combat) Battalion. Additional organic units were assigned to the Brigade. The 20th Engineer Battalion was transferred from Ft. Campbell, Kentucky and the 91st Engineer Battalion was activated to complete the alignment of the Brigade.

In November 1992, the unit designations and colors for the battalions from the former (Tiger) Brigade were returned to 2nd Armored Division in preparation for their redesignations at Ft. Hood on 02 December 1992. This action was done to realign the historical designations with their parent divisions.

Subsequently, on 29 November, the 1st Cavalry Division redesignated, realigned and restored the former (Tiger) Brigade, giving its members the titles of its historical units. These changes included:

- 3rd Battalion, 41st Infantry Regiment redesignated as 1st Battalion, 9th Cavalry Regiment,
- 1st Battalion, 67th Armor Regiment redesignated as 3rd Battalion, 8th Cavalry Regiment and
- 1st Battalion, 3rd Field Artillery redesignated as 2nd Battalion, 82nd Field Artillery.

On 16 December 1992, other 1st Cavalry Division units were redesignated and restored, giving them the titles of their historical units. These changes included:

- 1st Battalion, 32nd Armor Regiment redesignated as 2nd Battalion, 12th Cavalry Regiment,

- 3rd Battalion, 32rd Armor Regiment redesignated as 1st Battalion, 12th Cavalry Regiment and
- "A" Battery, 333rd Field Artillery redesignated as "B" Battery, 26th Field Artillery.

By 11 August 1993, the reflagging actions were completed when the 4th Battalion, 6th Infantry, 2nd Armored Division was reflagged the 2nd Battalion, 7th Cavalry Regiment which was assigned to the First Team's 3rd (Greywolf) Brigade. Following this last reorganization, the 1st Cavalry Division remained the largest division of the Army and the only armored contingency force, ready to deploy anywhere in the world at a moments notice.

Since the reorganization, elements of the First Team have returned to Kuwait no less than a dozen times, as part of a ten year training agreement between the US and Kuwait and also in a crisis situation when Iraq again threatened the borders of Kuwait.

While the First Team takes great strides to preserve its heritage and live up to the legend its forefathers created, training, maintaining readiness and staying on the leading edge of the warfare technology of today are top priorities. To support this initiative on a continuing basis, the 1st Cavalry Division deploys brigades to the National Training Center at Ft. Irwin, California. The National Training Center has a computerized, live-fire complex with sophisticated targetry, a full-time opposing force, a state-of-the-art range instrumentation system that monitors training battles, and full-time combat trainers who observe and control units during exercises.

Chapter 12: Desert Peacekeepers

1992-2000

US Army Forces Central Command - Kuwait

The United States Army Forces Central Command (ARCENT) - Kuwait, a major subordinate command of the United States ARCENT of Ft. McPherson, Georgia. and United States Central Command (CENTCOM) of Tampa, Florida, is the operational unit of the US Army in Kuwait. The mission of the United States CENTCOM is to a) support US and free-world interests by assuring access to Mideast oil resources, b) help friendly regional states maintain their own security and collective defense, c) maintain an effective and visible US military presence in the region, d) deter threats by hostile regional states and e) project a military force into the region if necessary.

The mission of US ARCENT, Kuwait is to acquire, maintain and protect a heavy brigade (reinforced) equipment set, to plan, direct and support all joint training exercises with the Kuwaiti Armed Forces and, in concert with the Government of Kuwait, to establish and maintain the contingency plans for the security of Kuwait. The center of Central Command operations is at Camp Doha, twenty miles north of Kuwait City. Doha is a large logistics base with a working population of over two thousand personnel - US soldiers and airmen, and both US and Kuwaiti contract personnel.

The performance of all support missions is critical in carrying out the training and tactical missions of the Central Command at Camp Doha. These key support missions are carried out by the 385th Signal Company which provides Echelons-Above-Corps Signal support for the USCENTCOM, Third US Army and all units that come to Kuwait to train. The maintenance support contractor, ITT Federal Services International Corporation has the primary mission for the performance of Organizational, Direct Support, General Support and Limited Depot Maintenance to ensure that all of the heavy armored brigade equipment sets prepositioned in Kuwait are always combat ready.

The following poem, written by a member of the 2nd Battalion, 12th Cavalry during one of their early "Intrinsic Action" assignments, summarizes the ultimate benefit gained by the training and eternal vigilance of our First Team.

"Ode to the Desert"

Upon scorching desert sands we have arrived.
Like a wind from the West we have come to chafe the settled grain.
Upon steel beasts we ride.
Through desert twisters and barren wastelands we shall travel,
wreaking havoc on those who oppose.
On the front lines we wait and wait and wait,
waiting for the enemy to attack . . . but they will not.
For they know what we would bestow upon them should they dare attempt to.
They will all drown in a sea of blood.
We shall hack them down with unimaginable force and fury,
and crush their meager army into oblivion!
So let it be said, "Upon steel beasts we sleep and eat
and sleep and eat until the time comes to fight
and then we shall unleash the Dogs of War!"

by Specialist Jacob A. Brooks
"A" Company, 2nd Battalion, 12th Cavalry Regiment

Mutual Defense Cooperation Agreement

Following the Gulf War, members of US Central Command's Army component and the armed forces of Kuwait agreed to participate in a series of Combined Exercises held within the framework of the Mutual Defense Cooperation Agreement between Kuwait and the United States. In these exercises, Army battalions rotate into Camp Doha for training with the Kuwaiti while brigade command elements rotate into various locations in the country to gain familiarity with the terrain to develop and practice a mutual set of defensive postures. These exercises represented an opportunity for US Army forces to work with Kuwaiti armed forces in country while at the same time demonstrating US capability and commitment to the region.

"Upon Steel Beasts, We Ride"

These four-month exercises are conducted utilizing the pre-positioned Army War Reserves Set 5 (AWR-5) at Camp Doha, Kuwait, to minimize the time/distance challenge of CONUS deployments. AWR-5 is a full, heavy brigade set of equipment that is ready to fight as fast as troops can be flown into theater. ARCENT-Kuwait is able to issue at least a battalion set of that equipment every 24 hours. The Army routinely exercises the equipment at least twice a year during the exercise series with brigade command posts and battalion Task Forces (TFs) from the 1st Cavalry Division, Fort Hood, Texas, and the 3rd Infantry Division (ID)(M), Fort Stewart, Georgia, the 4th ID(M), Fort Hood, and the 1st Armored Division, Bad Kreuznach, Germany.

Although termed an exercise, these deployments actually constitute a continuous operational presence of American ground forces in Kuwait, with different units rotating in for periods of four months at a time. These exercises are designed to provide continued US ground presence in Kuwait, improve interoperability and battle staff proficiency between US and Kuwaiti armed forces, and enhance US military force capabilities to rapidly deploy to the region. A typical exercise involves approximately 1,250 - 1,500 US Army soldiers from elements of Headquarters, US Army Forces Central Command, Ft. McPherson, Georgia. combined with elements of the rotating Task Force (TF). The deployment typically included a battalion task force, combat support units and combat service support units. The task force, deployed without equipment, draws from the prepositioned equipment in Kuwait.

Since Desert Storm, the 1st Cavalry Division has responded several times to contingency requirements to participate in joint desert training, and deploy in maneuvering exercises which support the mission objectives of the US Central Command. Each operation has underscored the need for vigilance and quick response and reinforced the value of pre-positioned equipment and limited forward presence in offsetting the strategic time/distance challenges inherent in winning the "race for Kuwait."

Each training mission starts with a complete familiarization program of Reception, Staging, Onward Movement and Integration (RSOI) which helps ease the transition from the hill country of Ft, Hood, Texas to the desert environment and operations of Camp Doha within a short period of time.

Reception

Reception starts when soldiers or equipment land at one of the ports of debarkation of Kuwait. Equipment, bags and supplies are palletized and prepared for convoy to Camp Doha's drawing lot.

Staging

At the drawing lot, soldiers draw their equipment from pre-positioned stores. The available inventory includes

M1A1 Abrams tanks and M2 Bradley fighting vehicles as well as other armor, artillery and engineer sets. This stage is not finished until all units have completed the draw and uploaded all combat equipment. All equipment receives preventive maintenance checks and services prior to departing.

Onward Movement

Onward movement begins with convoy operations to the marshaling area. Supplies and equipment are loaded onto trucks, and tracked vehicles are loaded onto heavy equipment transports. The convoys then move to their positions in the desert to begin training or contingency operations. This phase ends with the downloading of assets and tactical road march to the Kabal.

Integration

The final phase, integration, is the tactical fusion of the arriving units into the Kuwait defensive plan. Training - at the individual and unit level - takes place at the multi-service and multi-national level. These training exercises have four main purposes:

- Force Protection - Training to better protect against hazards or threats.
- Coalition Training with Kuwaiti Forces.
- Operations Plans Refinement.
- Mission Essential Task Training - Including live fire exercises.

The significant elements of each of tour of duty of the Division during the last decade of the century follows:

Operation INTRINSIC ACTION 92-3, 1992

Fresh from organizational restructuring, which restored the 3rd Brigade to the 1st Cavalry Division, plans began to return units to Kuwait in compliance to the Mutual Defense Cooperation Agreements. The first units selected for these exercises were elements of the newly activated 3rd Brigade. Commanders and staffs deployed to the Area of Responsibility (AOR) for the first time since Operation DESERT STORM to provide local command and control of the forces and to emphasize US resolve to counter any Iraqi aggression.

On 20 October 1992, TF 3-41 Infantry, comprised of soldiers from the newly reactivated 3rd (Greywolf) Brigade came to Camp Doha for extensive ground training. Iraq remained quiet before and during the operation, but the soldiers of the First Team were back at Camp Doha, Kuwait. TF 3-41 Infantry was composed of the following elements:

"HHC", 3rd Battalion, 41st Infantry Regiment
"A" Company, 3rd Battalion, 41st Infantry Regiment
"B" Company, 1st Battalion, 67th Armor Regiment
"C" Company, 3rd Battalion, 67th Armor Regiment
"B" Battery, 3rd Battalion, 82nd Field Artillery
"B" Battery, 4th Battalion, 5th Air Defense Artillery
"B" Company, 8th Engineer Battalion
15th Personnel Support Company
215th Forward Support Battalion

They deployed on short notice and upon arrival, drew their allotment of prepositioned equipment and were ready to either train or fight. Phase one of the exercise consisted of extensive desert training. Phase two, which consisted of joint training and maneuver exercises with the Kuwait armed forces, was completed in December allowing the TF to return to Ft. Hood 15 December.

Operation IRIS GOLD 93-1, 1993

Reorganization of the Division continued throughout 1992 when the unit designations for the battalions remaining from the former "Tiger" Brigade were returned to the 2nd Armored Division. This action was done to realign the designations to their parent divisions. The result being that the 1st Battalion, 41st Infantry Regiment being redesignated 1st Battalion, 9th Cavalry Regiment just before their redeployment to Kuwait in January for participation in "Iris Gold", the joint Task-Force Kuwait.

Operation IRIS GOLD, a deployment in January 1993 to support a Special Forces Operation, was not a scheduled Intrinsic Action Rotation. Iraq had begun conducting training exercises close to the Kuwait border. Inasmuch as the elements of TF 3-41 Infantry had just completed an Intrinsic Action Rotation it only made sense to send them back immediately as a reactionary force.

On 15 January 1993, TF 1-9 Cavalry, comprised of soldiers from the newly redesignated elements of the 3rd (Greywolf) Brigade returned to Camp Doha, Kuwait to prepare for a defensive posture against Iraq, if necessary. TF 1-9 Cavalry was composed of the following elements:

"HHC", 1st Battalion, 9th Cavalry Regiment
"A" Company, 1st Battalion, 9th Cavalry Regiment
"C" Company, 1st Battalion, 9th Cavalry Regiment
"B" Company, 3rd Battalion, 8th Cavalry Regiment
"C" Company, 3rd Battalion, 8th Cavalry Regiment
"C" Company, 3rd Battalion, 67th Armor Regiment
"B" Battery, 3rd Battery, 82nd Field Artillery
"B" Battery, 4th Battalion, 5th Air Defense Artillery
"B" Company, 8th Engineer Battalion
15th Personnel Support Company
215th Forward Support Battalion

Their training included coalition work with the 151st Armored Battalion of Kuwait including Military Operations on Urbanized Terrain (MOUT) training with Bradley fighting vehicles and the practice of underground concealment of M1A1 tanks. In all, their training provided another opportunity to strengthen the ties with Kuwait showing a resolve to support the Mutual Defense Agreements and enhance the deployment capability of the 1st Cavalry Division. The Task Force returned to Ft. Hood on 15 March 1993.

Operation VIGILANT WARRIOR, 1994

In early October 1994, Operation VIGILANT WARRIOR was initiated. Forces were deployed to the Central Region in response to the threat of aggression by the Iraq troop movements which amassed large bodies of troops in southern Iraq along the Kuwait border in November.

As Iraq began to pull back its Republican Guard units north of the 32nd parallel, coalition forces conducted joint and combined maneuver training designed to further enhance operations and interoperability.

Operation VIGILANT SENTINEL, 1995

By August 1995, Iraq was continuing a five-week-old pattern of "unusual" troop movements of training in and around garrisons, at surface-to-air missile launcher sites and at air bases. The United States had not detected any signs of planning or preparation for invasion of any of its neighbors by Iraq.

The Pentagon dubbed its reply to the Iraqi movements Operation VIGILANT SENTINEL, an operation that included sending forces and warships closer to the Gulf as well as initiating war games in Jordan.

From August to December 1995 during VIGILANT SENTINEL, the 1st Cavalry Division trained with the equipment, showing they are fully prepared to fight alongside our Kuwaiti allies. Well maintained, rapidly issued and versatile, AWR-5 puts a fighting force on the ground swiftly. In addition the scheduled exercise of Operation INTRINSIC ACTION was moved up in Kuwait.

Operation INTRINSIC ACTION 95-3, 1995

On 22 August, two Air Mobility Command C-141 Starlifters and one C-5 Galaxy arrived with an advance team of 93 soldiers from TF 1-5, "Black Knights," from the 1st Cavalry Division's 2nd (Blackjack) Brigade, based at Ft. Hood, Texas. They were carrying 150,000 pounds of cargo and support equipment. Two additional flights were scheduled through the night, with eight aircraft scheduled for 23 August.

The main body of 1,300 soldiers, fully operational within four hours after landing, came highly motivated and ready for some of the most intense, real-world, combined-arms training the world has to offer. Becoming a part of the ongoing Operation VIGILANT SENTINEL, INTRINSIC ACTION was originally scheduled for October, but the unexplained troop movements in Iraq prompted an earlier starting date for the exercise.

The mission of TF 1-5 was to conduct combined arms training and force protection operations with the Kuwaiti Army at the Udairi Range. The training began with platoon gunnery, then progressed to company-sized operations. The mission culminated with a Combined Arms Live Fire Exercise.

TF 1-5, part of the 2nd Brigade Combat Team, was composed of the following elements:

1st Battalion, 5th Cavalry Regiment
2nd Battalion, 12th Cavalry Regiment
1st Battalion, 8th Cavalry Regiment
3rd Battalion, 82nd Field Artillery
15th Forward Support Battalion
27th Main Support Battalion
3rd Battalion, 29th Field Artillery
2nd Battalion, 77th Armor
91st Engineer Battalion
"A" Battery, 21st Field Artillery;
"C" Battery, 4th Battalion, 5th Air Defense Artillery
545th Military Police Company

The Udairi Range is one of those unique operational areas because it is not only a training area, but also a defensive perimeter in case of war. With sand berms and concertina wire blockades permanently in place, the range is perfect for many training exercises.

Navigating over an ocean of sand, the TF 1-5 thundered across the Kuwaiti desert under a blanket of darkness. Suddenly, the boom of howitzers firing pierced the night. mortar crews lit the night up with rounds of illuminating ground bursts to mark sectors of fire. After transitioning into a delay, TF 1-5 spent the remainder of the night battling a motorized rifle regiment. As the sun began to rise over the horizon, the tanks and Bradley vehicles started to pull back, conducting a rearward passage of lines.

In November 1995, TF 1-5 left Kuwait after an accelerated Intrinsic Action to deter Iraqi aggression.

Operation INTRINSIC ACTION 96-3, 1996

On 26 August 1996, TF 1-9 comprised of soldiers from the 3rd (Greywolf) Brigade came to Camp Doha for extensive ground training. Iraq remained quiet before and during the operation, but the soldiers of the First Team were back at Camp Doha, Kuwait. TF 1-9, part of the 3rd (Greywolf) Brigade, composed of the following elements:

"HHC", 1st Battalion, 9th Cavalry Regiment
"D" Company, 1st Battalion, 9th Cavalry Regiment
"B" Company, 1st Battalion, 12th Cavalry Regiment
"D" Company, 1st Battalion, 12th Cavalry Regiment
"A" Battery, 2nd Battalion, 82nd Field Artillery
"A" Company, 8th Engineer Battalion
215th Forward Support Battalion
15th Finance Battalion

Immediately after the jet-engines of the transport airplanes were silenced, TF 1-9 hit the ground running. Following a bus ride to Camp Doha, the 3rd (Greywolf) Brigade put a carefully planned operation into action.

A row of Bradley Fighting Vehicles (BFVs) lay in wait for their crews at the Tactical Assembly Area (TAA) TUESDAY. Once there, the troops picked up force protection ammunition and moved to the marshaling area for staging and reorganization. Tracked vehicles were loaded onto heavy equipment transports and lowboys for transport to the Logistical Release Point One at TAA HEADHUNTER.

The 1st Cavalry Division earned its nickname, America's First Team, by being the first unit to do so many things. First in Tokyo, First in Korea, First in Vietnam, First heavy armored division into Iraq and now Artillerymen from

"A" Battery, 2nd Battalion, 82nd Field Artillery were the First to use the new M-109A6 Paladin Howitzers on Udaira Range in Kuwait. This new weapon provided an excellent opportunity for the TF 1-9 Team to demonstrate their ability to project combat power anywhere in the world, showing their allies and others the resolve of the United States to maintain peace.

The hot air at Udairi Range can be like a blast furnace. But for the mortarmen of TF 1-9, the heat becomes a great advantage in that you can actually see downrange where the rounds impact. As hot as it is, the air is less dense, providing less aerodynamic drag, allowing the rounds go farther with greater accuracy. The mortar platoon from Headquarters Company, 1st Battalion, 9th Cavalry adapted to the new conditions and quickly used this advantage to get some unbeatable training.

Designing ranges was just one of the tasks the engineers engaged in during Intrinsic Action 96-3. Every field site was surrounded by a berm. The engineers had constructed ten to twelve miles of sand walls during the first week of training. Upon completion of the tactical training, the engineers concentrated on site improvements and their own role in the TF 1-9 training for the next three months before they returned home in December. Yet, even though they left, the results of their work remained for the next units that would come to Kuwait, an enduring piece of craftsmanship for future gunners to use.

As the soldiers of TF 1-9 engaged in their training exercises, word came of the alert sounded for their comrades at Ft. Hood, Texas. Tensions in the Gulf areas increased when the Iraqi troops moved into a "safe haven" at Irbil in northern Iraq.

On 17 September after several days of preparation, III Corps was given the green light to deploy additional forces to Kuwait as part of Exercise Intrinsic Action. To support the directive, a US Army Brigade Combat Team headquarters and two battalions, consisting of approximately 3,500 soldiers, from the 3rd Brigade, 1st Cavalry Division deployed from Robert Gray Army Airfield. The deploying units were the 3rd Battalion, 8th Cavalry, the 1st Battalion, 5th Cavalry, the remainder of the 82nd Field Artillery and support personnel from "A" Battery, 21st Multiple Rocket System, 68th Chemical Company and the 13th Signal Battalion to augment the elements already with TF 1-9.

The deploying 1st Cavalry Division forces, commanded by 3rd Brigade Commander Colonel Eric T. Olson, reinforced the 1,200 soldiers already in Kuwait training with the Kuwaiti armed forces as part of exercise Intrinsic Action 96-3 since early August. Troops deployed with individual arms and equipment by air from Robert Gray Army Airfield at Ft. Hood, Texas and used equipment such as M1A1 Abrams tanks and M2A2 Bradley Fighting Vehicles prepositioned in the Kuwait theater.

On 19 September, the 3rd Brigade Combat Team of the 1st Cavalry Division arrived at Kuwait City International Airport. They and the nearly 1,100 soldiers already here with TF 1-9 became the main body of TF KUWAIT, commanded by the Third US Army, Ft. McPherson, Ga.

The 3rd Brigade was no stranger to Kuwait or the hash environment of desert training. It was here during operations DESERT SHIELD and DESERT STORM. Immediately after landing, the 3rd Brigade began drawing prepositioned equipment and after only six hours in country, the soldiers were moving out to locations just twenty five kilometers south of the Iraq border.

Operation INTRINSIC ACTION 97-1, 1997

The 1st Squadron, 7th Cavalry deployed to Kuwait in mid-February 1997 for Intrinsic Action 97-01. The TF 1-7, of approximately 1,100 soldiers, was composed of the following elements:

"HHT", 1st Squadron, 7th Cavalry
"A", "B", and "C" Troops, 1st Squadron, 7th Cavalry
"B" Battery, 1st Battalion, 82nd Field Artillery
"C" Company, 20th Engineer Battalion
"A" Battery, 4th Battalion, 5th Air Defense Artillery
545th Military Police Company
312th Military Intelligence Battalion

With all of its added slice elements, the squadron-turned-task force continued to lead the division. TF 1-7 Cavalry, the first division cavalry squadron to do an Intrinsic Action rotation, set the record for reception, staging, onward movement, and integration. Apache Troop rolled out, fully uploaded, from its staging base in Doha less than four hours after it touched down at Kuwait City. TF 1-7 was the first unit to set up a consolidated assembly area. In the past, units had split up into separate camps, but the single assembly area has greatly improved quality of life compared to earlier rotations. During the rotation, the task force artillery battery conducted a Paladin gunnery cycle and showed the Kuwaitis how to train in live-fire and simulated exercises, culminating in a highly successful force-on-force exercise and combined arms live-fire exercise.

During April, the TF 1-7 fired its best tank and Bradley gunneries ever. Thirteen crews distinguished their tanks with a squadron average of 884, including a perfect 1000-point crew. Thirty five crews distinguished their Bradleys with a squadron average of 939. In May, the squadron executed Cavalry Table XXII (a maneuver and live-fire exercise combining both tanks and Bradleys), and in finishing the rotation, the entire TF 1-7 deployed to the field to conduct a zone reconnaissance and execute screen and delay operations.

Operation INTRINSIC ACTION 97-2, 1997

The advance party for Operation TF 2-12 deployed to Kuwait on 03 June 1997. The remainder of the organization deployed in mid June. Drawing their equipment, they quickly moved out to the desert to TAA Charger which and the troopers settled in beginning training in live fire exercises at the platoon, company and task force level and, individual and crew-served weapon ranges. TF 2-12, of approximately 1,200 troops, was composed of the following elements:

"HHC", 2nd Battalion, 12th Cavalry Regiment
"A" Company, 2nd Battalion, 12th Cavalry Regiment
"D" Company, 2nd Battalion, 12th Cavalry Regiment
"B" Company, 2nd Battalion, 5th Cavalry Regiment
"D" Company, 2nd Battalion, 5th Cavalry Regiment
"A" Battery, 3rd Battalion, 82nd Field Artillery
"A" Company, 91st Engineer Battalion
15th Forward Support Battalion
4th Platoon, 545th Military Police Company
2nd Platoon, "B" Battery, 4th Battalion, 5th Air Defense Artillery
312th Military Intelligence Detachment

Following 120 days of extensive training and live fire exercises, TF 2-12 returned to Ft. Hood in October.

Operation INTRINSIC ACTION 98-2, 1998

The 3rd Brigade deployed TF 2-7 to Kuwait in May 1998 for Operation INTRINSIC ACTION 98-02. This deployment was part of the 1st Brigade Combat Team, 3rd Infantry Division, in response to increased tensions along the Iraq-Kuwait border. The Task Force of approximately 1,100 soldiers was composed of the following elements:

"HHC", 2nd Battalion, 7th Cavalry Regiment
"A" Company, 2nd Battalion, 7th Cavalry Regiment
"D" Company, 2nd Battalion, 7th Cavalry Regiment
"A" Company, 3rd Battalion, 8th Cavalry Regiment
"D" Company, 3rd Battalion, 8th Cavalry Regiment
"B" Company, 8th Engineer Battalion
"C" Battery, 2nd Battalion, 82nd Field Artillery
215th Forward Support Battalion

Upon arrival, TF 2-7 drew all of their prepositioned equipment in less than three hours, and were in position at the forward area in less than fifteen hours and ready for combat operations. The 1,100 soldier TF 2-7 executed Level I Gunnery, Gunnery Situational Tactical Exercise (STX) Lanes and a TF Movement to Contact/Live File. Following completion of training, the TF 2-7 turned in all equipment and redeployed to Ft. Hood in August.

Operation DESERT SPRING 00-1, 2000

The TF 1-9 deployed to Kuwait in April 2000 for Operation DESERT SPRING 00-01 that gave the soldiers an opportunity to sharpen their fighting skills. TF 1-9 was composed of the following elements:

"HHC", 1st Battalion, 9th Cavalry Regiment
"A" Company, 1st Battalion, 9th Cavalry
"B" Company, 1st Battalion, 9th Cavalry
"B" Company, 2nd Battalion, 8th Cavalry
"C" Company, 2nd Battalion, 8th Cavalry
"B" Battery, 2nd Battalion, 82nd Field Artillery
"A" Company, 20th Engineer Battalion
3rd Platoon, 68th Chemical Company (DECON)
5th Platoon. 545th Military Police Company
4th Platoon, "D" Company, 312th Military Intelligence Battalion (GSR)

Operation INTRINSIC ACTION 01-1, 2001

TF 2-5, consisting of about 1,000 Troopers including attached support units from other 1st Cavalry Division units, deployed to Kuwait in December 2000 for INTRINSIC ACTION 01-01. The advance party arrived on 07 December to prepare for the garrison of TF 2-5 in a well fortified Kabal (an enclave or base camp) in the Kuwait desert only a few miles from the Iraq border, but many miles from the nearest Kuwait town or other defense forces. TF 2-5 was composed of the following elements:

"HHC" 2nd Battalion, 5th Cavalry
"A" Company, 2nd Battalion, 5th Cavalry
"B" Company, 2nd Battalion, 5th Cavalry
"D" Company, 2nd Battalion, 5th Cavalry
"D" Company, 2nd Battalion, 8th Cavalry
"A" Company, 20th Engineer Battalion
2nd Platoon, "A" Battery, 4th Battalion, 5th Air Defense Artillery
Decon Platoon, 68th Chemical Company
"A" Battery. 1st Battalion, 82nd Field Artillery
115th Forward Support Battalion (-)
"A" Company, 312th Military Intelligence Battalion (-)
6th Platoon, 545th Military Police Company
Enlisted Terminal Attack Controllers (ETAC), 1st Cavalry Division

In addition to the normal training exercises, February began with Coalition Training with the 35th Mechanized Infantry Battalion, 10th Land Forces (Kuwait). Additional special training in medical technicial skills and chemical detection were conducted. Following completion of training, TF 2-5 returned to the base camp and turned in all equipment and began preparations for redeployment to Ft. Hood in late April.

Chapter 13: Bosnia Peacekeepers

1998-1999

Bosnia-Herzegovina is a crossroads country. Straddling the Dinaric Alps and sandwiched between Croatia and Serbia, it has long been a zone of multi-cultural contention. In 1992 a three-way civil war pitted Muslim Slavs, Orthodox Serbs and Catholic Croats - all formerly neighbors and all of the same South Slavic stock - against one another. It devastated the infrastructure of the country, left refugees numbering in the millions and gave its partisans the ignominious distinction of having introduced the phrase "ethnic cleansing" into modern parlance. The war is now officially over, but it will be many years yet before the new wounds have healed and the country again boasts a significant tourist draw. In the meantime, around 34,000 NATO troops remain stationed in Bosnia-Hercegovina to stabilize inter-ethnic fighting, and a large international civilian presence is working to rebuild the country.

History of Task Force Eagle

On 14 December 1995, a NATO-imposed cease-fire halted the destructive four year Balkan conflict by the issuance of the General Framework Agreement for Peace (GFAP). The US 1st Armored Division, as part of NATO's Allied Command Europe, Rapid Reaction Corps, established Task Force Eagle (TFE) which was ordered to Bosnia-Herzegovina as part of Operation JOINT ENDEAVOR. On 20 December 1995, TFE assumed control of its Area Of Responsibility (AO) during a Transfer of Authority (TA) ceremony with United Nations forces at Eagle Base, Tuzla.

After the historic bridging of the Sava River on 31 December 1995, the 1st Armored (Old Ironsides) Division, with supporting forces from V (US) Corps, was joined by armed forces of twelve other nations that included Estonia, Latvia, Finland, Poland, Denmark, Lithuania, Norway, Iceland, Sweden, Russia, Turkey and the United States. Together, these nations comprised MultiNational Division - North (MND-N), one of three MultiNational Divisions in NATO Stabilization Force (SFOR). TFE represented the Unites States contribution to SFOR.

One of the most powerful UN formations ever fielded, MND-N, enforced the cease-fire, supervised the marking of boundaries, and monitored the zone of separation between the former warring factions, Entity Armed Forces (EAF) which consisted of the Bosnian-Croat Entity (the Federation) and the Bosnian Serb Entity (the Republika Srpska), enforced the withdrawal of the combatants and the movement of the heavy weapons to designated storage sites. TFE also supported the efforts of the Organization for Security and Cooperation in Europe (OSCE) to administer the country's first ever democratic national elections.

On 10 November 1996, the 1st Armored Division transferred authority for command and control of TFE to the 1st Infantry Division. The 1st Infantry Division deployed as a covering force to allow the safe return of the 1st Armored Division units to Germany. Upon completion of the covering force mission, the 1st Infantry Division continued ensuring that all military aspects of the Dayton Peace Accord were accomplished firmly and fairly.

On 20 December 1996, the implementation force mission came to successful conclusion and the 1st Infantry Division was directed to continue serving in Bosnia as part of the new NATO Stabilization FORce (SFOR). This decision brought to close the mission of Operation JOINT ENDEAVOR and the beginning for Operation JOINT GUARD.

On 22 October 1997, the 1st Armored Division conducted a relief in place of the 1st Infantry Division, and reassumed command of MND-N. Soldiers from America's tank division, accustomed to the mission, quickly adjusted to their familiar role and challenges of maintaining a safe and secure and peaceful environment in the MND-N sector of SFOR.

Through careful planning and the skillful execution of every mission, the soldiers of the 1st Armored Division and those of our multi-national allies, continued to monitor the militaries of the former combatants and provide a climate of stability in the war-torn land of Bosnia-Herzegovina.

1st Cavalry Division Deployment Order

On 17 April 1998, a Pentagon spokesman announced that the 1st Cavalry Division troops from Ft. Hood, Texas, would deploy in late summer to replace the US peace keeping force in Bosnia, the 1st Armored Division. It was an historic move. It was the first time a Continental United States (CONUS) based contingency unit would assume the Bosnian peacekeeping mission.

While not obvious to the layman, the challenge of moving a division-size Army task force from its home station in Texas to a deployment zone far away, was great. The task force of approximately five thousand soldiers required over seven hundred vehicles, aircraft, and shipping containers of supplies. The shipment had to travel to a far shore across many time zones, mountains, and kilometers of salt water. The route had to be secure and the movement had to be safe.

Planning for the move began in the spring and early summer of 1998. The equipment of the 1st Cavalry Division task force would be transported, staged and concentrated at the shipping terminal of the 596th Transportation Group in Beaumont, Texas, then be moved to its final destination by motor convoy, rail, and air delivery. Additional cargo of the 3rd Battalion, 229th Aviation Regiment, 82nd Airborne Division, which was to be attached to the 1st Cavalry Division for the Bosnian Operation, would be staged and loaded at Wilmington, North Carolina.

Spring turned into early summer. Military Traffic Management Command (MTMC) planners began to hear word of a change of plans. The original planed seaport of embarkation was to be Bremerhaven, Germany, but Pentagon planners were considering a port much closer to the destination. Aviators were concerned about maintenance and safety issues associated with a long air transit across the heart (and mountains) of Europe from northern Germany. Leaders in the 1st Cavalry Division wanted to hurry the movement of their heavy equipment to facilitate the mission transfer. Although a port closer to their destination such as the Adriatic Sea port of Rijeka, Croatia might create infrastructure and security problems, it was finally chosen because its location would speed the entry of the task force's heavy equipment into the theater.

In May, the 3rd Battalion, 229th Aviation Regiment, which has its origin in the "WidowMakers", "C" Company, 229th Aviation Regiment as an organic unit of the 1st

Task Force Eagle (TFE) was established as part of the NATO Allied Command Europe Rapid Reaction Forces when the cease-fire halted the destructive four year Balkan conflict. (14 December 1995)

The advance party of the 1st Cavalry Division arrived at Krk International Airport in Rijeka, Croatia to support the 21st TAACOM's logistical task force at the port of Rijeka with the offloading equipment, which will be used by the division.

On 23 October, a Humvee is the first vehicle to roll off the Military Sealift Command ship USNS Soderman at Rijeka, Croatia. Soderman, one of the large, medium-speed, roll-on/ roll-off ships, is off-loading helicopters and hundreds of vehicles and containers in support of the deployment of the 1st Cavalry Division to Bosnia and Herzegovina for Operation Joint Forge.

Cavalry Division in Vietnam, received orders to deploy to Bosnia as peacekeepers in support of Operation JOINT FORGE. In June, in preparation for deployment, its intensive training criteria included an Aviation Training Exercise at Ft. Rucker, Alabama; in July, a Mission Readiness Exercise at Fort Polk, Louisiana and in August, a Joint Air Attack Team live fire coordination exercise at Ft. Bragg, North Carolina.

On 20 June, the mission of the NATO SFOR was restated to deter hostilities and stabilize the peace, contribute to a secure environment by providing a continued military presence in the area of operations, target and coordinate SFOR support to key areas and primary civil implementation organizations and progress towards a lasting consolidation of peace without further need for NATO-led forces in Bosnia-Herzegovina.

As a result of the mission restatement, the transition of the current NATO led SFOR in Bosnia-Herzegovina changed over to a slightly smaller follow-on force. The United States agreed to provide a force of approximately 6,900 US military personnel to help maintain a capable military force in Bosnia-Herzegovina.

Concurrently, Operation JOINT GUARD ended and Operation JOINT FORGE began. Operation JOINT FORGE continued to build on the successes of Operations JOINT GUARD and JOINT ENDEAVOR. No timeline for the duration of Operation JOINT FORGE was established from the onset. Rather than establish a deadline for the termination of the mission, its accomplishments were assessed periodically and the level of force commitment adjusted, as circumstances required.

With the deployment to Bosnia inching closer, soldiers from the advance party, Headquarters and Headquarters Company, 1st Cavalry Division had already packed their military vans in preparation for their year-long deployment.

1st Cavalry Division Deployment

Different sections from the Division Headquarters loaded the military vans for about four days. The loading went fast because everyone was assigned a specific task. Once the military vans were packed, they were loaded onto vehicles, weighed and moved to the Deployment Readiness Reaction Field (DRRF) to be marshaled, and taken to the railhead in one group, In addition to the Headquarters equipment, High Mobility Multi-purpose Wheeled Vehicles (HMMWV), five-ton trucks and expando-vans from the Headquarters Company motorpool were being loaded on the trains for shipment to Beaumont, Texas.

Long lines of vehicles were concentrated at the Beaumont docks. The cargo was mostly from Ft. Hood. However, some cargo was from Ft. Carson, Colorado; Ft. Riley, Kansas; Ft. Sam Houston, Texas; and Ft. Polk, Louisiana. Helicopters were flown into the port, some from as far away as a National Guard unit in California. Army Reservists from the 1192nd Transportation Terminal Brigade, from New Orleans, Louisiana, began loading on 04 and 05 August. In all, the cargo covered 192,000 square feet.

On 04 August, the *USNS Soderman*, a roll-on, roll-off ship, belonging to the Military Sealift Command, three-football fields long and fifteen stories in height from keel to bridge, was loaded in Beaumont, Texas. The 1st Cavalry Division was on the way to Bosnian peacekeeping duties ahead of schedule, with help from MTMC who arranged to move the equipment for the Division via the Adriatic Sea - instead of a traditional approach through Northern Europe. The bold move through the Croatian port of Rijeka saved two weeks of transit time for the "First Team" soldiers and equipment.

For the *USNS Soderman*, the distant port of Rijeka drew closer by the hour. As the ship approached the lands of the former country of Yugoslavia, the personnel of MTMC were joined by some new and valuable partners, military and civilian. The military partners included the 21st Theater Area Army Command (TAACOM) of Kaiserslautern, Germany. who were charged with the responsibility of setting up a small city to speed the transition from ship unloading to onward movement to Bosnia.

On 20 August 1998, it was "wheels up" for the troopers of the 1st Cavalry Division as it became the first Continental United States (CONUS) based division to support the ongoing peace-keeping mission in Bosnia and assume command of Task Force Eagle and MND-N. Members of the division departed Robert Gray Army Airfield, Ft. Hood to set the stage for more than six thousand soldiers of the 1st and 2nd Brigade that would follow.

In a ceremony held at Tuzla Airfield, Major General Kevin P. Byrnes, Commander, 1st Cavalry Division, accepted authority of the NATO Multi-National Division and Task Force Eagle from outgoing commander Major General Larry R. Ellis, 1st Armored Division. (07 October 1998)

On the same day, the 3rd Battalion, 229th Aviation Regiment along with the 301st Attack Squadron, Royal Netherlands Air Force and "I/K" Company, 4th Battalion. 159th Aviation that supplements the Battalion's organic aviation maintenance, was attached to the 1st Cavalry Division as Task Force ANGEL. Concurrently, the first element of Task Force ANGEL (the 3rd Battalion, 229th Aviation Regiment) conducted an incident free deployment from Fort Bragg, North Carolina and arrived at the Camp Comanche airfield. Task Force Angel immediately assumed the attack aviation mission for the 4th Brigade, 1st Cavalry Division as if it were organic to the division. Their defined mission was to conduct reconnaissance, security and attack helicopter operations across the SFOR Area of Responsibility (AOR); to deter EAF hostility and enforce the GFAP; on order, conduct mission handover and safely redeploy to CONUS.

On 21 August the first group of the 1st Cavalry Division advance party arrived at Krk International Airport in Rijeka, Croatia. The troops would assist the 21st TAACOM's logistical task force at the port of Rijeka with the offloading, inspection, preparing and transporting more than eight hundred pieces of equipment, which would be needed by the Division in October, when it would take operational control of the SFOR mission in Bosnia from the 1st Armored Division.

The operations of the port of Rijeka, a new one for MTMC presented many challenges. There were civilian contractors to work with as well. MTMC contracted with Jadroagent, of Rijeka, for port and stevedore services. The stevedores, with another language and culture, were required to undergo extensive training. American contractors included veteran overseas operators Brown & Root Services Corporation and DynCorp. Brown & Root operated a dining facility and shower point. DynCorp maintenance specialists, many of whom were Army veterans, were assigned to assist in the assembly and preparation of the helicopters for flight. In short order, a complete task force staff, from command sergeant major to public affairs officer, was functioning and ready for the arrival of the *USNS Soderman*.

After a stop in Wilmington, North Carolina to load the equipment of the 229th Aviation Regiment, the US Navy Military Sealift Command *USNS Soderman*, reached Rijeka on 23 August, a day early. The early arrival of the ship presented no problem as the support teams came together and the dockside operations were started. Unloading the equipment was accomplished in a timely manner.

From 23 to 25 August, stevedores at the Rijeka dockside unloaded 377 wheeled and tracked vehicles, 287 shipping containers, 45 Apache and Black Hawk helicopters, numerous support vehicles, and assorted unit supplies. The cargo was prepared immediately for onward movement. From 25 to 30 August, some twelve trains were loaded with vehicles and shipping containers. Some nonstandard pieces of equipment went by road convoy to the Bosnia-Herzegovina 1st Cavalry Division area near Tuzla.

The shipment into Croatia was a plus for the aviators and saved hundreds of hours of helicopter time for the aircraft of the Division. From a hastily formed dockside aviation compound, helicopters were reassembled, tested, and prepared for liftoff. Almost hourly, another helicopter completed safety and maintenance checks and lifted off high above the sun-drenched docks and out over the azure waters and blue skies of the Adriatic. Pilots flew the helicopters directly to the Camp Comanche airfield, deep in Bosnia.

On 24 August the main body of the Division began deploying from Robert Gray Army Airfield. Flights continued through early October. In preparation for the mission, the training soldiers had conducted prior to deploying had been very demanding and the whole experience had pulled the Division Staff together. Morale was quite high.

In settling into Eagle Base, Major General Kevin P. Byrnes observed that the surroundings were a little bland and lonely. To correct the situation, he commissioned a group of resourceful troopers to survey the conditions, recommend and implement a solution. The results of their first, of many, labors of love were illustrated by many of the colorful 1st Cavalry Division insignia and signs that quickly appeared.

To the citizens of Bosnia, stability is associated with the presence of peacekeepers who conducted approximately 100 patrols daily throughout the Area of Responsibility (ARO).

Although units of the 1st Cavalry Division are deployed routinely around the world, the Division Headquarters had not deployed since DESERT STORM. The headquarters command staff of the division would be stationed at Camp Eagle near Tuzla, Bosnia during the entire year-long deployment. It was planned that the 1st Brigade Combat Team would be relieved by the 2nd Brigade Combat Team in March 1999. In August 1999, the 10th Mountain Division (and other units) would replace the 1st Cavalry Division Headquarters, 4th Brigade, 312th Military Intelligence Battalion, and 13th Signal Battalion. The 2nd Brigade Combat Team would remain in Bosnia under the control of the 10th Mountain Division until October 1999 when, as the last unit of the 1st Cavalry Division to return, it would redeploy to Ft. Hood.

On 07 October, with the change-over in command, the units of the 1st Cavalry Division became responsible for enforcing the Dayton Peace Accords and maintaining a stable and secure environment in war-torn Bosnia-Herzegovina.

General Eric M. Shinseki, NATO SFOR Commander was the reviewing officer for the afternoon ceremony. The ceremony began with an inspection of the honor company. After a performance of the Turkish band playing the "Star Spangled Banner" the colors of the 1st Armored Division were cased and the colors of the 1st Cavalry Division were uncased and presented. The ceremony closed with a benediction by the 1st Cavalry Division Chaplain and the band playing "Hit the Leather", "Iron Soldier March", and "The Army Song".

On 12 October the last flight of the 1st Cavalry Division soldiers arrived in Bosnia. The 1st Cavalry Division had completed deployment of Division Headquarters, two brigade task forces and a number of supporting units to Bosnia. This flight marked the first commercial flight, carrying military soldiers, to fly into the Tuzla International Airport, which officially opened on 10 September 1998. The capability shortened deployment and redeployment times for soldiers assigned to Task Force Eagle.

First, working in a coordinated, preplanned effort, each unit of the Combat Team was assigned tasks or undertook specific missions, that fulfilled the overall mission of the 1st Cavalry Division. The missions included the following tasks:

- Patrolling the 1,400 km Zone of Separation (ZOS)
- Monitoring 766 cantonment sites
- Conducting site inspections and confiscation of unauthorized weapons
- Monitoring Entity Armed Forces (EAF) training and movement activities
- Monitoring of de-mining efforts
- Enforcing restrictions on EAF training
- Removing unauthorized checkpoints
- Maintaining a secure environment for civil implementation
- Improving local infrastructure of bridges, roads and railroads
- Maintaining local airport facilities for civil air traffic

To effectively cover their theater of operations, the 1st Cavalry Division units were functionally deployed and stationed at various base camps in Bosnia. The units deployed for the first six months of duty were garrisoned in the following camp areas: *SEE NEXT PAGE*

Following deployment to Bosnia, 1st Cavalry Division troopers underwent a post deployment training and transition program with the soldiers of 1st Armored Division that was built on the concept of a "right-seat-ride". During this time, leaders, staffs and soldiers trained alongside their counterparts with the 1st Cavalry Division soldiers accompanying 1st Armored Division soldiers as they accomplished daily missions in Bosnia. Each soldier and officer rode "in the right seat" for a minimum of two weeks. 1st Cavalry Division soldiers assumed more responsibility for missions as 1st Armored Division soldiers were slowly phased out from daily operations.

Supervision and monitoring of EAF, consisting of the Bosnian-Croat Entity (the Federation) and the Bosnian Serb Entity (the Republika Srpska), was a primary mission. Training and movements of the EAF were monitored daily to ensure that their missions were conducted safely and

Force Protection is a prime objective of the mission of the Task Force. Even on Christmas eve, additional security was provided by generator-fed floodlights that spilled a halogen shimmer on the icy road and frosted spirals of concetina wire strung along the perimeter of Eagle Base.

1st BRIGADE COMBAT TEAM (20 August 1998 to 15 Mar 1999)		
EAGLE BASE	**CAMP McGOVERN**	**CAMP COMANCHE**
"HHC", 1st Cav Div "HHC", 1st Bde Cmbt Tm "HHB", 1-82nd FA Bn "B" Btry, 1-82 FA "C" Btry, 1-82nd FA "C" Btry, 1-21st FA "HHC", 13th Signal Bn "C" Co, 13th Signal Bn "D" Co, 13th Signal Bn 15th Finance Det 15th Pers Serv Det "HHC", 312th MI Bn "B" Co, 312th MI Bn 545th MP Company 410th MP Company	"HHC", 1-8th Cavalry "A" Co, 2-5th Cavalry "B" Co, 1-8th Cavalry "C" Co, 1-8th Cavalry "D" Co, 1-8th Cavalry "B" Co, 20th Engineer Bn	"HHC", 4th Bde Cmbt Tm "HSC", 2-227th Avn Regt "A" Co, 2-227th Avn Regt "B" Co, 2-227th Avn Regt "C" Co, 2-227th Avn Regt "D" Co, 2-227th Avn Regt "HHC", 615th Avn Spt Bn "HHC", 115th Fwd Spt Bn "A" Co, 2-8th Cavalry "C" Co, 2-5th Cavalry "HHC", 3-229th Avn Regt ** "A" Co, 3-229th Avn Regt ** "B" Co, 3-229th Avn Regt ** "C" Co, 3-229th Avn Regt ** "D" Co, 3-229th Avn Regt ** "I/K" Co, 4-159th Avn Regt ** 301st Atk Sqn, RNAF **
CAMP DOBOL	**CAMP BEDROCK**	**CAMP DEMI**
"HHC", 2-5th Cavalry (-) "D" Co, 2-5th Cavalry "A" Co, 1-8th Cavalry "B" Co, 2-8th Cavalry "A" Co, 20th Engineer Bn "A" Co, 4-5th ADA (-)	"HHC", 2-8th Cavalry "C" Co, 2-8th Cavalry "D" Co, 2-8th Cavalry "C" Co, 20th Engineer Bn "A" Co, 52nd Engineer Bn	"HHC", 2-5th Cavalry "B" Co, 2-5th Cavalry "A" Btry, 1-82nd FA
** Units attached 20 Aug 1998 and not organic to 1st Cavalry Division. Relieved from attachment 30 Mar 1999		

within the stipulations of the Dayton Peace Accord. To the citizens of Bosnia, stability was associated with the presence of peacekeepers, who on a daily basis, conducted eight to ten inspections of EAF arms and ammunition storage.

13th Signal Battalion

Units of the 13th Signal Battalion were deployed to Bosnia in support of Operation JOINT FORGE. They were stationed at various camps sites to upgrade or replace existing networks and make radio, data link, land line and internet communication for commanders and service personnel deployed to the area, simple and efficient. The objective of the 13th Signal Battalion was to facilitate communicating throughout the rough terrain of the MND-N sector as easy as in the states. Another major responsibility of the 13th Signal Battalion was to install and maintain secure data links for communicating to the various NATO commanders and back to Ft. Hood, Texas.

On 15 October the first of many video teleconferences between the Command Division staff and the Family Readiness Group leaders at Ft. Hood occurred. Task Force Eagle updated the leaders about the status of current operations in Bosnia and the leaders were given the opportunity to provide information on any critical family issues and ask questions. Following the Family Readiness conference, the Task Force Hood staff briefed the Commanding General on the situations at Fort Hood.

227th Aviation Regiment

On 24 October an unidentified laser device was directed at a UH-60 helicopter and resulted in minor eye injuries for two crewmembers of "B" Company, 2nd Battalion, 227th Aviation. The incident occurred near the town of Zenica. The flight crew completed their mission and returned to base where they were both immediately examined and treated. The pilot and a crew member had sustained minor burns to the outer tissue of their eyes. The optometrist who treated the pilots indicated that the injuries would not be permanent. Subsequent examinations showed improvement and the crew returned to full flight status within the next few days. By 01 November all UH-60 and AH-64 helicopter crews flying in support of the SFOR in the Bosnia-Herzegovina areas began using newly issued laser protective equipment in the wake of the recent incident.

Immediately following the confirmation of the incident, coalition military units and units from the EAF were notified and a full investigation was initiated. This was the first verified laser incident in which medical treatment was required. There had been three other reports of suspected pen-type laser devices being aimed at US aircraft. However, there had been no injuries in any of these cases.

In November 1989, a detachment of the Cavalry Division Band from Ft. Hood, Texas, deployed to Bosnia to provide entertainment for the troopers of the Division and local citizens at the Bosnian Cultural Center in Tuzla. The three-month tour of the band in Bosnia included a series of concerts for SFOR soldiers and citizens. The thirty nine member band delighted its audiences with a musical sampling of rock-n-roll, dixie land, country western and classical music. In addition, the holiday concerts included combined performances with Turkish and Russian military bands.

At Christmas, the brass players performed a tree lighting ceremony at the MND-N Headquarters, complete with Santa Claus in a HMMWV and a bag full of gifts. After the ceremony, the band members, trading their battle dress uniforms for "dress blues", held a concert for Major General Byrnes and the staff. Throughout the tour and at all the concerts, the band served the Division well by promoting esprit de corps throughout Bosnia.

As 1998 came to a close at Camp Eagle, a HMMWV driven by the commander of the guard mount, crunched over the snow-packed perimeter road as it made the rounds for the night, checking on soldiers at the gatehouses and on guard towers along the airstrip side of the main American base.

Security, at all times, was the prime mission for the Division. New Year's Eve passed much like any winter night of pulling guard duty at Eagle Base except for the 15-minute fusillade that was touched off at the stroke of midnight by Bosnians in nearby communities, who celebrated by firing pistols, rifles, and machine guns into the sky along with several small explosions for good measure.

Engineer Brigade

The most important and dangerous mission performed by units of the engineer battalions deployed to Bosnia was monitoring demining operations. Their mission was: "To collect, manage and provide mine information for the protection of the force and support the removal, dismantling, and destruction of all mines, unexploded ordnance, explosive devices, and demolitions in accordance with the Dayton Accords."

Bosnia and Herzegovina has the most severe landmine problem in Europe, where it is estimated that one million land mines were emplaced during the three year civil war. To prepare themselves for the mission, the engineers conducted an extensive training program that consisted of: 1) deminer refresher training; 2) mechanical deminer training; 3) survey training; 4) mine awareness and 5) Explosive Ordnance Disposal (EOD) training.

Two types of remote controlled mechanical crawlers were employed to ensure the safe movement of soldiers through areas suspected to be mined. The first, the Mini-Flail, was designed to clear a footpath for troops to move through suspected mine areas. A larger sized mechanical version, the M-60A3 Panther Mine Proofing Vehicle, was designed for use to clear a safe path wide enough to accommodate vehicles.

Significant improvements in demining operations were developed in Bosnia. The significant change was made when the methodology changed from "mine lifting" to "mine clearing". Mine lifting consists of identifying and removing known mines in accordance with the quantities shown on existing minefield records and any additional devices identified during the lifting operation. This method does not account for inaccurate maps or mines that had been added at later dates. Mine clearing consists in a concentrated clearance of mines from a designated area and physical proofing of the ground. This process results in a 99.6% certainty that the area will be safe and can be returned for public use.

In addition to the dangerous task of monitoring mine clearing, units of the Engineer Brigade were busy constructing and reshaping the landscape with their heavy equipment. Among the most critical high priority tasks, was the restoring, remodeling and new construction of the various camps and specialized training facilities. A major effort was undertaken to improve the local infrastructure system of roads and bridges around each of the camps and patrol areas. Working with the Navy Mobile Construction Battalion (NMCB-40) Seabees, the engineers rebuilt the drainage system and added new access roads and several Forward Area Refueling Points (FARP) pads at the Comanche airfield.

One of the assignments of the Division was to routinely set up checkpoints to exercise the interoperability of various forces during day and night operations and to spot check vehicles that might be carrying illegal cargo. On the evening of 24 February, troopers of "A" Company, 2nd Battalion, 5th Cavalry, operating out of Camp McGovern, found a large quantity of contraband munitions in two trucks during checkpoint operations near Brcko. Acting within their mandate, they seized the two trucks and all military weapons and ammunition on board.

The seized weapons, hidden behind piles of firewood, were not the types of weapons that ordinary citizens would have in their possession. They included multiple rocket launch systems, SA-7 air defense systems, and other sophisticated and expensive weapons that one would expect to see only in the possession of military forces.

Following up on the results on their investigation, troopers supported by Russian troops moved to a warehouse suspected of holding illegal weapons in Bijeljina. Entering the warehouse, they found hundreds of cases of military equipment, plastic explosives, grenades, arms and ammunition. The equipment was seized and removed from the site along with paperwork and records.

Subsequent investigations revealed that several personnel assigned to the 311th Infantry Brigade, III Corps Bosnian Serb Army Voijska Rep Srpska (VRS) were directly involved in supporting this illegal act. As punishment for the movement of counterband arms, the 311th Infantry Brigade was disbanded in its entirety. The headquarters, barracks, weapons storage sites, ammunition depots, ranges and other facilities associated with the 311th Infantry Brigade were barred from any military function.

3rd Battalion, 229th Aviation Regiment

The 3rd Battalion, 229th Aviation Regiment, under Operational Control (OPCON) to the 1st Cavalry Division, accomplished its mission and each of its Mission Essential Tasks (MET), as well as exceeding every standard by which a unit can be measured.

In the Bosnian AOR, Task Force ANGEL conducted multinational operations with more than ten foreign countries, from NATO and the former Warsaw Pact. In doing so, it executed its traditional attack helicopter missions as well as non-traditional roles such as hostage rescue exercises, border surveillance, weapon storage site reconnaissance, escort and combined arms operations with infantry and armor.

In February, "C" Company conducted an aerial gunnery qualification and live fire exercise at Glamoc Range in southwestern Bosnia-Herzegovina, most notably firing live Hellfire missiles. Each crew passed its certification requirements by firing 30mm rounds and 2.75 folding-fin practice rockets, in preparation for the live fire exercise. The live fire operation was a capstone training event that allowed the company commander to control the fires of multiple aircraft in a single engagement. Successful training such as the Glamoc exercise gave the Company confidence in its ability to shoot, communicate and move out.

After it arrived in Bosnia, TF ANGEL flew over five hundred hours per month. This Operating Tempo (OPTEMPO) reflected an average of 32.5 hours per month per aircraft. In carrying out its mission, TF ANGEL surpassed the Field Mission Capability (FMC) standard for both aircraft (84% FMC) and vehicle (88% FMC) readiness each and every month by 14% and 8% respectively.

Nuclear, Biological and Chemical Training

In March, seven countries participated in Nuclear, Biological and Chemical (NBC) training hosted by the 1st Battalion, 8th Cavalry at Camp McGovern. US troops participated along with Russian, Danish, Turkish, Swedish, Finish, Polish, and Norwegian chemical soldiers in a joint static display of NBC defense equipment. The US soldiers also held a decontamination exercise demonstrated by soldiers from Camp Bedrock.

The multinational forces converged to observe the US soldiers perform an operational decontamination demonstration. The demonstration consisted of a vehicle wash down to simulate cleansing vehicles of contamination. The exercise also involved a Mission Oriented Protective Posture (MOPP) gear exchange, where soldiers traded out simulated contaminated protective suits for clean garments.

In early March, the 1st Brigade with its attached units began the process of deploying to Ft. Hood and was relieved in place by the 2nd Brigade with its corresponding attached units. The Division Headquarters, the 4th Brigade, the 312th MI and 13th Signal Battalions, scheduled as "Holdover Units", remained deployed in theater until redeployment in August.

On 20 March the first train loaded with military equipment, vehicles and containers, belonging to the 1st Brigade, 1st Cavalry Division, arrived late at night in the port of Rijeka, Croatia. This was the first staging for the unit in its redeployment to CONUS following its tour of duty as part of the NATO Stabilization Forces in the Balkans. The equipment and containers were placed in a sterile area, adjacent to the docks, before being loaded. Croatian longshoremen, working under the direction of the MTMC

Military equipment, vehicles and containers, belonging to the 1st Brigade, 1st Cavalry Division, was loaded on the USNS Antares *(T-AKR-294), a "Roll On/Roll Off" Transport bound for the return voyage to Wilmington, North Carolina and the Port of Corpus Christi, Texas. (30 March 1999)*

NATO F-15s aircraft shot down two of three Federal Republic of Yugoslavia (FRY) MiG 29 air superiority fighter aircraft that entered the airspace of Bosnia-Herzegovina. One of the downed aircraft crashed near Task Force Eagle. (26 March 1999)

<table>
<tr><th colspan="3">2nd BRIGADE COMBAT TEAM (15 March 1999 to 01 October 1999)</th></tr>
<tr><th>EAGLE BASE</th><th>CAMP McGOVERN</th><th>CAMP COMANCHE</th></tr>
<tr><td>"HHC", 1st Cav Div *
"HHC", 2nd Bde Combat Tm
"HHB", 3-82nd FA
"A" Btry, 3-82nd FA
"B" Btry, 3-82nd FA
"C" Btry, 1-21st FA *
"HHC", Engineer Bde
"HHC", 91st Engineer Bn (-)
"C" Co, 91st Engineer Bn
545th MP Co (-) *
"HHC", 312th MI Bn *
"B" Co, 312th MI Bn *
"HHC", 13th Signal Bn *
"B" Co, 13th Signal Bn
15th Finance Det *
15th Pers Serv Det *
401st MP Co (-) **</td><td>HHC", 1-12th Cavalry
'A" Co, 1-12th Cavalry
'C" Co, 1-12th Cavalry
'D" Co, 1-12th Cavalry
'C" Co, 1-5th Cavalry
1st Plt, "C" Btry, 3-82nd FA
'A" Co, 91st Engineer Bn
2nd Plt, 401st MP Co **</td><td>"HSD", 4th Bde
"HHC", 1-4th Avn Regt **
"A" Co, 1-4th Avn Regt **
"C" Co, 1-4th Avn Regt **
"D" Co, 1-4th Avn Regt **
"D" Trp, 1-7th Cavalry
"E" Trp, 1-7th Cavalry
"F" Trp, 1-7th Cavalry
"HHC", 1-159th Avn Regt **
"A" Co, 1-159th Avn Regt **
"B" Co, 1-159th Avn Regt **
"D" Co, 1-159th Avn Regt **
"HHC", 15th Fwd Spt Bn
"A" Co, 15th Fwd Spt Bn
"B" Co, 15th Fwd Spt Bn
"C" Co, 15th Fwd Spt Bn
"A" Co, 62nd Engineer Bn **
"HHC", 404th Avn Spt Bn **
"C" Btry, 4-5th ADA Bn</td></tr>
<tr><th>CAMP DOBOL</th><th>CAMP BEDROCK</th><th>CAMP DEMI</th></tr>
<tr><td>"HHC", 1-5th Cavalry
"B" Co, 1-5th Cavalry
"D" Co, 1-5th Cavalry
"B" Co, 1-12th Cavalry
"B" Co, 2-12th Cavalry
"C" Co, 2-12th Cavalry
"B" Co, 91st Engineer Bn
1st Plt, 401st MP Co **</td><td>** CLOSED **</td><td>"HHC", 2-12th Cavalry
"A" Co, 2-12th Cavalry
"D" Co, 2-12th Cavalry
"A" Co, 1-5th Cavalry
2nd Plt, "C" Btry, 3-82nd FA
1st Plt, "C" Co, 91st Engineer Bn</td></tr>
<tr><td colspan="3">* Holdover units from fall 1998 deployment
** Units attached and not organic to 1st Cavalry Division</td></tr>
</table>

839th Transportation Battalion, loaded the cargo of vehicles, shrink wrapped helicopters, and specialized equipment on the *USNS Antares* (T-AKR-294), a "Roll On/Roll Off" Transport. On 30 March, the *USNS Antares* left Rijeka, Croatia, bound for Wilmington, North Carolina and the Port of Corpus Christi, Texas.

On 24 March 1999, NATO launched a major offensive against Kosovo, a neighboring Balkan province of the Federal Republic of Yugoslavia. The objective of the strike was to force the Yugoslavian leader, Slobodan Milosevic to end his "ethnic" cleansing of Albanians who constituted approximately eighty-five percent of the population in the Serb province.

The threat from the warring factions in the neighboring country of Kosovo became more evident on 26 March, when three Federal Republic of Yugoslavia (FRY) MiG 29 air superiority fighter aircraft entered the sovereign airspace of Bosnia-Herzegovina with hostile intent. They were immediately engaged by F-15s aircraft flying defensive air patrols in support of the UN sanctioned, NATO Forces. Two of the MiGs were shot down, the third retreated to FRY airspace before being intercepted. One of the downed MiGs crashed near the town of Teocak, about 10 kilometers inside the Bosnia-Herzegovina border and the other downed aircraft crashed a little farther south. Access to the crash sites was hampered by concern for the many land mines which had not yet been cleared.

Although wreckage from the two downed aircraft landed near the US Sector. troopers of the 1st Cavalry Division remained safe and secure as they continued with their mission. As always, force protection was the top priority. Immediately, access to all camps became limited to mission essential personnel only, with normal operations resuming as soon as was practical. Troops were back in flak jackets

In a Transfer of Authority ceremony, Colonel Bernard S. Champoux, commander of 2nd Brigade, 10th Mountain Division and his Brigade took over responsibility of the American sector of the MND-N, previously controlled by 2nd Brigade, 1st Cavalry Division under the command of Colonel George E. Bowers. (01 October 1999)

and helmets and were eating Meals, Ready to Eat (MREs) supplemented with fresh fruit, breads and simply prepared foods. Contractor personnel only returned when the potential of any threat to the troops was eliminated.

On 27 March, a second violation of Bosnia-Herzegovina airspace occurred when two FRY helicopters crossed the border south of Bijeljina. The helicopters were tracked by NATO aircraft and retreated to FRY airspace before they could be intercepted.

In light of possible air retaliation for the stepped up NATO attack against the neighboring country of the FRY, force protection measures were strengthened and accelerated. On 01 April a team from the 4th Battalion, 5th Air Defense Artillery, 1st Cavalry Division at Ft. Hood, Texas deployed in support of Operation JOINT FORGE. The unit, equipped with the AN/MPQ-64 "Sentinel" radar, was assigned to MND-N in Bosnia. The deployment of the unit enhanced an already robust, protection program of force that ensured the safety of the 1st Cavalry Division and other attached forces supporting Operation JOINT FORGE.

The Sentinel radar, a towed, ground-based unit is assigned to short range air defense battalions. It is capable of automatically detecting, tracking, identifying and classifying threat targets in all weather conditions up to forty kilometers away. Its low altitude system is capable of alerting and cueing air defense gunners to aerial targets approaching their engagement area.

Amid the flurry and background of a war occurring in a neighboring country, the members of the 2nd Brigade Combat Team readied themselves to relieve the 1st Brigade Combat Team in Bosnia. Upon deployment for the second round, the 2nd Brigade, along with its support units, including the non-divisional attached units as well as the "Holdover Units" were stationed at the following five camp areas. (See Previous Page.)

On 01 April, the Division initiated the formal implementation of the second and final mission phase as the 2nd Brigade Combat Team relieved the 1st Brigade Combat Team. On 02 April Major General Byrnes, considering the morale and dedication of all the MND-N troops necessary to keep Bosnia on a peaceful track, made arrangements to get the troops back on one hot meal a day, and they eagerly awaited the day of the return of "home cooking" provided by the contractor, Brown and Root.

Division Support Command

The Division Support Command (DISCOM) "Wagonmasters" provided TFE with the "life line" of "fuel-fix-arm-move-sustain" spectrum of support for the operations of the 1st Cavalry Division in Bosnia. Previously, in the first mission phase, this function was provided by the 115th Forward Support Battalion (FSB) and the 615th Aviation Support Battalion (ASB). Now the colors of the 15th FSB provided the focal point of the Logistics Task Force (LTF) which included aviation maintenance elements of the "A" Company, 404th ASB, 4th Infantry Division and higher-echelon maintenance from the 183rd Maintenance Support Team (MST), III Corps from Ft. Carson, Colorado.

The LTF was also home to a platoon of Heavy Equipment Transports (HETs), as well as a section of Reverse Osmosis Water Purification Units (ROWPUs) from the division's own 27th Main Support Battalion (MSB), the largest and most powerful battalion of its kind in the entire United States Army. The HETs add the capability for transporting armored vehicles quickly and efficiently over long distances, and the ROWPUs provides a standby fresh water "on call" capability, should the local contracting efforts break down or become overloaded. The 27th MSB also provided personnel backfill to the 15th FSB prior to the deployment to Bosnia in order to fill critical slots in a variety of Military Occupational Specialties (MOS).

Although the 15th FSB "Gamblers" of the LTF operated mainly from Comanche Base, they also had soldiers and equipment spread throughout the MND-N area of responsibility, providing support where needed. MSTs were attached to each battalion TF to ensure that weapon system repairs were done at the lowest possible echelon of support, and the HET drivers of the 27th MSB operated constantly, shifting combat power where needed in the sector.

On 16 April, the Navy ship, *USNS Antares* (T-AKR-294), a "Roll On/Roll Off" Transport, arrived at the Port of Corpus Christi, Texas. The docks became a flurry of military activity as Blackhawk helicopters, HMMWVs and supplies were loaded onto trains and trucks after their return from Bosnia. The equipment, coming from Rijeka, Croatia, was moving to the home base of the 1st Cavalry Division, Ft. Hood, Texas.

The MTMC planned the highly coordinated endeavor. It involved synchronizing seventy rail cars, forty-six trucks and more than two-hundred people, including one-hundred local stevedores, who descended on the cargo dock and began unloading the ship. Among the contents of the ship were twenty-four helicopters, which were shipped disassembled. The pieces of each of the $14 million helicopters would be reassembled on the dock, inspected, refueled and then flown on to Ft. Hood.

On 23 July, initiating the 6th rotation of US Troops into Bosnia, an advanced party of what would eventually be 3,200 soldiers from 10th Mountain Division, Ft. Drum, New York arrived to begin the staffing build-up for the Change of Command of MND-N. The 10th Mountain Division had been in training for the past year, preparing for their mission assignment to Bosnia. The 2nd Brigade, 10th Mountain Division began deploying later in the month to fall in on equipment which sailed from the Howland Hook Marine Terminal, Staten Island, New York on 04 August.

On Wednesday 04 August at 10:00 hours, the MND-N hosted a Change of Command ceremony at the Eagle Sports Complex, during which the 10th Mountain Division Headquarters from Ft. Drum, New York, formally took over the duties of the MND-N mission replacing the 1st Cavalry Division Headquarters, 4th Brigade, 312th Military Intelligence Battalion, and 13th Signal Battalion. Major General James L. Campbell assumed command of MND-N from Major General Kevin P. Byrnes. Carrying out the TA was a gradual, incrementally executed operation that was planned in a manner to ensure that there would be no disruption in the mission of MND-N, which was to provide a safe and secure environment throughout its area of responsibility. The 1st Cavalry Division soldiers would continue to serve the citizens of Bosnia and Herzegovina as the 2nd Brigade Combat Team remained in place under the command of the 10th Mountain Division, completing the full mission transition in October

In the continuation of its assignment, members of the 2nd Brigade stayed busy conducting their regular patrolling activities and participating in, and monitoring the variety of EAF training and movement events throughout the sector. The missions included presence patrols, weapons search and inspections, Operation HARVEST '99, the summer demining campaign, Joint Assessment Team investigations and displaced persons and refuges returns.

By 01 October, serving the final six months rotation of the 1st Cavalry Division in Bosnia, the 2nd Cavalry Brigade had brought immeasurable credit to its proud lineage and tradition in being a part of the struggle of Bosnia to pull itself out of the aftermath of four years of war. The 1st Cavalry Division had served as our nation's ambassadors. The day to day presence of its soldiers and their commitment to the citizens of Bosnia helped prove and ensure that a lasting and self-sustaining peace is possible.

This transition represented more than a brigade level transition of authority. It signaled the mission accomplishment of the military provisions set forth by the Dayton Peace Accords and the end of the magnificent tour of the First Team in Bosnia. Following the TA Ceremony, the final efforts of the 2nd Brigade were spent organizing personnel, cleaning, packing and loading equipment. Upon arrival at Ft. Hood, a Division Patch Welcome Home Ceremony was held on 18 October, at Cooper Field.

"BOSNIA ROCK" - Operation Joint Forge

1st Cavalry Division Headquarters Eagle Base, Tulsa, Bosnia 04 October 1998 to 04 August 1999 SAPPERS FORWARD !

Chapter 14: Today's Cavalry

2000

The 1st Cavalry Division, a 17,000-soldier heavy armored division, consists of three maneuver brigades, an aviation brigade, engineer brigade, division artillery and division support command plus an air defense battalion, a signal battalion, a military intelligence battalion and a military police company. The First Team remains a combat proven force ready to fight anywhere, anytime and win. The organizational structure for the 1st Cavalry Division of today is shown below:

Commanding General

- Chief of Staff
- Assistant Division Commander (Maneuver)
- Assistant Division Commander (Support)
- Division Command Sergeant Major

General Staff
- G1 Adjutant General
- G2 Intelligence
- G3 Operations/Plans
- G4 Logistics
- G5 Civil Affairs
- G6 Communications

Special Staff
- Provost Marshal
- Surgeon
- Comptroller
- Chemical Officer
- Equal Opportunity Officer
- Saftey Officer

Personal Staff
- Inspector General
- Staff Judge Advocate
- Chaplain
- Public Affairs Officer

Major Subordinate Commands

1st Brigade
- HQ&HQ Company
- "C" Trp, 10th Cavalry
- 2nd Bn, 5th Cavalry
- 2nd Bn, 8th Cavalry
- 1st Bn, 12th Cavalry

2nd Brigade
- HQ&HQ Company
- 1st Bn, 5th Cavalry
- 1st Bn, 8th Cavalry
- 2nd Bn, 12th Cavalry

3rd Brigade
- HQ&HQ Company
- 2nd Bn, 7th Cavalry
- 3rd Bn, 8th Cavalry
- 1st Bn, 9th Cavalry

Aviation Brigade
- HQ&HQ Company
- 1st Sqdn, 7th Cavalry
- 1st Bn, 227th Aviation
- 2nd Bn, 227th Aviation
- "F" Co, 1st Bn, 58th Avn

Division Artillery
- HQ&HQ Battery
- 1st Bn, 21st Fld Arty
- 1st Bn, 82nd Fld Arty
- 2nd Bn, 82nd Fld Arty
- 3rd Bn, 82nd Fld Arty
- 68th Chemical Company

Engineer Brigade
- HQ&HQ Detachment
- 8th Engineer Bn
- 20th Engineer Bn
- 91st Engineer Bn

Division Support Command
- HQ&HQ Company
- 15th Support Bn (FWD)
- 27th Support Bn (MAIN)
- 115th Support Bn (FWD)
- 215th Support Bn (FWD)
- 615th Support Bn (AVN)
- Div Mat'l Mgm't Center

Separate Units
- HQ&HQ Co. 1st Cav Div
- 1st Cavalry Div Band
- 1st Cavalry Horse Det
- 4th Bn, 5th Air Defense
- 13th Signal Bn
- 312th M.I. Bn
- 545th M.P. Company

Direct Support
- 9th Air Spt Operations
- 15th Finance Bn
- 15th Personnel Svc Bn

Museum

Chapter 15: The Next Generation

2000-2020

The Challenge of a New Century

The Army plays a critical role in National Military Strategy. Its job is to remain the preeminent land war fighting force in the world to be capable of successfully conducting a major theater war while remaining always ready to undertake such other missions - from fire fighting to peace keeping as may come its way.

Responding to changes in warfare and how it executes its core missions, the Army has outlined the nature of future conflicts and what must be done to win. The fundamental concept of winning is based upon the control and dominance of information across the spectrum of all military operations. This concept is an extension of the intelligence gathering and field execution exhibited in the Gulf War.

Currently, forces of the Army are not optimally designed and organized to fully support these missions. The heavy forces of today, while decisive in a variety of combat and non-combat environments, require tremendous resources to deploy quickly and sustain themselves. In contrast, light forces can strike quickly but lack survivability, lethality, and tactical mobility. Both types of forces lack capabilities that will assume increasing importance in future decades. The requirements for greater lethality, survivability, and deployability across the entire force, resulting in greater versatility and agility for full-spectrum operations, point to the need for fundamental transformation and a new vision.

The Army Vision

For the 21st Century, the Army envisions a strategically responsive force that is dominant across the entire spectrum of operations and is responsive, deployable, agile, versatile, lethal, survivable, and sustainable.

- Strategic responsiveness requires forward-deployed forces, forward positioned capabilities, and force projection from the continental United States or any other location where needed capabilities reside. Specifically, the Army will be able to deploy a combat-capable brigade anywhere in the world within 96 hours, a combat-capable division within 120 hours, and five combat capable divisions within 30 days.
- Agility and versatility is the ability to move forces quickly from stability and support operations to war fighting and back. It requires platforms that are smaller and lighter, yet lethal and survivable.
- Lethality calls for every element in the war fighting formation to be capable of generating combat power and contributing decisively to the fight.
- Survivability means maximum protection to forces at the individual soldier level, whether mounted or dismounted.
- Sustainability is achieved by reducing the logistics footprint and replenishment demands of the Army, by controlling the number of vehicles deployed, leveraging reach-back capabilities, and investing in a system-of-systems approach to the design of weapons and equipment.

Achieving the Army Vision

To achieve its vision, the Army has implemented a three-phase strategy: sustainment of the Legacy Force, development of an Interim Force, and evolution to an Objective Force, known today as FORCE XXI.

Legacy Force, Today's Army

The process of transforming the Army into the FORCE XXI will span decades. During this time, the Legacy Force will be recapitalized and selectively modernized. The objectives of this Legacy Force include the following:

- Retain the present day level of combat overmatch over all potential adversaries;
- Sustain combat power and survivability, with less weight and bulk;
- Achieve a commonality among platform, chassis, caliber, component, and battlefield operating systems that responds to a much broader range of operations;
- Reduce the deployed logistical footprint; and
- Enhance strategic responsiveness.

Systems now being fielded, and systems that will be fielded in the near term, may not be part of the ultimate Objective Force, though they could be used for up to thirty years or longer. An advanced field artillery system and the Heavy Equipment Transport (HET) are examples. An advanced field artillery system to replace the current palladin system would restore combat overmatch in cannon artillery to the United States. Cost-effective enhancements to the HET will reduce future Operations and Support (O&S) costs.

For the immediate future and well into the mid-term, the Legacy Force of today, complemented by the Interim Force as it is fielded over the next few years, will be the primary means by which the Army fulfills its readiness responsibilities to the Nation. Beginning with the Legacy Force, followed by the introduction of new equipments to the Interim Force, the transformation to the Objective Force will eventually encompass the entire Army.

Interim Force - Army After Next (AAN)

To bridge the gap between capabilities of today and the Objective Force, the Army is fielding an Interim Force. This force will be combined arms in design and organized as a rapidly deployable, full-spectrum force. Its core qualities include operational and tactical mobility, enhanced situational understanding, combined arms integration down to and including company level, and increased dismounted strength for close combat in urban and complex terrain.

The Interim Force includes systems that will not be a part of the ultimate Objective Force, though they may be in use for an indefinite period before they are phased out of service. The primary Interim Force combat platform will be a family of Interim Armored Vehicles (IAVs), capable of deployment anywhere in the world in a combat-ready configuration. By 2010, the Army will have outfitted most of its active component divisions with these systems.

Objective Force - FORCE XXI

The Objective Force, currently being tested as FORCE XXI, is the ultimate transformation goal which will become the future force that achieves all the characteristics described in the Army Vision. Based on data gathered during a series of advanced warfighting experiments conducted at Ft. Hood, Texas, and Ft. Irwin, California, The implementation of the Objective Force requires the adoption of a new division which will still be the most lethal combat force in the world even though it will have fewer soldiers, armored vehicles and weapon systems. Many of those systems include systems available today, as well as those planned and being developed for production in the future.

Future Combat Systems (FCS) will be the centerpiece of the Objective Force. The Objective Force concept and its enabling technologies are currently under development. Its enabling capabilities will require continued investments in science and technology. FCS is envisioned as a digitized system-of-systems land combat capability with multi-mission functionality. Its main characteristics will include:

- A common design, allowing inter-changeable full-spectrum capability;
- Increased responsiveness, accuracy and lethality for a full range of fire-support missions;
- Improved operational force autonomy with reduced demands for fuel, spare parts, and munitions;
- Improved networked command, control, communications, computers, intelligence, surveillance, and reconnaissance (C4ISR) links;
- Use of unmanned air and ground systems performing as reconnaissance, surveillance, or attack systems, as well as providing other battlefield functions;
- Current or enhanced levels of survivability;
- Enhanced lift capability, to enable rapid deployment; and
- Improved early warning and intercept of enemy ground and air launched conventional and smart weapons.

Transition of Force XXI Concepts to the 1st Cavalry Division

In Fiscal Year (FY) 01 the 1st Cavalry Division began its transition to the Force XXI design. "Force XXI is the comprehensive process of the Army for modernizing and preparing for the challenges of the 21st Century. Force XXI projects our soldiers into the 21st century and provides them with the necessary doctrine and organizations, the most realistic training, and the best equipment and weapon systems that our nation can provide.

When the Division is equipped with Force XXI equipment that is fully digitized, it will create a "collaborative virtual environment" that, among other advantages, will give commanders separated by great distances the ability to communicate through various digital systems to provide situational awareness across the battlefield.

Implementation of Force XXI Concepts to the 1st Cavalry Division encompasses the following strategies or reorganization:

1. Seamless Integration -

- Multi-component Unit with a single Table of Organization.
- Reorganization downsizes active-duty personnel from the 01 May 2001 authorization of 17,000 to 15,526.
- Integration of 236 Reserve Component personnel throughout the Division consisting of 3 US Air Reserve and 233 Army Reserve National Guard.

2. Command and Control -

- Information exchange capability increased to control a larger battlespace.
- Division Main and Rear Command Post collocated.
- Signal planning performed in the Division Communications (G6).
- Added an Information Operations (IO) Cell that coordinates and synchronizes IO assets across the Division's Battlespace.

3. Maneuver -

- Added a Reconnaissance Troop to each Maneuvering Brigade.
- Utilization of 44 Weapon System Platforms per battalion, 3 companies per battalion and 4 vehicles per platoon.
- Infantry companies are composed of 3 squads x 3 platoons.
- Reduced and standardized scouting operations.
- Standardized mortar organizations to 4 tubes per battalion.
- Centralized Combat Services Support (CSS) functions,
- Added a Forward Repair System (FRS) that provides improved maintenance readiness and responsiveness at the front where it is needed.

4. Fire Support -

- Maneuver brigades supported with the addition of Bradley Fire Support Team Vehicles (FISTV) and STRIKER HMMWV Vehicles.
- Eliminated Fire Support below company level.
- Downsized Paladin Crew from 9 to 8.
- Merged Service and Headquarters Batteries into one Headquarters Service Battery.
- Eliminated the 2nd Fire Direction Controller.

5. Mobility, Countermobility and Survivability -

- Engineer Brigade Planning Function is relocated to Division Headquarters in FY03.
- Engineer Armored Personnel Carrier is replaced with the Bradley Tracked Armored Fighting Vehicle.
- Engineer Armored Combat Earthmover. (ACE) is replaced by the GRIZZLY, an Obstacle Breaching Vehicle, and the Heavy Assault Bridge (AVLB) is replaced with the Wolverine as they are fielded.
- Combat Support Services (CSS) for engineers is centralized in Forward Support Battalions.
- The 68th Chemical Company is deactivated in FY03. The NBC function migrates to III Corps and the reconnaissance mission is transferred to the 1st Squadron, 7th Cavalry.

6. Combat Service Support -

- Incorporated a distributed, maneuver level based supply system.
- Centralized Combat Support Services (CSS) for Maneuver and Engineer Battalions that reduces stock inventories.
- Expanded maintenance training.
- Some functions "passbacked" to III Corps.

7. Air Defense Artillery -

- The 4th Battalion, 5th Air Defense eliminates all Manportable Air Defense System (MANPADS) and dismount STINGERS.
- Added an AN/MPQ64, Search and Track Radar (SENTINEL) Platoon.
- Added twelve Low Level Air Defense (AVENGER) Systems, to bring the total to thirty-six vehicles, each employing MANPADS Under Armor (MUA).

8. Intelligence -

- Increased strength and staffing of Analysis Control Team for Aviation or Artillery use.
- Increased strength and staffing of Intelligence (S2) in Aviation Regiments and Cavalry Squadron.
- Increased interface with assets external to the Division.
- Added a Tactical Unmanned Aerial Vehicle (TUAV) capability that provides Reconnaissance, Surveillance, and Target Acquisition (RSTA) information in limited adverse weather conditions.

9. Communications -

- Added enhanced company, battalion, division and joint defense communication systems which enable the transfer of voice and digital information in a secure, common usage environment.

10. Force Protection and Security -

- Phase in of "up armored" vehicles.

Unit Designation	Unit Type	Force XXI Equipment	Notes
1st Brigade			
Brigade Recon Troop	Reconnaissance	14 M1025/26 HMMWV SCOUT	
2nd Bn, 8th Cavalry	Armored	44 M1A2 SEP MBT	
1st Bn, 12th Cavalry	Armored	44 M1A2 SEP MBT	
2nd Bn, 5th Cavalry	Mechanized Infantry	44 M2A3 BFVS	
2nd Brigade			
Brigade Recon Troop	Reconnaissance	14 M1025/26 HMMWV SCOUT	
1st Bn, 8th Cavalry	Armored	44 M1A2 SEP MBT	
2nd Bn, 12th Cavalry	Armored	44 M1A2 SEP MBT	
1st Bn, 5th Cavalry	Mechanized Infantry	44 M2A3 BFVS	
3rd Brigade			
Brigade Recon Troop	Reconnaissance	14 M1025/26 HMMWV SCOUT	
2nd Bn, 7th Cavalry	Mechanized Infantry	44 M2A3 BFVS	
1st Bn, 9th Cavalry	Mechanized Infantry	44 M2A3 BFVS	
3rd Bn, 8th Cavalry	Armored	44 M1A2 SEP MBT	

Unit Designation	Unit Type	Force XXI Equipment	Notes
Aviation Brigade			
1st Bn, 227th Avn Rgt	Attack Combat	24 AH64D LOT IV	FY03 AH64D LOT VII
2nd Bn, 227th Avn Rgt	Combat Support	24 UH60L	
"B" Co, 2nd Bn, 149th Avn	Combat Support	8 UH60L	TX ARNG Unit, 49th AD
1st Sqdrn, 7th Cavalry	Armored & Air Cavalry	16 OH58D Kiowa	UpgradeRAH66
		27 M1A2 SEP MBT	W/Available
		43 M3A3 BFBS	
		6 M93A1 FOX NBCRS	Scheduled FY03
Division Artillery			
1st Bn, 82nd Field Arty	Medium Artillery	18 M109A6 Paladin (N1)	Supts 1st Bde
2nd Bn, 82nd Field Arty	Medium Artillery	18 M109A6 Paladin (N1)	Supts 2nd Bde
3rd Bn, 82nd Field Arty	Medium Artillery	18 M109A6 Paladin (N1)	Supts 3rd Bde
1st Bn, 21st Field Arty	Rocket Artillery	12 M270 MLRS	General Supt
"C" Btry, 131st Field Arty	Rocket Artillery	6 M270 MLRS	TX ARNG 49th AD
68th Chemical Company	NBC Reconnaissance	6 M93A1 FOX NBCRS	FY03 Deactivated
Engineer Brigade			FY03 Deactivated
8th Engineer Bn	Combat Engineer	Each Bn Eqp'd With	Supts 3rd Bde
		29 M2A2ODS-E	
20th Engineer Bn	Combat Engineer	21 M9 ACE (N2)	Supts 1st Bde
		12 M60 AVLB (N3)	
91st Engineer Bn	Combat Engineer		Supts 2nd Bde
Divisional Support (DISCOM)			
15th Supt Battalion (FWD)			Supts 2nd Bde
27th Supt Battalion (MAIN)	Heavy Equipment Transport (HET)		2nd Level Supt
115th Supt Battalion (FWD)	Palletized Load System (PAL)		Supts 1st Bde
215th Supt Battalion (FWD)	Forward Repair System (FRS)		Supts 3rd Bde
615th Supt Battalion (AVN)	Load Handling System (HEMTT_LHS)		Supts Avn Bde
Material Mgm't Center	Movement Tracking System (MTS)		
Division Assets			
4th Bn, 5th ADA	Forward Air Defense	1 Sentinel AN/MPQ64	Aerial Acquisition
		36 Avenger	Aerial Defense
		24 M6 Linebacker	Each W/6 MUR
13th Signal Bn	Communications	Warfighter Information Network (WIN)	
		Global Broadcasting System (GBS)	
		AN/PSC-5 EMUT Spitfire	
		SMARTT Secure Terminal	
		Joint Tactical Radio System (JTRS)	
312th Intelligence Bn	Intelligence & EW	Tactical Unmanned Aerial Vehicle (TUAV)	
		Common Ground Station (CGS) AN/TSQ-179	
545 MP Company	Security & Protection	14 M1025/26 HMMWV	
		29 M1114 HMMWV W/Up Armor	

Note 1:Replaced W/Advanced Field Artillery System
Note 2:Replaced W/Obstacle Breaching Vehicle (Grizzly)
Note 3:Replaced W/Heavy Assault Bridge (Wolverine)

Objective Force XXI Equipment Acquisition Summary

The initial equipment selected for the units of the 1st Cavalry Division is tabulated on page 151. However, note that the final configuration of the Objective Force XXI will be modified based on simulations, field exercises and analysis of the Division design. In its end configuration, Force XXI equipment will be ideally suited for joint operations and will be fully compatible with the operational systems of the other services. Seamless information connectivity with the other elements of the joint force will be its primary characteristic and is essential for the success of joint operations.

Because the Force XXI organizational and equipment concept will be modular in nature, it can easily tailored to the necessary force package to meet the needs of a joint force commander. This modularity will allow the generation, projection, and sustaining force packages for any type of military situations and operations.

The mission of the Army is to equip a capability based upon the achievement of a full-spectrum dominance against any potential enemy across the entire field of military operations by 2015 - 2020. Appendix "L" details the descriptions and features of the next generation weapon system replacements that have been tested and will be fielded to support the reorganization of the new 1st Cavalry Division. It is important to note that even though each of the systems described can deliver a deadly force, they are not isolated pieces of individual equipment. Rather, they are the building blocks needed to create the integrated capabilities that are the defining capabilities of AAN and FORCE XXI.

Appendix A: Commanding Generals

The ability of the 1st Cavalry Division to transition through the seamless integration of the methods of war fighting and improvements in armament technology is founded upon the evolution of subordinate units with experienced noncommissioned and commissioned personnel who continuously upgrade their management of warfighting skills. On 10 October 2001, the Noncommissioned Officers and soldiers of the 1st Cavalry Division participated in a colorful Change of Command Ceremony, at Cooper Field, Ft. Hood, Texas in which Major General David D. McKiernan relinquished his command to Major General Joseph F.H. Peterson. The flag was flown at "half staff" in memory to the victims of the 9/11 attack.

The Honor Guard presents the Division's Colors to the Outgoing and Incoming Commanders of the 1st Cavalry Division.

The next assignment for Major General McKiernan, who had assumed the command of the 1st Cavalry Division in October 1999, was as the Deputy Chief of Staff for Operations, Headquarters, Department of the Army, Washington, DC. Major General Joe Peterson came from Ft. Stewart, Georgia and in his last assignment served as Commanding General, 3rd Infantry Division (Mechanized).

First team Troopers prepare to pass in review.

COMMANDING GENERALS

Major General Robert L. Howze	Sep 1921 - Jun 1925
Brigadier General Joseph C. Castner	Jun 1925 - Jan 1926
Major General Edwin B. Winans	Jan 1926 - Oct 1927
Brigadier General Samuel D. Rockenback	Oct 1927 - Nov 1927
Major General Van Horn Moseley	Nov 1927 - Sep 1929
Brigadier General Charles J. Symmonds	Sep 1929 - Oct 1930
Brigadier General George C. Barnhardt	Oct 1930 - Dec 1930
Major General Ewing E. Booth	Dec 1930 - Mar 1932
Brigadier General Walter C. Short	Mar 1932 - Mar 1933
Major General Frank R. McCoy	Mar 1933 - Oct 1933
Brigadier General Walter C. Short	Oct 1933 - Apr 1934
Brigadier General Hamilton S. Hawkins	Apr 1934 - Sep 1936
Brigadier General Francis LeJ. Parker	Sep 1936 - Oct 1936
Major General Ben Lear	Oct 1936 - Nov 1938
Major General Kenyon A. Joyce	Nov 1938 - Oct 1940
Major General Robert C. Richardson Jr.	Oct 1940 - Feb 1941
Major General Innis Palmer Swift	Feb 1941 - Aug 1944
Major General Verne D. Mudge	Aug 1944 - Feb 1945
Brigadier General Hugh F. T. Hoffman	Feb 1945 - Jul 1945
Major General William C. Chase	Jul 1945 - Feb 1949
Brigadier General William B. Bradford	Feb 1949 - Feb 1949
Major General John M. Devine	Feb 1949 - Aug 1949
Brigadier General Henry I. Hodes	Aug 1949 - Sep 1949
Major General Hobart R. Gay	Sep 1949 - Feb 1951
Major General Charles D. Palmer	Feb 1951 - Jul 1951
Major General Thomas L. Harrold	Jul 1951 - Mar 1952
Major General Arthur G. Trudeau	Mar 1952 - Mar 1953
Brigadier General William J. Bradley	Mar 1953 - Apr 1953
Major General Joseph P. Cleland	May 1953 - Jun 1953
Major General Armistead D. Mead	Jun 1953 - Dec 1954
Brigadier General Orlando C. Troxel Jr.	Dec 1954 - May 1955
Major General Edward J. McGaw	May 1955 - Nov 1956
Major General Edwin H. J. Carns	Nov 1956 - Aug 1957
Major General Ralph W. Zwicker	Oct 1957 - Jan 1958
Major General George E. Bush	Jan 1958 - Apr 1959
Major General Charles E. Beauchamp	Apr 1959 - May 1960
Major General Charles C. Dodge	May 1960 - Dec 1960
Major General Frank H. Britton	Dec 1960 - Jul 1961
Major General James K. Woolnough	Jul 1961 - Sep 1962
Brigadier General D. C. Clayman	Sep 1962 - Oct 1962
Major General Clifton F. Von Kann	Oct 1962 - Jun 1963
Brigadier General Charles P. Brown	Jun 1963 - Aug 1963

Major General Charles F. Leonard Jr.	Aug 1963 - Oct 1964
Major General Hugh Exton	Oct 1964 - June 1965
Major General Harry W. O. Kinnard	Jul 1965 - May 1966
Major General John Norton	May 1966 - Mar 1967
Major General John J. Tolson	Mar 1967 - Aug 1968
Brigadier General Richard L. Irby	Aug 1968 - Aug 1968
Major General George I. Forsythe	Aug 1968 - Apr 1969
Major General E. B. Roberts	May 1969 - May 1970
Major General George W. Casey	May 1970 - Jul 1970
Major General George W. Putnam	Aug 1970 - May 1971
Major General James C. Smith	May 1971 - Jan 1973
Major General Robert M. Shoemaker	Jan 1973 - Feb 1975
Major General Julius W. Becton Jr.	Feb 1975 - Nov 1976
Major General W. Russell Todd	Nov 1976 - Nov 1978
Major General Paul S. Williams Jr.	Nov 1978 - Nov 1980
Major General Richard D. Lawrence	Nov 1980 - Jul 1982
Major General Andrew P. Chambers	Jul 1982 - Jun 1984
Major General Michael J. Conrad	Jun 1984 - Jun 1986
Major General John J. Yeosock	Jun 1986 - May 1988
Major General William F. Streeter	May 1988 - Jul 1990
Major General John H. Tilelli Jr.	Jul 1990 - Jul 1992
Major General Wesley K. Clark	Jul 1992 - Mar 1994
Major General Eric K. Shinseki	Mar 1994 - Jul 1995
Major General Leon J. LaPorte	Jul 1995 - Jul 1997
Major General Kevin P. Byrnes	Jul 1997 - Oct 1999
Major General David D. McKiernan	Oct 1999 - Oct 2001
Major General Joseph F. H. Peterson	Oct 2001 -

Appendix B: Medal of Honor

The Medal of Honor is the highest medal awarded by the United States. The following is an excerpt of the Army regulation that describes the eligibility criteria for award of the Medal of Honor...

The Medal of Honor [Army], section 3741 title 10, United States Code (10 USC 3741), was established by Joint Resolution of Congress, 12 July 1862 (amended by acts 9 July 1918 and 25 July 1963).

The Medal of Honor is awarded by the President in the name of Congress to a person who, while a member of the Army, distinguishes himself or herself conspicuously by gallantry and intrepidity at the risk of his life or her life above and beyond the call of duty while engaged in an action against an enemy of the United States; while engaged in military operations involving conflict with an opposing foreign force; or while serving with friendly foreign forces engaged in an armed conflict against an opposing armed force in which the United States is not a belligerent party. The deed performed must have been one of personal bravery or self-sacrifice so conspicuous as to clearly distinguish the individual above his comrades and must have involved risk of life. Incontestable proof of the performance of the service will be exacted and each recommendation for the award of this decoration will be considered on the standard of extraordinary merit.

From chapter 3-6, Army Regulation 600-8-22 (Military Awards) dated 25 February 1995

The Medal of Honor has been awarded to thirty-six members of the 1st Cavalry Division. Exact copies of the full text citation describing the battle situation and heroic actions taken by each member of the 1st Cavalry Division are listed below:

Medal of Honor

1944

Sergeant Troy A. McGill, G-5 Cavalry earns the Medal of Honor. (04 March)

Rank and organization: Sergeant, U.S. Army, Troop G, 5th Cavalry Regiment, 1st Cavalry Division. Place and date: Los Negros Islands, Admiralty Group, 4 March 1944. Entered service at: Ada, Okla. Birth: Knoxville, Tenn. G.O. No.: 74, 11 September 1944. Citation: For conspicuous gallantry and intrepidity above and beyond the call of duty in action with the enemy at Los Negros Island, Admiralty Group, on 4 March 1944. In the early morning hours Sergeant McGill, with a squad of 8 men, occupied a revetment which bore the brunt of a furious attack by approximately 200 drinkcrazed enemy troops. Although covered by crossfire from machineguns on the right and left flank, he could receive no support from the remainder of our troops stationed at his rear. All members of the squad were killed or wounded except Sergeant McGill and another man, whom he ordered to return to the next revetment. Courageously resolved to hold his position at all cost, he fired his weapon until it ceased to function. Then, with the enemy only 5 yards away, he charged from his foxhole in the face of certain death and clubbed the enemy with his rifle in hand-to-hand combat until he was killed. At dawn, 105 enemy dead were found around his position. Sergeant McGill's intrepid stand was an inspiration to his comrades and a decisive factor in the defeat of a fanatical enemy.

1945

Private First Class Private First Class William J. Grabiarz, E-5 Cavalry earns the Medal of Honor. (23 February)

Rank and organization: Private First Class, U.S. Army. Troop E, 5th Cavalry, 1st Cavalry Division. Place and date: Manila, Luzon, Philippine Islands, 23 February 1945. Entered service at: Buffalo, N.Y. Birth: Buffalo, N.Y. G.O. No.: 115, 8 December 1945. Citation: He was a scout when the unit advanced with tanks along a street in Manila, Luzon,

Philippine Islands. Without warning, enemy machinegun and rifle fire from concealed positions in the Customs building swept the street, striking down the troop commander and driving his men to cover. As the officer lay in the open road, unable to move and completely exposed to the pointblank enemy fire, Private First Class Grabiarz voluntarily ran from behind a tank to carry him to safety, but was himself wounded in the shoulder. Ignoring both the pain in his injured, useless arm and his comrades' shouts to seek the cover which was only a few yards distant, the valiant rescuer continued his efforts to drag his commander out of range. Finding this impossible, he rejected the opportunity to save himself and deliberately covered the officer with his own body to form a human shield, calling as he did so for a tank to maneuver into position between him and the hostile emplacement. The enemy riddled him with concentrated fire before the tank could interpose itself. Our troops found that he had been successful in preventing bullets from striking his leader, who survived. Through his magnificent sacrifice in gallantly giving his life to save that of his commander, Private First Class Grabiarz provided an outstanding and lasting inspiration to his fellow soldiers.

1950

Private First Class Melvin L. Brown, D-8 Engineers earns the Medal of Honor. (04 September)

Rank and organization: Private First Class, U.S. Army, Company D, 8th Engineer Combat Battalion. Place and date: Near Kasan, Korea, 4 September 1950. Entered service at: Erie, Pa. Birth: Mahaffey, Pa. G.O. No.: 11, 16 February 1951. Citation. Private First Class Brown, Company D distinguished himself by conspicuous gallantry and intrepidity above and beyond the call of duty in action against the enemy. While his platoon was securing Hill 755 (the Walled City), the enemy, using heavy automatic weapons and small arms, counterattacked. Taking a position on a fifty-foot-high wall he delivered heavy rifle fire on the enemy. His ammunition was soon expended and although wounded, he remained at his post and threw his few grenades into the attackers causing many casualties. When his supply of grenades was exhausted his comrades from nearby foxholes tossed others to him and he left his position, braving a hail of fire, to retrieve and throw them at the enemy. The attackers continued to assault his position and Private First Class Brown, weaponless, drew his entrenching tool from his pack and calmly waited until they one by one peered over the wall, delivering each a crushing blow upon the head. Knocking ten or twelve enemy from the wall, his daring action so inspired his platoon that they repelled the attack and held their position. Private First Class Brown's extraordinary heroism, gallantry, and intrepidity reflect the highest credit upon himself and was in keeping with the honored traditions of the military service. Reportedly missing in action and officially killed in action, September 5, 1950.

1950

Corporal Gorden M. Craig, 16th Reconnaissance earns the Medal of Honor. (10 September)

Rank and organization: Corporal, U.S. Army, Reconnaissance Company, 1st Cavalry Division. Place and date: Near Kasan, Korea 10 September 1950. Entered service at. Brockton, Mass. Born: 1 August 1929, Brockton, Mass. G.O. No.: 23, 25 April 1951. Citation: Corporal Craig, 16th Reconnaissance Company, distinguished himself by conspicuous gallantry and intrepidity above and beyond the call of duty in action against the enemy. During the attack on a strategic enemy-held hill his company's advance was subjected to intense hostile grenade, mortar, and small-arms fire. Corporal Craig and four comrades moved forward to eliminate an enemy machine gun nest that was hampering the company's advance. At that instance an enemy machine gunner hurled a hand grenade at the advancing men. Without hesitating or attempting to seek cover for himself, Corporal Craig threw himself on the grenade and smothered its burst with his body. His intrepid and selfless act, in which he unhesitantly gave his life for his comrades, inspired them to attack with such ferocity that they annihilated the enemy machine gun crew, enabling the company to continue its attack. Corporal Craig's noble self-sacrifice reflects the highest credit upon himself and upholds the esteemed traditions of the military service.

1950

Private First Class Robert H. Young, E-8 Cavalry earns the Medal of Honor. (09 October)

Rank and organization: Private First Class, U.S. Army, Company E, 8th Cavalry Regiment, 1st Cavalry Division. Place and date: North of Kaesong, Korea, 9 October 1950. Entered service at: Vallejo, Calif. Born: 4 March 1929, Oroville. Calif. G.O. No.: 65, 2 August 1951. Citation: Private First Class Young distinguished himself by conspicuous gallantry and intrepidity above and beyond the call of duty in action. His company, spearheading a battalion drive deep in enemy territory, suddenly came under a devastating barrage of enemy mortar and automatic weapons crossfire which inflicted heavy casualties among his comrades and wounded him in the face and shoulder. Refusing to be evacuated, Private First Class Young remained in position and continued to fire at the enemy until wounded a second time. As he awaited first aid near the company command post the enemy attempted an enveloping movement. Disregarding medical treatment he took an exposed position and firing with deadly accuracy killed five of the enemy. During this action he was again hit by hostile fire which knocked him to the ground and destroyed his helmet. Later when supporting tanks moved forward, Private First Class Young, his wounds still unattended, directed tank fire which destroyed three enemy gun positions and enabled the company to advance. Wounded again by an enemy mortar burst, and while aiding several of his injured comrades, he demanded that all others be evacuated first. Throughout the course of this action the leadership and combative instinct displayed by Private First Class Young exerted a profound influence on the conduct of the company. His aggressive example affected the whole course of the action and was responsible for its success. Private First Class Young's dauntless courage and intrepidity reflect the highest credit upon himself and uphold the esteemed traditions of the U.S. Army.

1950

First Lieutenant Samuel Coursen, C-5 Cavalry earns the Medal of Honor. (12 October)

Rank and organization: First Lieutenant, U.S. Army, Company C 5th Cavalry Regiment. Place and date: Near Kaesong, Korea, 12 October 1950. Entered service at: Madison, N.J. Born: 4 August 1926 Madison, N.J. G.O. No.: 57, 2 August 1951. Citation: First Lieutenant Coursen distinguished himself by conspicuous gallantry and intrepidity above and beyond the call of duty in action. While Company C, was attacking Hill 174 under heavy enemy small-arms fire, his platoon received enemy fire from close range. The platoon returned the fire and continued to advance. During this phase one of his men moved into a well-camouflaged emplacement, which was thought to be unoccupied, and was wounded by the enemy who were hidden within the emplacement. Seeing the soldier in difficulty he rushed to the man's aid and, without regard for his personal safety, engaged the enemy in hand-to-hand combat in an effort to protect his wounded comrade until he himself was killed. When his body was recovered after the battle seven enemy dead were found in the emplacement. As the result of First Lieutenant Coursen's violent struggle several of the enemies' heads had been crushed with his rifle. His aggressive and intrepid actions saved the life of the wounded man, eliminated the main position of the enemy roadblock, and greatly inspired the men in his command. First Lieutenant Coursen's extraordinary heroism and intrepidity reflect the highest credit on himself and are in keeping with the honored traditions of the military service.

1951

First Lieutenant Robert M. McGovern, A-5 Cavalry earns the Medal of Honor. (30 January)

Rank and organization: First Lieutenant, U.S. Army, Company A, 5th Cavalry Regiment, 1st Cavalry Division. Place and date: Near Kamyangjan-ni, Korea, 30 January 1951. Entered service at: Washington, D.C. Birth: Washington, D.C. G.O. No.: 2, 8 January 1952. Citation: First Lieutenant McGovern, a member of Company A, distinguished himself by conspicuous gallantry and intrepidity at the risk of life above and beyond the call of duty in action against an armed enemy of the United Nations. As First Lieutenant McGovern led his platoon up a slope to engage hostile troops emplaced in bunker-type pillboxes with connecting trenches, the unit came under heavy machine gun and rifle fire from the crest of the hill, approximately seventy-five yards distant. Despite a wound sustained in this initial burst of withering fire, First Lieutenant McGovern, assured the men of his ability to continue on and urged them forward. Forging up the rocky incline, he fearlessly led the platoon to within several yards of its objective when the ruthless foe threw and rolled a vicious barrage of hand grenades on the group and halted the advance. Enemy fire increased in volume and intensity and First Lieutenant McGovern, realizing that casualties were rapidly increasing and the morale of his men badly shaken, hurled back several grenades before they exploded. Then, disregarding his painful wound and weakened condition he charged a machine gun emplacement which was raking his position with flanking fire. When he was within ten yards of the position a burst of fire ripped the carbine from his hands, but, undaunted, he continued his lone-man assault and, firing his pistol and throwing grenades, killed seven hostile soldiers before falling mortally wounded in front of the gun he had silenced. First Lieutenant McGovern's incredible display of valor imbued his men with indomitable resolution to avenge his death. Fixing bayonets and throwing grenades, they charged with such ferocity that hostile positions were overrun and the enemy routed from the hill. The inspirational leadership, unflinching courage, and intrepid actions of First Lieutenant McGovern reflected utmost glory on himself and the honored tradition of the military services.

1951

First Lieutenant Lloyd L. Burke, G-5 Cavalry earns the Medal of Honor. (28 October)

Rank and organization: First Lieutenant, U.S. Army, Company G, 5th Cavalry Regiment, 1st Cavalry Division. Place and date: Near Chong-dong, Korea, 28 October 1951. Entered service at: Stuttgart, Ark. Born: 29 September 1924, Tichnor, Ark. G.O. No.: 43. Citation: First Lieutenant Burke, distinguished himself by conspicuous gallantry and outstanding courage above and beyond the call of duty in action against the enemy. Intense enemy fire had pinned down leading elements of his company committed to secure commanding ground when First Lieutenant Burke left the command post to rally and urge the men to follow him toward three bunkers impeding the advance. Dashing to an exposed vantage point he threw several grenades at the bunkers, then, returning for an Ml rifle and adapter, he made a lone assault, wiping out the position and killing the crew. Closing on the center bunker he lobbed grenades through the opening and, with his pistol, killed three of its occupants attempting to surround him. Ordering his men forward he charged the third emplacement, catching several grenades in midair and hurling them back at the enemy. Inspired by his display of valor his men stormed forward, overran the hostile position, but were again pinned down by increased fire. Securing a light machine gun and three boxes of ammunition, First Lieutenant Burke dashed through the impact area to an open knoll, set up his gun and poured a crippling fire into the ranks of the enemy, killing approximately seventy-five. Although wounded, he ordered more ammunition, reloading and destroying two mortar emplacements and a machine gun position with his accurate fire. Cradling the weapon in his arms he then led his men forward, killing some twenty-five more of the retreating enemy and securing the objective. First Lieutenant Burke's heroic action and daring exploits inspired his small force of thirty-five troops. His unflinching courage and outstanding leadership reflect the highest credit upon himself, the infantry, and the U.S. Army.

1951

First Lieutenant James L. Stone, F-8 Cavalry earns the Medal of Honor. (21/22 November)

Rank and organization: First Lieutenant, U.S. Army, Company F, 8th Cavalry Regiment, 1st Cavalry Division. Place and date: Near Sokkogae, Korea, 21 and 22 November 1951. Entered service at: Houston Tex. Born: 27 December 1922, Pine Bluff, Ark. G.O. No: 82, 20 October 1953. Citation: First Lieutenant Stone, distinguished himself by conspicuous gallantry and indomitable courage above and beyond the call of duty in action against the enemy. When his platoon, holding a vital outpost position, was attacked by overwhelming Chinese forces, First Lieutenant Stone stood erect and exposed to the terrific enemy fire calmly directed his men in the defense. A defensive flame-thrower failing to function, he personally moved to its location, further exposing himself, and personally repaired the weapon. Throughout a second attack, First Lieutenant Stone; though painfully wounded, personally carried the only remaining light machine gun from place to place in the position in order to bring fire upon the Chinese advancing from two directions. Throughout he continued to encourage and direct his depleted platoon in its hopeless defense. Although again wounded, he continued the fight with his carbine, still exposing himself as an example to his men. When this final overwhelming assault swept over the platoon's position his voice could still be heard faintly urging his men to carry on, until he lost consciousness. Only because of this officer's driving spirit and heroic action was the platoon emboldened to make its brave but hopeless last ditch stand.

1965

Captain Ed W. Freeman A-229th Assault Helicopter battalion, earns the Medal of Honor. (14 November)

Rank and organization: Captain, US Army, Co. A, 229th Assault Helicopter Battalion, First Cavalry Division (Airmobile). Place and date: Ia Drang Valley, RVN, 14 Nov 1965. Born: 20 Nov 1927. Entered service at: Hattiesburg, MS. Citation: For conspicuous gallantry and intrepidity at the risk of his life above and beyond the call of duty. Captain Ed W. Freeman, United States Army, distinguished himself by numerous acts of conspicuous gallantry and extraordinary intrepidity on 14 Nov 1965, while serving with Co. A, 229th Assault Helicopter Battalion, First Cavalry Division (Airmobile). As a flight leader and second in command of a 16-helicopter lift unit, he supported a heavily engaged American infantry battalion at landing zone X-ray in the Ia Drang Valley, Republic of Vietnam. The infantry unit was almost out of ammunition, after taking some of the heaviest casualties of the war, fighting off a relentless attack from a highly motivated, heavily armed enemy force. When the infantry commander closed the helicopter landing zone, due to intense direct enemy fire, Captain Freeman risked his own life by flying his unarmed helicopter through a gauntlet of enemy fire, time after time, delivering critically needed ammunition, water and medical supplies to the undersiege battalion. His flights had a direct impact on the battle's outcome by providing the engaged units with timely supplies of ammunition critical to their survival without which they would almost surely have experienced a much greater loss of life. After medical evacuation helicopters refused to fly into the area, due to intense enemy fire, Captain Freeman flew fourteen separate rescue missions, providing life-saving evacuation of an estimated thirty seriously wounded soldiers, some of whom would not have survived, had he not acted. All flights were made into a small emergency landing zone within one hundred to two hundred meters of the defensive perimeter where heavily committed units were perilously holding off the attacking elements. Captain Freeman's selfless acts of great valor, extraordinary perseverance and intrepidity were far above and beyond the call of duty or mission and set a superb example of leadership and courage for all of his peers. Captain Freeman's extraordinary heroism and devotion to duty are in keeping with the highest traditions of military service and reflect great credit upon himself, his unit and the United States Army.

1965

Second Lieutenant Walter J. Marm, A-1/7 Cavalry earns the Medal of Honor. (14 November)

Rank and organization: First Lieutenant (then Second Lieutenant), U.S. Army, Company A, 1st Battalion, 7th Cavalry, 1st Cavalry Division (Airmobile). Place and date: Vicinity of Ia Drang Valley, Republic of Vietnam, 14 November 1965. Entered service at: Pittsburgh, Pa. Born: 20 November 1941, Washington, Pa. G.O. No: 7, 15 February 1967. Citation: For conspicuous gallantry and intrepidity at the risk of life above and beyond the call of duty. As a platoon leader in the 1st Cavalry Division (Airmobile), First Lieutenant Marm demonstrated indomitable courage during a combat operation. His company was moving through the valley to relieve a friendly unit surrounded by an enemy force of estimated regimental size. First Lieutenant Marm led his platoon through withering fire until they were finally forced to take cover. Realizing that his platoon could not hold very long, and seeing four enemy soldiers moving into his position, he moved quickly under heavy fire and annihilated all four. Then, seeing that his platoon was receiving intense fire from a concealed machine gun, he deliberately exposed himself to draw its fire. Thus locating its position, he attempted to destroy it with an antitank weapon. Although he inflicted casualties, the weapon did not silence the enemy fire. Quickly, disregarding the intense fire directed on him and his platoon, he charged thirty meters across open ground, and hurled grenades into the enemy position, killing some of the eight insurgents manning it. Although severely wounded, when his grenades were expended, armed with only a rifle, he continued the momentum of his assault on the position and killed the remainder of the enemy. First Lieutenant Marm's selfless actions reduced the fire on his platoon, broke the enemy assault, and rallied his unit to continue toward the accomplishment of this mission. First Lieutenant Marm's gallantry on the battlefield and his extraordinary intrepidity at the risk of his life are in the highest traditions of the U.S. Army and reflect great credit upon himself and the Armed Forces of his country.

1966

Staff Sergeant Jimmy G. Stewart, B-2/12 Cavalry earns the Medal of Honor. (18 May)

Rank and organization: Staff Sergeant, U.S. Army, Company B, 2d Battalion, 12th Cavalry, 1st Cavalry Division (Airmobile). Place and date: Republic of Vietnam, 18 May 1966. Entered service at: Ashland, Ky. Born: 25 December 1942, West Columbia, W. Va. Citation: For conspicuous gallantry and intrepidity in action at the risk of his life above and beyond the call of duty. Early in the morning a reinforced North Vietnamese company attacked Company B, which was manning a defensive perimeter in Vietnam. The surprise onslaught wounded five members of a six-man squad caught in the direct path of the enemy's thrust. Staff Sergeant Stewart became a lone defender of vital terrain—virtually one man against a hostile platoon. Refusing to take advantage of a lull in the firing which would have permitted him to withdraw, Staff Sergeant Stewart elected to hold his ground to protect his fallen comrades and prevent an enemy penetration of the company perimeter. As the full force of the platoon-sized main attack struck his lone position, he fought like a man possessed; emptying magazine after magazine at the determined, on-charging enemy. The enemy drove almost to his position and hurled grenades, but Staff Sergeant Stewart decimated them by retrieving and throwing the grenades back. Exhausting his ammunition, he crawled under intense fire to his wounded team members and collected ammunition that they were unable to use. Far past the normal point of exhaustion, he held his position for four harrowing hours and through three assaults, annihilating the enemy as they approached and before they could get a foothold. As a result of his defense, the company position held until the arrival of a reinforcing platoon which counterattacked the enemy, now occupying foxholes to the left of Staff Sergeant Stewart's position. After the counterattack, his body was found in a shallow enemy hole where he had advanced in order to add his fire to that of the counterattacking platoon. Eight enemy dead were found around his immediate position, with evidence that 15 others had been dragged away. The wounded whom he gave his life to protect, were recovered and evacuated. Staff Sergeant Stewart's indomitable courage, in the face of overwhelming odds, stands as a tribute to himself and an inspiration to all men of his unit. His actions were in the highest traditions of the U.S. Army and the Armed Forces of his country.

1966

Sergeant David C. Dolby, B-1/8 Cavalry earns the Medal of Honor. (21 May)

Rank and organization. Sergeant (then Specialist Fourth Class), U.S. Army, Company B, 1st Battalion (Airborne), 8th Cavalry, 1st Cavalry Division (Airmobile). Place and date. Republic of Vietnam, 21 May 1966. Entered service at: Philadelphia, Pa. Born: 14 May 1946, Norristown, Pa. G.O. No.: 45, 20 October 1967. Citation: For conspicuous gallantry and intrepidity at the risk of life above and beyond the call of duty, when his platoon, while advancing tactically, suddenly came under intense fire from the enemy located on a ridge immediately to the front. Six members of the platoon were killed instantly and a number were wounded, including the platoon leader. Sergeant Dolby's every move brought fire from the enemy. However, aware that the platoon leader was critically wounded, and that the platoon was in a precarious situation, Sergeant Dolby moved the wounded men to safety and deployed the remainder of the platoon to engage the enemy. Subsequently, his dying platoon leader ordered Sergeant Dolby to withdraw the forward elements to rejoin the platoon. Despite the continuing intense enemy fire and with utter disregard for his own safety, Sergeant Dolby positioned able-bodied men to cover the withdrawal of the forward elements, assisted the wounded to the new position, and he, alone, attacked enemy positions until his ammunition was expended. Replenishing his ammunition, he returned to the area of most intense action, single-handedly killed three enemy machine gunners and neutralized the enemy fire, thus enabling friendly elements on the flank to advance on the enemy redoubt. He defied the enemy fire to personally carry a seriously wounded soldier to safety where he could be treated and, returning to the forward area, he crawled through withering fire to within fifty meters of the enemy bunkers and threw smoke grenades to mark them for air strikes. Although repeatedly under fire at close range from enemy snipers and automatic weapons, Sergeant Dolby directed artillery fire on the enemy and succeeded in silencing several enemy weapons. He remained in his exposed location until his comrades had displaced to more secure positions. His actions of unsurpassed valor during four hours of intense combat were a source of inspiration to his entire company, contributed significantly to the success of the overall assault on the enemy position, and were directly responsible for saving the lives of a number of his fellow soldiers. Sergeant Dolby's heroism was in the highest tradition of the U.S. Army.

1966

Private First Class Billy L. Lauffer, C-2/5 Cavalry earns the Medal of Honor. (21 September)

Rank and organization: Private First Class, U.S. Army, Company C, 2d Battalion, 5th Cavalry, 1st Air Cavalry Division. Place and date: Near Bon Son in Binh Dinh province, Republic of Vietnam, 21 September 1966. Entered service at: Phoenix, Ariz. Born: 20 October 1945, Murray, Ky. Citation: For conspicuous gallantry and intrepidity in action at the risk of his life above and beyond the call of duty. Private First Class Lauffer's squad, a part of Company C, was suddenly struck at close range by an intense machine gun crossfire from 2 concealed bunkers astride the squad's route. Private First Class Lauffer, the second man in the column, saw the lead man fall and noted that the remainder of the squad was unable to move. Two comrades, previously wounded and being carried on litters, were lying helpless in the beaten zone of the enemy fire. Reacting instinctively, Private First Class Lauffer quickly engaged both bunkers with fire from his rifle, but when the other squad members attempted to maneuver under his covering fire, the enemy fusillade increased in volume and thwarted every attempt to move. Seeing this and his wounded comrades helpless in the open, Private First Class Lauffer rose to his feet and charged the enemy machine gun positions, firing his weapon and drawing the enemy's attention. Keeping the enemy confused and off balance, his one-man assault provided the crucial moments for the wounded point man to crawl to a covered position, the squad to move the exposed litter patients to safety, and his comrades to gain more advantageous positions. Private First Class Lauffer was fatally wounded during his selfless act of courage and devotion to his fellow soldiers. His gallantry at the cost of his life served as an inspiration to his comrades and saved the lives of an untold number of his companions. His actions are in keeping with the highest traditions of military service and reflect great credit upon himself, his unit, and the U.S. Army.

1966

Private First Class Lewis Albanese, B-5/7 Cavalry earns the Medal of Honor. (01 December)

Rank and organization: Private First Class, U.S. Army, Company B, 5th Battalion (Airmobile), 7th Cavalry, 1st Cavalry Division. Place and date: Republic of Vietnam, 1 December 1966. Entered service at: Seattle, Wash. Born: 27 April 1946, Venice, Italy. G.O. No.: 12, 3 April 1968. Citation: For conspicuous gallantry and intrepidity in action at the risk of his life and beyond the call of duty. Private First Class Albanese's platoon, while advancing through densely covered terrain to establish a blocking position, received intense automatic weapons fire from close range. As other members maneuvered to assault the enemy position, Private First Class Albanese was ordered to provide security for the left flank of the platoon. Suddenly, the left flank received fire from enemy located in a well-concealed ditch. Realizing the imminent danger to his comrades from this fire, Private First Class Albanese fixed his bayonet and moved aggressively into the ditch. His action silenced the sniper fire, enabling the platoon to resume movement toward the main enemy position. As the platoon continued to advance, the sound of heavy firing emanated from the left flank from a pitched battle that ensued in the ditch which Private First Class Albanese had entered. The ditch was actually a well-organized complex of enemy defenses designed to bring devastating flanking fire on the forces attacking the main position. Private First Class Albanese, disregarding the danger to himself, advanced one hundred meters along the trench and killed six of the snipers, who were armed with automatic weapons. Having exhausted his ammunition, Private First Class Albanese was mortally wounded when he engaged and killed two more enemy soldiers in fierce hand-to-hand combat. His unparalleled actions saved the lives of many members of his platoon who otherwise would have fallen to the sniper fire from the ditch, and enabled his platoon to successfully advance against an enemy force of overwhelming numerical superiority. Private First Class Albanese's extraordinary heroism and supreme dedication to his comrades were commensurate with the finest traditions of the military service and remain a tribute to himself, his unit, and the U.S. Army.

1966

Staff Sergeant Delbert O. Jennings, C-1/12 Cavalry earns the Medal of Honor. (27 December)

Rank and organization: Staff Sergeant, U.S. Army, Company C, 1st Battalion (Airborne), 12th Cavalry, 1st Air Cavalry Division. Place and date: Kim Song Valley, Republic of Vietnam, 27 December 1966. Entered service at: San Francisco, Calif. Born: 23 July 1936, Silver City, N. Mex. Citation: For conspicuous gallantry and intrepidity at the risk of life above and beyond the call of duty. Part of Company C was defending an artillery position when attacked by a North Vietnamese Army regiment supported by mortar, recoilless-rifle, and machine gun fire. At the outset, Staff Sergeant Jennings sprang to his bunker, astride the main attack route, and slowed the on-coming enemy wave with highly effective machine gun fire. Despite a tenacious defense in which he killed at least twelve of the enemy, his squad was forced to the rear. After covering the withdrawal of the squad, he rejoined his men, destroyed an enemy demolition crew about to blow up a nearby howitzer, and killed three enemy soldiers at his initial bunker position. Ordering his men back into a secondary position, he again covered their withdrawal, killing one enemy with the butt of his weapon. Observing that some of the defenders were unaware of an enemy force in their rear, he raced through a fire-swept area to warn the men, turn their fire on the enemy, and lead them into the secondary perimeter. Assisting in the defense of the new position, he aided the air-landing of reinforcements by throwing white phosphorous grenades on the landing zone despite dangerously silhouetting himself with the light. After helping to repulse the final enemy assaults, he led a group of volunteers well beyond friendly lines to an area where eight seriously wounded men lay. Braving enemy sniper fire and ignoring the presence of booby traps in the area, they recovered the eight men who would have probably perished without early medical treatment. Staff Sergeant Jenning's extraordinary heroism and inspirational leadership saved the lives of many of his comrades and contributed greatly to the defeat of a superior enemy force. His actions stand with the highest traditions of the military profession and reflect great credit upon himself, his unit, and the U.S. Army.

1967

Private First Class James H. Monroe, C-1/8 Cavalry earns the Medal of Honor. (16 February)

Rank and organization: Private First Class, U.S. Army, Headquarters and Headquarters Company, 1st Battalion, 8th Cavalry, 1st Cavalry Division (Airmobile). Place and date: Bong Son, Hoai Nhon Province, Republic of Vietnam, 16 February 1967. Entered service at: Chicago, Ill. Born: 17 October 1944, Aurora, Ill. Citation: For conspicuous gallantry and intrepidity in action at the risk of his life above and beyond the call of duty. His platoon was deployed in a night ambush when the position was suddenly subjected to an intense and accurate grenade attack, and one foxhole was hit immediately. Responding without hesitation to the calls for help from the wounded men Private First Class Monroe moved forward through heavy small-arms fire to the foxhole but found that all of the men had expired. He turned immediately and crawled back through the deadly hail of fire toward other calls for aid. He moved to the platoon sergeant's position where he found the radio operator bleeding profusely from fragmentation and bullet wounds. Ignoring the continuing enemy attack, Private First Class Monroe began treating the wounded man when he saw a live grenade fall directly in front of the position. He shouted a warning to all those nearby, pushed the wounded radio operator and the platoon sergeant to one side, and lunged forward to smother the grenade's blast with his body. Through his valorous actions, performed in a flash of inspired selflessness, Private First Class Monroe saved the lives of two of his comrades and prevented the probable injury of several others. His gallantry and intrepidity were in the highest traditions of the U.S. Army, and reflect great credit upon himself and the Armed Forces of his country.

1967

Specialist Fifth Class Charles C. Hagemeister, A-1/5 earns the Medal of Honor. (20 March)

Rank and organization: Specialist Fifth Class (then Specialist Fourth Class) U .S. Army, Headquarters and Headquarters Company, 1st Battalion, 5th Cavalry, 1st Cavalry Division (Airmobile). Place and date: Binh Dinh Province, Republic of Vietnam, 20 March 1967. Entered service at: Lincoln, Nebr. Born: 21 August 1946, Lincoln, Nebr. Citation: For conspicuous gallantry and intrepidity in action at the risk of his life above and beyond the call of duty. While conducting combat operations against a hostile force, Specialist Fifth Class Hagemeister's platoon suddenly came under heavy attack from three sides by an enemy force occupying well concealed, fortified positions and supported by machine guns and mortars. Seeing two of his comrades seriously wounded in the initial action, Specialist Fifth Class Hagemeister unhesitatingly and with total disregard for his safety, raced through the deadly hail of enemy fire to provide them medical aid. Upon learning that the platoon leader and several other soldiers also had been wounded, Specialist Fifth Class Hagemeister continued to brave the withering enemy fire and crawled forward to render lifesaving treatment and to offer words of encouragement. Attempting to evacuate the seriously wounded soldiers, Specialist Fifth Class Hagemeister was taken under fire at close range by an enemy sniper. Realizing that the lives of his fellow soldiers depended on his actions, Specialist Fifth Class Hagemeister seized a rifle from a fallen comrade, killed the sniper, three other enemy soldiers who were attempting to encircle his position and silenced an enemy machine gun that covered the area with deadly fire. Unable to remove the wounded to a less exposed location and aware of the enemy's efforts to isolate his unit, he dashed through the fusillade of fire to secure help from a nearby platoon. Returning with help, he placed men in positions to cover his advance as he moved to evacuate the wounded forward of his location. These efforts successfully completed, he then moved to the other flank and evacuated additional wounded men despite the fact that his every move drew fire from the enemy. Specialist Fifth Class Hagemeister's repeated heroic and selfless actions at the risk of his life saved the lives of many of his comrades and inspired their actions in repelling the enemy assault. Specialist Fifth Class Hagemeister's indomitable courage was in the highest traditions of the U.S. Armed Forces and reflect great credit upon himself.

1967

Specialist Fourth Class George A. Ingalls, A-2/5 Cavalry earns the Medal of Honor. (16 April)

Rank and organization: Specialist Fourth Class, U.S. Army, Company A, 2d Battalion, 5th Cavalry, 1st Cavalry Division (Airmobile). Place and date: Near Duc Pho, Republic of Vietnam, 16 April 1967. Entered service at: Los Angeles, Calif. Born: 9 March 1946, Hanford, Calif. Citation: For conspicuous gallantry and intrepidity in action at the risk of his life above and beyond the call of duty. Specialist Fourth Class Ingalls, a member of Company A, accompanied his squad on a night ambush mission. Shortly after the ambush was established, an enemy soldier entered the killing zone and was shot when he tried to evade capture. Other enemy soldiers were expected to enter the area, and the ambush was maintained in the same location. Two quiet hours passed without incident, then suddenly a hand grenade was thrown from the nearby dense undergrowth into the center of the squad's position. The grenade did not explode, but shortly thereafter a second grenade landed directly between Specialist Fourth Class Ingalls and a nearby comrade. Although he could have jumped to a safe position, Specialist Fourth Class Ingalls, in a spontaneous act of great courage, threw himself on the grenade and absorbed its full blast. The explosion mortally wounded Specialist Fourth Class Ingalls, but his heroic action saved the lives of the remaining members of his squad. His gallantry and selfless devotion to his comrades are in keeping with the highest traditions of the military service and reflects great credit upon Specialist Fourth Class Ingalls, his unit, and the U.S. Army.

1967

Specialist Fifth Class Edgar L. McWethy Jr., B-1/5 Cavalry earns the Medal of Honor. (21 June)

Rank and organization: Specialist Fifth Class, U.S. Army, Company B, 1st Battalion, 5th Cavalry, 1st Cavalry Division (Airmobile). Place and date: Binh Dinh province, Republic of Vietnam, 21 June 1967. Entered service at: Denver, Colo. Born: 22 November 1944, Leadville, Colo. Citation: For conspicuous gallantry and intrepidity in action at the risk of his life above and beyond the call of duty. Serving as a medical aidman with Company B, Specialist Fifth Class McWethy accompanied his platoon to the site of a downed helicopter. Shortly after the platoon established a defensive perimeter around the aircraft, a large enemy force attacked the position from three sides with a heavy volume of automatic weapons fire and grenades. The platoon leader and his radio operator were wounded almost immediately, and Specialist Fifth Class McWethy rushed across the fire-swept area to their assistance. Although he could not help the mortally wounded radio operator, Specialist Fifth Class McWethy's timely first aid enabled the platoon leader to retain command during this critical period. Hearing a call for aid, Specialist Fifth Class McWethy started across the open toward the injured men, but was wounded in the head and knocked to the ground. He regained his feet and continued on but was hit again, this time in the leg. Struggling onward despite his wounds, he gained the side of his comrades and treated their injuries. Observing another fallen rifleman lying in an exposed position raked by enemy fire, Specialist Fifth Class McWethy moved toward him without hesitation. Although the enemy fire wounded him a third time, Specialist Fifth Class McWethy reached his fallen companion. Though weakened and in extreme pain, Specialist Fifth Class McWethy gave the wounded man artificial respiration but suffered a fourth and fatal wound. Through his indomitable courage, complete disregard for his safety, and demonstrated concern for his fellow soldiers, Specialist Fifth Class McWethy inspired the members of his platoon and contributed in great measure to their successful defense of the position and the ultimate rout of the enemy force. Specialist Fifth Class McWethy's profound sense of duty, bravery, and his willingness to accept extraordinary risks in order to help the men of his unit are characteristic of the highest traditions of the military service and reflect great credit upon himself and the U.S. Army.

1967

Specialist Fourth Class Carmel B. Harvey Jr., B-1/5 Cavalry earns the Medal of Honor. (21 June)

Rank and organization: Specialist Fourth Class, U.S. Army, Company B, 1st Battalion, 5th Cavalry, 1st Cavalry Division (Airmobile). Place and date: Binh Dinh Province, Republic of Vietnam, 21 June 1967. Entered service at: Chicago, Ill. Born: 6 October 1946, Montgomery, W. Va. Citation: For conspicuous gallantry and intrepidity in action at the risk of his life above and beyond the call of duty. Specialist Fourth Class Harvey distinguished himself as a fire team leader with Company B, during combat operations. Ordered to secure a downed helicopter, his platoon established a defensive perimeter around the aircraft, but shortly thereafter a large enemy force attacked the position from three sides. Specialist Fourth Class Harvey and two members of his squad were in a position directly in the path of the enemy onslaught, and their location received the brunt of the fire from an enemy machine gun. In short order, both of his companions were wounded, but Specialist Fourth Class Harvey covered this loss by increasing his deliberate rifle fire at the foe. The enemy machine gun seemed to concentrate on him and the bullets struck the ground all around his position. One round hit and armed a grenade attached to his belt. Quickly, he tried to remove the grenade but was unsuccessful. Realizing the danger to his comrades if he remained and despite the hail of enemy fire, he jumped to his feet, shouted a challenge at the enemy, and raced toward the deadly machine gun. He nearly reached the enemy position when the grenade on his belt exploded, mortally wounding Specialist Fourth Class Harvey, and stunning the enemy machine gun crew. His final act caused a pause in the enemy fire, and the wounded men were moved from the danger area. Specialist Fourth Class Harvey's dedication to duty, high sense of responsibility, and heroic actions inspired the others in his platoon to decisively beat back the enemy attack. His acts are in keeping with the highest traditions of the military service and reflect great credit upon himself and the U.S. Army.

1967

Sergeant Allen J. Lynch, D-1/12 Cavalry earns the Medal of Honor. (15 December)

Rank and organization: Sergeant, U.S. Army, Company D, 1st Battalion (Airmobile), 12th Cavalry, 1st Cavalry

Division (Airmobile). Place and date: Near My An (2), Binh Dinh province, Republic of Vietnam, 15 December 1967. Entered service at: Chicago, Ill. Born: 28 October 1945, Chicago, Ill. Citation: For conspicuous gallantry and intrepidity in action at the risk of his life above and beyond the call of duty. Sergeant Lynch (then Specialist Fourth Class) distinguished himself while serving as a radio telephone operator with Company D. While serving in the forward element on an operation near the village of My An, his unit became heavily engaged with a numerically superior enemy force. Quickly and accurately assessing the situation, Sergeant Lynch provided his commander with information which subsequently proved essential to the unit's successful actions. Observing three wounded comrades lying exposed to enemy fire, Sergeant Lynch dashed across fifty meters of open ground through a withering hail of enemy fire to administer aid. Reconnoitering a nearby trench for a covered position to protect the wounded from intense hostile fire, he killed two enemy soldiers at point blank range. With the trench cleared, he unhesitatingly returned to the fire-swept area three times to carry the wounded men to safety. When his company was forced to withdraw by the superior firepower of the enemy, Sergeant Lynch remained to aid his comrades at the risk of his life rather than abandon them. Alone, he defended his isolated position for two hours against the advancing enemy. Using only his rifle and a grenade, he stopped them just short of his trench, killing five. Again, disregarding his safety in the face of withering hostile fire, he crossed seventy meters of exposed terrain five times to carry his wounded comrades to a more secure area. Once he had assured their comfort and safety, Sergeant Lynch located the counterattacking friendly company to assist in directing the attack and evacuating the three casualties. His gallantry at the risk of his life is in the highest traditions of the military service, Sergeant Lynch has reflected great credit on himself, the 12th Cavalry, and the U.S. Army.

1968

Sergeant William D. Port, C-5/7 earns the Medal of Honor. (12 January)

Rank and organization: Sergeant (then Private First Class), U.S. Army, Company C, 5th Battalion, 7th Cavalry, 1st Air Cavalry Division. Place and date: Que Son Valley, Heip Duc Province, Republic of Vietnam, 12 January 1968. Entered service at: Harrisburg, Pa. Born: 13 October 1941, Petersburg, Pa. Citation: For conspicuous gallantry and intrepidity at the risk of his life above and beyond the call of duty. Sergeant Port distinguished himself while serving as a rifleman with Company C, which was conducting combat operations against an enemy force in the Que Son Valley. As Sergeant Port's platoon was moving to cut off a reported movement of enemy soldiers, the platoon came under heavy fire from an entrenched enemy force. The platoon was forced to withdraw due to the intensity and ferocity of the fire. Although wounded in the hand as the withdrawal began, Sergeant Port, with complete disregard for his safety, ran through the heavy fire to assist a wounded comrade back to the safety of the platoon perimeter. As the enemy forces assaulted in the perimeter, Sergeant Port and three comrades were in position behind an embankment when an enemy grenade landed in their midst. Sergeant Port, realizing the danger to his fellow soldiers, shouted the warning, "Grenade," and unhesitatingly hurled himself towards the grenade to shield his comrades from the explosion. Through his exemplary courage and devotion he saved the lives of his fellow soldiers and gave the members of his platoon the inspiration needed to hold their position. Sergeant Port's selfless concern for his comrades, at the risk of his life above and beyond the call of duty are in keeping with the highest tradition of the military service and reflect great credit on himself, his unit, and the U.S. Army.

1968

Chief Warrant Officer Fredrick E. Ferguson, C-227 Aviation Battalion, earns the Medal of Honor. (31 January)

Rank and organization: Chief Warrant Officer, U.S. Army, Company C, 227th Aviation Battalion, 1st Cavalry Division (Airmobile). Place and date: Hue, Republic of Vietnam, 31 January 1968. Entered service at: Phoenix, Ariz. Born: 18 August 1939, Pilot Point, Tex. Citation: For conspicuous gallantry and intrepidity in action at the risk of his life above and beyond the call of duty. Chief Warrant Officer Ferguson, U.S. Army distinguished himself while serving with Company C. Chief Warrant Officer Ferguson, commander of a resupply helicopter monitoring an emergency call from wounded passengers and crewmen of a downed helicopter under heavy attack within the enemy controlled city of Hue, unhesitatingly volunteered to attempt evacuation. Despite warnings from all aircraft to stay clear of the area due to heavy antiaircraft fire, Chief Warrant Officer Ferguson began a low-level flight at maximum airspeed along the Perfume River toward the tiny, isolated South Vietnamese Army compound in which the crash survivors had taken refuge. Coolly and skillfully maintaining his course in the face of intense, short range fire from enemy occupied buildings and boats, he displayed superior flying skill and tenacity of purpose by landing his aircraft in an extremely confined area in a blinding dust cloud under heavy mortar and small-arms fire. Although the helicopter was severely damaged by mortar fragments during the loading of the wounded, Chief Warrant Officer Ferguson disregarded the damage and, taking off through the continuing hail of mortar fire, he flew his crippled aircraft on the return route through the rain of fire that he had experienced earlier and safely returned his wounded passengers to friendly control. Chief Warrant Officer Ferguson's extraordinary determination saved the lives of five of his comrades. His actions are in the highest traditions of the military service and reflect great credit on himself and the U.S. Army.

1968

Captain James M. Sprayberry, D-5/7 Cavalry earns the Medal of Honor. (25 April)

Rank and organization: Captain (then First Lieutenant), U.S. Army, Company D, 5th Battalion, 7th Cavalry , 1st Cavalry Division (Airmobile). Place and date: Republic of Vietnam, 25 April 1968. Entered service at: Montgomery, Ala. Born: 24 April 1947, LaGrange, Ga. Citation: For conspicuous gallantry and intrepidity in action at the risk of his life above and beyond the call of duty. Captain Sprayberry, Armor, U.S. Army, distinguished himself by exceptional bravery while serving as executive officer of Company D. His company commander and a great number of the men were wounded and separated from the main body of the company. A daylight attempt to rescue them was driven back by the well entrenched enemy's heavy fire. Captain Sprayberry then organized and led a volunteer night patrol to eliminate the intervening enemy bunkers and to relieve the surrounded element. The patrol soon began receiving enemy machinegun fire. Captain Sprayberry quickly moved the men to protective cover and without regard for his own safety, crawled within close range of the bunker from which the fire was coming. He silenced the machinegun with a hand grenade. Identifying several one-man enemy positions nearby, Captain Sprayberry immediately attacked them with the rest of his grenades. He crawled back for more grenades and when two grenades were thrown at his men from a position to the front, Captain Sprayberry, without hesitation, again exposed himself and charged the enemy-held bunker killing its occupants with a grenade. Placing two men to cover his advance, he crawled forward and neutralized three more bunkers with grenades. Immediately thereafter, Captain Sprayberry was surprised by an enemy soldier who charged from a concealed position. He killed the soldier with his pistol and with continuing disregard for the danger neutralized another enemy emplacement. Captain Sprayberry then established radio contact with the isolated men, directing them toward his position. When the two elements made contact he organized his men into litter parties to evacuate the wounded. As the evacuation was nearing completion, he observed an enemy machinegun position which he silenced with a grenade. Captain Sprayberry returned to the rescue party, established security, and moved to friendly lines with the wounded. This rescue operation, which lasted approximately seven and one-half hours, saved the lives of many of his fellow soldiers. Captain Sprayberry personally killed twelve enemy soldiers, eliminated two machineguns, and destroyed numerous enemy bunkers. Captain Sprayberry's indomitable spirit and gallant action at great personal risk to his life are in keeping with the highest traditions of the military service and reflect great credit upon himself, his unit, and the U.S. Army.

1968

First Lieutenant Douglas B. Fournet, B-1/7 Cavalry earns the Medal of Honor. (04 May)

Rank and organization: First Lieutenant, U.S. Army, Company B, 1st Battalion, 7th Cavalry, 1st Cavalry Division (Airmobile). Place and date: A Shau Valley, Republic of Vietnam, 4 May 1968. Entered service at: New Orleans, La. Born: 7 May 1943, Lake Charles, La. Citation: For conspicuous gallantry and intrepidity in action at the risk of his life above and beyond the call of duty. First Lieutenant Fournet, Infantry, distinguished himself in action while serving as rifle platoon leader of the 2d Platoon, Company B. While advancing uphill against fortified enemy positions in the A Shau Valley, the platoon encountered intense sniper fire, making movement very difficult. The right flank man suddenly discovered an enemy claymore mine covering the route of advance and shouted a warning to his comrades. Realizing that the enemy would also be alerted, First Lieutenant Fournet ordered his men to take cover and ran uphill toward the mine, drawing a sheath knife as he approached it. With complete disregard for his safety and realizing the imminent danger to members of his command, he used his body as a shield in front of the mine as he attempted to slash the control wires leading from the enemy positions to the mine. As he reached for the wire the mine was detonated, killing him instantly. Five men nearest the mine were slightly wounded, but First Lieutenant Fournet's heroic and unselfish act spared his men of serious injury or death. His gallantry and willing self-sacrifice are in keeping with the highest traditions of the military service and reflect great credit upon himself, his unit, and the U.S. Army.

1968

Specialist Fourth Class Hector Santiago-Colon, B-5/7 Cavalry earns th Medal of Honor. (28 June)

Rank and organization: Specialist Fourth Class, U.S. Army, Company B, 5th Battalion, 7th Cavalry, 1st Cavalry Division (Airmobile). Place and date: Quang Tri Province, Republic of Vietnam, 28 June 1968. Entered service at: New York, N.Y. Born: 20 December 1942, Salinas, Puerto Rico. Citation: For conspicuous gallantry and intrepidity in action at the risk of his life above and beyond the call of duty. Specialist Fourth Class Santiago-Colon distinguished himself at the cost of his life while serving as a gunner in the mortar platoon of Company B. While serving as a perimeter sentry, Specialist Fourth Class Santiago-Colon heard distinct movement in the heavily wooded area to his front and flanks. Immediately he alerted his fellow sentries in the area to move to their foxholes and remain alert for any enemy probing forces. From the wooded area around his position heavy enemy automatic weapons and small-arms fire suddenly broke out, but extreme darkness rendered difficult the precise location and identification of the hostile force. Only the muzzle flashes from enemy weapons indicated their position. Specialist Fourth Class Santiago-Colon and the other members of his position immediately began to repel the attackers, utilizing hand grenades, antipersonnel mines and small-arms fire. Due to the heavy volume of enemy fire and exploding grenades around them, a North Vietnamese soldier was able to crawl, undetected, to their position. Suddenly, the enemy soldier lobbed a hand grenade into Specialist Fourth Class Santiago-Colon's foxhole. Realizing that there was no time to throw the grenade out of his position, Specialist Fourth Class Santiago-Colon retrieved the grenade, tucked it in to his stomach and, turning away from his comrades, absorbed the full impact of the blast. His heroic self-sacrifice saved the lives of those who occupied the foxhole with him, and provided them with the inspiration to continue fighting until they had forced the enemy to retreat from the perimeter. By his gallantry at the cost of his life and in the highest traditions of the military service, Specialist Fourth Class Santiago-Colon has reflected great credit upon himself, his unit, and the U.S. Army.

1968

Sergeant John N. Holcomb, D-2/7 Cavalry earns the Medal of Honor. (03 December)

Rank and organization: Sergeant, U.S. Army, Company D, 2d Battalion, 7th Cavalry, 1st Cavalry Division. Place and date: Near Quan Loi, Republic of Vietnam, 3 December 1968. Entered service at: Corvallis, Oreg. Born: 11 June 1946, Baker, Oreg. Citation: For conspicuous gallantry and intrepidity in action at the risk of his life above and beyond the call of duty. Sergeant Holcomb distinguished himself while serving as a squad leader in Company D during a combat assault mission. Sergeant Holcomb's company assault had landed by helicopter and deployed into a hasty defensive position to organize for a reconnaissance-in-force mission when it was attacked from three sides by an estimated battalion-size enemy force. Sergeant Holcomb's squad was directly in the path of the main enemy attack. With complete disregard for the heavy fire, Sergeant Holcomb moved among his men giving encouragement and directing fire on the assaulting enemy. When his machine gunner was knocked out, Sergeant Holcomb seized the weapon, ran to a forward edge of the position, and placed withering fire on the enemy. His gallant actions caused the enemy to withdraw. Sergeant Holcomb treated and carried his wounded to a position of safety and reorganized his defensive sector despite a raging grass fire ignited by the incoming enemy mortar and rocket rounds. When the enemy assaulted the position a second time, Sergeant Holcomb again manned the forward machine gun, devastating the enemy attack and forcing the enemy to again break contact and withdraw. During the enemy withdrawal an enemy rocket hit Sergeant Holcomb's position, destroying his machine gun and severely wounding him. Despite his painful wounds, Sergeant Holcomb crawled through the grass fire and exploding mortar and rocket rounds to move the members of his squad, everyone of whom had been wounded, to more secure positions. Although grievously wounded and sustained solely by his indomitable will and courage, Sergeant Holcomb as the last surviving leader of his platoon organized his men to repel the enemy, crawled to the platoon radio and reported the third enemy assault on his position. His report brought friendly supporting fires on the charging enemy and broke the enemy attack. Sergeant Holcomb's inspiring leadership, fighting spirit, conspicuous gallantry and intrepidity in action at the cost of his life were in keeping with the highest traditions of the military service and reflect great credit on himself, his unit, and the U.S. Army.

1969

Specialist Fourth Class Donald R. Johnston, D-1/8 Cavalry earns the Medal of Honor. (21 March)

Rank and organization: Specialist Fourth Class, U.S. Army, Company D, 1st Battalion, 8th Cavalry, 1st Cavalry Division. Place and date: Tay Ninh Province, Republic of Vietnam, 21 March 1969. Entered service at: Columbus, Ga. Born: 19 November 1947, Columbus, Ga. Citation: For conspicuous gallantry and intrepidity in action at the risk of his life above and beyond the call of duty. Specialist Fourth Class Johnston distinguished himself while serving as a mortarman with Company D, at a fire support base in Tay Ninh Province. Specialist Fourth Class Johnston's company was in defensive positions when it came under a devastating rocket and mortar attack. Under cover of the bombardment, enemy sappers broke through the defensive perimeter and began hurling explosive charges into the main defensive bunkers. Specialist Fourth Class Johnston and six of his comrades had moved from their exposed positions to one of the bunkers to continue their fight against the enemy attackers. As they were firing from the bunker, an enemy soldier threw three explosive charges into their position. Sensing the danger to his comrades, Specialist Fourth Class Johnston, with complete disregard for his safety, hurled himself onto the explosive charges, smothering the detonations with his body and shielding his fellow soldiers from the blast. His heroic action saved the lives of six of his comrades. Specialist Fourth Class Johnston's concern for his fellow men at the cost of his life were in the highest traditions of the military service and reflect great credit upon himself, his unit, and the U.S. Army.

1969

First Lieutenant Robert L. Poxon, B-1/9 Cavalry earns the Medal of Honor. (02 June)

Rank and organization: First Lieutenant, U.S. Army, Troop B, 1st Squadron, 9th Cavalry, 1st Cavalry Division. Place and date: Tay Ninh Province, Republic of Vietnam, 2 June 1969. Entered service at: Detroit, Mich. Born: 3 January 1947, Detroit, Mich. Citation: For conspicuous gallantry and intrepidity in action at the risk of his life above and beyond the call of duty. First Lieutenant Poxon, Armor, Troop B, distinguished himself while serving as a platoon leader on a reconnaissance mission. Landing by helicopter in an area suspected of being occupied by the enemy, the platoon came under intense fire from enemy soldiers in concealed positions and fortifications around the landing zone. A soldier fell, hit by the first burst of fire. First Lieutenant Poxon dashed to his aid, drawing the majority of the enemy fire as he crossed 20 meters of open ground.

The fallen soldier was beyond help and First Lieutenant Poxon was seriously and painfully wounded. First Lieutenant Poxon, with indomitable courage, refused medical aid and evacuation and turned his attention to seizing the initiative from the enemy. With sure instinct he marked a central enemy bunker as the key to success. Quickly instructing his men to concentrate their fire on the bunker, and in spite of his wound, First Lieutenant Poxon crawled toward the bunker, readied a hand grenade and charged. He was hit again but continued his assault. After succeeding in silencing the enemy guns in the bunker he was struck once again by enemy fire and fell, mortally wounded. First Lieutenant Poxon's comrades followed their leader, pressed the attack and drove the enemy from their positions. First Lieutenant Poxon's gallantry, indomitable will, and courage are in keeping with the highest traditions of the military service and reflect great credit upon himself, his unit, and the U.S. Army.

1969

Sergeant Rodney J. Evans, D-1/12 earns the Medal of Honor. (18 July)

Rank and organization: Sergeant, U.S. Army, Company D, 1st Battalion, 12th Cavalry, 1st Cavalry Division. Place and date: Tay Ninh Province, Republic of Vietnam, 18 July 1969. Entered service at: Montgomery, Ala. Born: 17 July 1948, Chelsea, Mass. Citation: For conspicuous gallantry and intrepidity in action at the risk of his life above and beyond the call of duty. Sergeant Evans distinguished himself by extraordinary heroism while serving as a squad leader in a reconnaissance sweep through heavy vegetation to reconnoiter a strong enemy position. As the force approached a well-defined trail, the platoon scout warned that the trail was booby-trapped. Sergeant Evans led his squad on a route parallel to the trail. The force had started to move forward when a nearby squad was hit by the blast of a concealed mine. Looking to his right Sergeant Evans saw a second enemy device. With complete disregard for his safety he shouted a warning to his men, dived to the ground and crawled toward the mine. Just as he reached it an enemy soldier detonated the explosive and Sergeant Evans absorbed the full impact with his body. His gallant and selfless action saved his comrades from probable death or injury and served as an inspiration to his entire unit. Sergeant Evans' gallantry in action at the cost of his life were in keeping with the highest traditions of the military service and reflect great credit upon himself, his unit, and the U.S. Army.

1969

Sergeant Donald S. Skidgel, D-1/9 Cavalry earns the Medal of Honor. (14 September)

Rank and organization: Sergeant, U.S. Army, Troop D, 1st Squadron, 9th Cavalry, 1st Cavalry Division. Place and date: Near Song Be, Republic of Vietnam, 14 September 1969. Entered service at: Bangor, Maine. Born: 13 October 1948, Caribou, Maine. Citation: For conspicuous gallantry and intrepidity in action at the risk of his life above and beyond the call of duty. Sergeant Skidgel distinguished himself while serving as a reconnaissance section leader in Troop D. On a road near Song Be in Binh Long Province, Sergeant Skidgel and his section with other elements of his troop were acting as a convoy security and screening force when contact occurred with an estimated enemy battalion concealed in tall grass and in bunkers bordering the road. Sergeant Skidgel maneuvered off the road and began placing effective machinegun fire on the enemy automatic weapons and rocket-propelled grenade positions. After silencing at least one position, he ran with his machinegun across sixty meters of bullet-swept ground to another location from which he continued to rake the enemy positions. Running low on ammunition, he returned to his vehicle over the same terrain. Moments later he was alerted that the command element was receiving intense automatic weapons, rocket-propelled grenade and mortar fire. Although he knew the road was saturated with enemy fire, Sergeant Skidgel calmly mounted his vehicle and with his driver advanced toward the command group in an effort to draw the enemy fire onto himself. Despite the hostile fire concentrated on him, he succeeded in silencing several enemy positions with his machinegun. Moments later Sergeant Skidgel was knocked down onto the rear fender by the explosion of an enemy rocket-propelled grenade. Ignoring his extremely painful wounds, he staggered back to his feet and placed effective fire on several other enemy positions until he was mortally wounded by hostile small arms fire. His selfless actions enabled the command group to withdraw to a better position without casualties and inspired the rest of his fellow soldiers to gain fire superiority and defeat the enemy. Sergeant Skidgel's gallantry at the cost of his life were in keeping with the highest traditions of the military service and reflect great credit upon himself, his unit, and the U.S. Army.

1969

Second Lieutenant Robert R. Leisy, B-1/8 Cavalry earns the Medal of Honor. (02 December)

Rank and organization: Second Lieutenant, U.S. Army, Infantry, Company B, 1st Battalion, 8th Cavalry, 1st Cavalry Division. Place and date: Phuoc Long province, Republic of Vietnam, 2 December 1969. Entered service at: Seattle, Wash. Born: 1 March 1945, Stockton, Calif. Citation: For conspicuous gallantry and intrepidity in action at the risk of his life above and beyond the call of duty. Second Lieutenant Leisy, Infantry, Company B, distinguished himself while serving as platoon leader during a reconnaissance mission. One of his patrols became heavily engaged by fire from a numerically superior enemy force located in a well-entrenched bunker complex. As Second Lieutenant Leisy deployed the remainder of his platoon to rescue the beleaguered patrol, the platoon also came under intense enemy fire from the front and both flanks. In complete disregard for his safety, Second Lieutenant Leisy moved from position to position deploying his men to effectively engage the enemy. Accompanied by his radio operator he moved to the front and spotted an enemy sniper in a tree in the act of firing a rocket-propelled grenade at them. Realizing there was neither time to escape the grenade nor shout a warning, Second Lieutenant Leisy unhesitatingly, and with full knowledge of the consequences, shielded the radio operator with his body and absorbed the full impact of the explosion. This valorous act saved the life of the radio operator and protected other men of his platoon who were nearby from serious injury. Despite his mortal wounds, Second Lieutenant Leisy calmly and confidently continued to direct the platoon's fire. When medical aid arrived, Second Lieutenant Leisy valiantly refused attention until the other seriously wounded were treated. His display of extraordinary courage and exemplary devotion to duty provided the inspiration and leadership that enabled his platoon to successfully withdraw without further casualties. Second Lieutenant Leisy's gallantry at the cost of his life are in keeping with the highest traditions of the military service and reflect great credit on him, his unit, and the U.S. Army.

1970

Specialist Fourth Class John P. Baca, D-1/12 Cavalry earns the Medal of Honor. (10 February)

Rank and organization: Specialist Fourth Class, U.S. Army, Company D, 1st Battalion, 12th Cavalry, 1st Cavalry Division. Place and date: Phuoc Long Province, Republic of Vietnam, 10 February 1970. Entered service at: Fort Ord, Calif. Born: 10 January 1949, Providence, R.I.. Citation: For conspicuous gallantry and intrepidity in action at the risk of his life above and beyond the call of duty. Specialist Fourth Class Baca, Company D, distinguished himself while serving on a recoilless rifle team during a night ambush mission. A platoon from his company was sent to investigate the detonation of an automatic ambush device forward of his unit's main position and soon came under intense enemy fire from concealed positions along the trail. Hearing the heavy firing from the platoon position and realizing that his recoilless rifle team could assist the members of the besieged patrol, Specialist Fourth Class Baca led his team through the hail of enemy fire to a firing position within the patrol's defensive perimeter. As they prepared to engage the enemy, a fragmentation grenade was thrown into the midst of the patrol. Fully aware of the danger to his comrades, Specialist Fourth Class Baca unhesitatingly, and with complete disregard for his own safety, covered the grenade with his steel helmet and fell on it as the grenade exploded, thereby absorbing the lethal fragments and concussion with his body. His gallant action and total disregard for his personal well-being directly saved eight men from certain serious injury or death. The extraordinary courage and selflessness displayed by Specialist Fourth Class Baca, at the risk of his life, are in the highest traditions of the military service and reflect great credit on him, his unit, and the U.S. Army.

1970

Sergeant Peter C. Lemon, E-2/8 Cavalry earns the Medal of Honor. (01 April)

Rank and organization: Sergeant, U.S. Army, Company E, 2d Battalion, 8th Cavalry, 1st Cavalry Division. place and date: Tay Ninh province, Republic of Vietnam, 1 April 1970. Entered service at: Tawas City, Mich. Born: 5 June 1950, Toronto, Canada. Citation: For conspicuous gallantry and intrepidity in action at the risk of his life above and beyond the call of duty. Sergeant Lemon (then Specialist Fourth Class), Company E, distinguished himself while serving as an assistant machine gunner during the defense of Fire Support Base Illingworth. When the base came under heavy enemy attack, Sergeant Lemon engaged a numerically superior enemy with machine gun and rifle fire from his defensive position until both weapons malfunctioned. He then used hand grenades to fend off the intensified enemy attack launched in his direction. After eliminating all but one of the enemy soldiers in the immediate vicinity, he pursued and disposed of the remaining soldier in hand-to-hand combat. Despite fragment wounds from an exploding grenade, Sergeant Lemon regained his position, carried a more seriously wounded comrade to an aid station, and, as he returned, was wounded a second time by enemy fire. Disregarding his personal injuries, he moved to his position through a hail of small arms and grenade fire. Sergeant Lemon immediately realized that the defensive sector was in danger of being overrun by the enemy and unhesitatingly assaulted the enemy soldiers by throwing hand grenades and engaging in hand-to-hand combat. He was wounded yet a third time, but his determined efforts successfully drove the enemy from the position. Securing an operable machine gun, Sergeant Lemon stood atop an embankment fully exposed to enemy fire, and placed effective fire upon the enemy until he collapsed from his multiple wounds and exhaustion. After regaining consciousness at the aid station, he refused medical evacuation until his more seriously wounded comrades had been evacuated. For his valiant action, Sergeant Peter C. Lemon received the Medal of Honor.

1971

Captain Jon E. Swanson, B-1/9 Cavalry earns the Medal of Honor. (26 February)

Rank and Organization: Captain, U.S. Army, Troop B, 1st Squadron, 9th Cavalry, 1st Cavalry Division. Place and date: Kingdom of Cambodia, 26 February, 1971. Born: 01 May 1942 San Antonio, Texas.

Citation: For conspicuous gallantry and intrepidity in action at the risk of his life above and beyond the call of duty. Captain Jon E. Swanson distinguished himself by acts of bravery on February 26, 1971, while flying an OH-6A aircraft in support of Army of the Republic of Vietnam (ARVN) Task Force 333 in the Kingdom of Cambodia. With two well-equipped enemy regiments known to be in the area, Captain Swanson was tasked with pinpointing the enemy's precise positions. Captain Swanson flew at treetop level at a slow airspeed, making his aircraft a vulnerable target. The advancing ARVN unit came under heavy automatic weapons fire from enemy bunkers 100 meters to their front. Exposing his aircraft to enemy antiaircraft fire, Captain Swanson immediately engaged the enemy bunker with concussion grenades and machinegun fire. After destroying five bunkers and evading intense ground-to-air fire, he observed a .51-caliber machinegun position. With all his heavy ordnance expended on the bunkers, he did not have sufficient explosives to destroy the position. Consequently, he marked the position with a smoke grenade and directed a Cobra gunship attack. After completion of the attack, Captain Swanson found the weapon still intact and an enemy soldier crawling over to man it. He immediately engaged the individual and killed him. During this time, his aircraft sustained several hits from another .51-caliber machinegun. Captain Swanson engaged the position with his aircraft's weapons, marked the target, and directed a second Cobra gunship attack. He volunteered to continue the mission, despite the fact that he was now critically low on ammunition and his aircraft was crippled by enemy fire. As Captain Swanson attempted to fly toward another .51-caliber machinegun position, his aircraft exploded in the air and crashed to the ground, causing his death. Captain Swanson's extraordinary heroism and devotion to duty are in keeping with the highest traditions of military service and reflect great credit upon himself, his unit, and the United States Army.

Appendix C: Lineage

While it would take a book to recount the rich history brought about by the changes of each unit in the 1st Cavalry Division, an overall look at the major reorganizational elements of the Divisional lineage presents a picture of an organization that brought strength from change and was accustomed to fighting

1st Cavalry Division
"The First Team"

22 January 1921; Constituted as 1st Cavalry Division

13 September 1921; Activated at Fort Bliss, Texas

03 January 1933; 1st Cavalry Regiment relieved from assignment to the 1st Cavalry Division

04 December 1943; Redesignated 1st Cavalry Division Special and reorganized partly under the Cavalry and partly under the Infantry Tables of Organization and Equipment

20 July 1945; The Regiments were organized wholly as Infantry, but retained their Cavalry designations

25 March 1949; Reorganized as a Triangular Division. The 5th, 7th and 8th Cavalry Regiments were retained and the 12th Cavalry Regiment was deactivated

01 November 1957; The 1st Cavalry Division was converted to the Pentomic structure with 5 Battle Groups. This change brought back the 12th Cavalry Regiment and added the 4th Cavalry Regiment as Battle Groups

15 July 1963; The new Army Division called, Reorganization Objective Army Division (ROAD) brought back the triangular Division and from the point of the 1st Cavalry, resulted in the loss of the 12th Cavalry Regiment once again as well as the 4th Cavalry Regiment

01 July 1965; On this date the 1st Cavalry Division was reorganized as an Airmobile Division. The 12th Cavalry and the 9th Cavalry Regiments became part of the Division

05 May 1971; The Division received an experimental designation as the TRICAP (Triple Capability) Division. It was organized with Armored, Infantry and Air Cavalry Brigades

21 February 1975; On this date, while retaining the famous name of the 1st Cavalry Division, it was designated an Armored Division

Headquarters and Headquarters Company
1st Cavalry Division
"The First Team"

20 August 1921; Constituted in the Regular Army as Headquarters, 1st Cavalry Division

13 September 1921; Activated at Fort Bliss, Texas

04 December 1943; Reorganized and redesignated as Headquarters, 1st Cavalry Division, Special

25 March 1949; Reorganized and redesignated as Headquarters, 1st Cavalry Division (Infantry)

01 July 1960; Reorganized and redesignated as Headquarters and Headquarters Company, 1st Cavalry Division

Headquarters and Headquarters Company
1st Brigade, 1st Cavalry Division
"Ironhorse"

29 August 1917; Constituted in the Regular Army as Headquarters, 1st Cavalry Brigade

February 1918; Organized at Fort Sam Houston, Texas, as an element of the 15th Cavalry Division

12 May 1918; Relieved from assignment to the 15th Cavalry Division

14 July 1919; Demobilized at Brownsville, Texas

20 August 1920; Reconstituted in the Regular Army as Headquarters and Headquarters Troop, 1st Cavalry Brigade, an element of the 1st Cavalry Division (later redesignated as the 1st Cavalry Division Special)

01 September 1921; Organized at Camp Harry J. Jones, Arizona

25 March 1949; Inactivated in Japan and relieved from assignment to the 1st Cavalry Division, Special

20 May 1949; Converted and redesignated as Headquarters, 1st Constabulary Brigade, assigned to the United States Constabulary, and activated in Germany

24 November 1950; Relieved from assignment to the United States Constabulary

15 August 1951; Inactivated in Germany

05 December 1951; Disbanded

15 July 1963; Reconstituted in the Regular Army as Headquarters and Headquarters Company, 1st Brigade, 1st Cavalry Division

01 September 1963; Activated in Korea

Headquarters and Headquarters Company 2nd Brigade, 1st Cavalry Division "Blackjack"

29 August 1917; Constituted in the Regular Army as Headquarters, 2nd Cavalry Brigade

27 December 1917; Organized at Fort Bliss, Texas, as an element of the 15th Cavalry Division

12 May 1918; Relieved from assignment to the 15th Cavalry Division

09 July 1919; Demobilized at Fort Bliss, Texas

20 August 1921; Reconstituted in the Regular Army as Headquarters and Headquarters Troop, 2nd Cavalry Brigade, an element of the 1st Cavalry Division (later redesignated as the 1st Cavalry Division, Special)

14 September 1921; Organized at Fort Bliss, Texas

25 March 1949; Inactivated in Japan and relieved from assignment to the 1st Cavalry Division, Special

20 May 1949; Converted and redesignated as Headquarters, 2nd Constabulary Brigade, assigned to the United States Constabulary, and activated in Germany

24 November 1950; Relieved from assignment to the United States Constabulary

15 December 1951; Inactivated in Germany

15 July 1963; Converted and redesignated as Headquarters and Headquarters Company, 2nd Brigade, 1st Cavalry Division

01 September 1963; Activated in Korea

Headquarters and Headquarters Company 3rd Brigade, 1st Cavalry Division "Greywolf"

29 August 1917; Constituted in the Regular Army as Headquarters, 3rd Cavalry Brigade

December 1917; Organized at Camp Harry J. Jones, Arizona, as an element of the 15th Cavalry Division

12 May 1918; Relieved from assignment to the 15th Cavalry Division

15 July 1919; Demobilized at Camp Harry J. Jones, Arizona

20 August 1921; Reconstituted in the Regular Army as Headquarters and Headquarters Troop, 3rd Cavalry Brigade, an element of the 2nd Cavalry Division

15 October 1940; Headquarters activated at Fort Riley, Kansas

21 February 1941; Headquarters Troop activated at Fort Riley, Kansas

15 July 1942; Relieved from assignment to the 2d Cavalry Division; concurrently, converted and redesignated as Headquarters and Headquarters Company, 9th Armored Division Trains

13 October 1945; Inactivated at Camp Patrick Henry, Virginia

15 July 1963; Relieved from assignment to the 9th Armored Division; concurrently, converted and redesignated as Headquarters and Headquarters Company, 3d Brigade, 1st Cavalry Division

01 September 1963; Activated in Korea

15 September 1980; Inactivated at Fort Hood, Texas

21 May 1991; Activated at Fort Hood, Texas

Headquarters and Headquarters Company Aviation Brigade, 1st Cavalry Division "Warriors"

01 September 1984; Constituted in the Regular Army as Headquarters and Headquarters Company, Aviation

Brigade, 1st Cavalry Division and activated at Fort Hood, Texas

Headquarters and Headquarters Battery
1st Cavalry Division Artillery
"The Red Team"

01 November 1940; Constituted as Headquarters and Headquarters Battery, 1st Cavalry Division Artillery

03 January 1941; Activated at Fort Bliss, Texas

04 December 1943; Redesignated Headquarters and Headquarters Battery, 1st Cavalry Division Artillery (Special) and reorganized partly under the Cavalry and partly under the Infantry Tables of Organization and Equipment

25 March 1949; Reorganized as a Triangular Division; designation as Headquarters and Headquarters Battery, 1st Cavalry Division remained the same

Headquarters and Headquarters Detachment
Engineer Brigade, 1st Cavalry Division
"Sapper Forward"

16 October 1991; Constituted in the Regular Army as Headquarters and Headquarters Detachment, Engineer Brigade, 1st Cavalry Division

28 April 1992; Provisionally organized under the Engineer Restructure Initiative of the US Army as Headquarters and Headquarters Detachment, Engineer Brigade, 1st Cavalry Division

16 October 1992; Headquarters and Headquarters Detachment, Engineer Brigade, 1st Cavalry Division activated at Fort Hood, Texas

Headquarters and Headquarters Company
1st Cavalry Division Support Command
"Wagon Masters"

22 August 1957; Constituted in the Regular Army as Headquarters and Headquarters Detachment, 1st Cavalry Division Trains

01 November 1957; Activated in Korea

01 September 1963; Consolidated with the 1st Cavalry Division Band (organized in 1856) and consolidated unit reorganized and redesignated as Headquarters, Headquarters Company and Band, 1st Cavalry Division Support Command

05 May 1971; Reorganized and redesignated as Headquarters and Headquarters Company, 1st Cavalry Division Support Command (Band concurrently withdrawn from the 1st Cavalry Division Support Command)

Appendix D: Date Line

"History will always tell of your predecessors' and of your own exploits. It is a special laudit to you, Cavalrymen, members of the First Team." John R. Tolson - Major General, USA Commanding

1921
1st Cavalry Regiment assigned to the 1st Cavalry Division. (20 August)

1921
7th, 8th and 10th Cavalry Regiments assigned to the 1st Cavalry Division. (13 September)

1st Cavalry Division Organization Day. (13 September)

1922
5th Cavalry Regiment assigned to the 1st Cavalry Division, relieving the 10th Cavalry Regiment. (18 December)

1923
Participated in the first 1st Cavalry Division maneuvers.

1927
Participated in the second 1st Cavalry Division maneuvers.

1929
Participated in the third 1st Cavalry Division maneuvers.

1930
Participated in the Fourth Corps Area Infantry/Cavalry maneuvers.

1933
12th Cavalry Regiment assigned to the 1st Cavalry Division, relieving the 1st Cavalry Regiment. (03 January)

1934
By order of The Secretary of War, /s/ E. T. Conley, Brigadier General, the saber was discontinued as an item of issue. (18 April)

1937
Participated in the Provisional Infantry Tests, Ft. Sam Houston, TX.

1938
Participated in the fourth 1st Cavalry Division maneuvers, Balmorhea, TX.

1939
Participated in the fifth 1st Cavalry Division maneuvers. (01 September)

1940
Camp Polk, Louisiana
Participated in the first 3rd Army Louisiana Maneuvers. (April/May & 13/24 August)

1941
Camp Polk, Louisiana
Participated in the second 3rd Army Louisiana Maneuvers. (05 August/08 October)

Japan attacks Pearl Harbor. (07 December)

Congress declares war on Japan. (08 December)

1942
Bataan Death March begins. (10 April)

Participated in the VIII Corps Louisiana Maneuvers near Mansfield, Louisiana (04 August/19 September)

1943
1st Cavalry Division transitions from mounted to dismounted troops. (01 February)

The first echelon of the 1st Cavalry Division, the 5th Cavalry Regiment and the 8th Engineer Combat Squadron, departs Camp Stoneman, California for the Pacific theater. (28 May)

The second echelon of the 1st Cavalry Division begins movement from Ft. Bliss to Camp Stoneman, California. (16 June)

The second echelon of the1st Cavalry Division departs Camp Stoneman, California for the Pacific theater. (03 June)

The second echelon of the 1st Cavalry Division closed on Strathpine, Queensland, Australia. (26 July)

Subordinate elements of the 9th Cavalry activated in Australia and assigned to the 1st Cavalry Division. (04 December)

1st Cavalry Division reorganized and redesignated as 1st Cavalry Division (Special). (04 December)

1944
Task Force Brewer makes an assault landing at White Beach, Los Negros, Admiralty Islands. (29 February)

SGT Troy A. McGill, G-5th Cavalry earns the Medal of Honor. (04 March)

Momote Airdrome, Los Negros secured. (04 March)

First amphibious landing made with out boats at southern shore of Los Negros by "G" Troop, 5th Cavalry. (24 March)

Admiralty Islands campaign (Task Force Brewer) concluded. (18 May)

1st Cavalry Division Association formed. (17 July)

1st Cavalry Division assault elements landed at Marasbaras, near the area of San Pedro Bay, Leyte. (20 October)

1945
Leyte-Samar campaign concluded. (04 January)

1st Division elements lands on Mabilao, near the area of Lignayan Gulf, Luzon, Philippine Islands. (27 January)

The "Flying Column", elements of the 1st Cavalry Division, initiates a forced march to Manila. (01 February)

The "Flying Column", elements of the 1st Cavalry Division, enters Manila, Philippine Islands and liberates Santo Tomas Prisoners. (03 February)

PFC William J. Grabiarz, E-5th Cavalry earns the Medal of Honor. (23 February)

Organized resistance in Manila ends. (03 March)

Kapatalan sawmill captured. (07 May)

Luzon campaign completed. (30 June)

The first atomic bomb, "Little Boy", was dropped at 8:15 AM over Hiroshima, Japan. (06 August)

The second atomic bomb, "Fat Man", was dropped at 11:02 AM over Nagasaki, Japan. (09 August)

Japan surrenders. (02 September)

In the early morning hours a long convey of ships, carrying the 1st Cavalry Division, steered into Yokohama Harbor and past the battleship Missouri where General MacArthur would later receive the Japanese surrender party. (02 September)

MG William C. Chase and 1st Cavalry Division Troopers enter Tokyo, Japan. (08 September)

1947
Department of Defense established by Congress. (25 July)

1950
Korean War begins. (25 June)

1st Cavalry Division alerted for deployment to Korea. (06 July)

1st Cavalry Division deploys to Korea. (14 July)

1st Cavalry Division amphibious landing at Pohang-Dong. (18 July)

1st combat of Korean War for 1st Division Cavalry Troopers. (23 July)

1st Cavalry Division sets up defensive line behind Naktong River. (03 August)

3rd Battalions, 5th, 7th and 8th Cavalry arrive on the USS General John Pope at Pusan. (25 August)

1st Cavalry Division occupies positions in the "bowling alley". (29 August)

PFC Melvin L. Brown, D-8 Engineer Bn earns the Medal of Honor. (04 September)

CPL Gorden M. Graig, 16th Reconnaissance Troop earns the Medal of Honor. (10 September)

2nd Battalion, 7 Cavalry captures Hill 314, earns Distinguished Unit Citation. (12 September)

General MacArthur launches Operation CHROMITE at Inchon. (15 September)

7th Cavalry initiates Task Forces 777, Lynch and Witherspoon to breakout of the Pusan Perimeter. (20 September)

Task Force 777, Lynch and Witherspoon complete 196 mile penetration through Pusan Perimeter, consolidating gains at Osan. (27 September)

1st Cavalry Division crosses the 38th parallel. (09 October)

PFC Robert H. Young, E-8th Cavalry earns the Medal of Honor. (09 October)

1st LT Samuel Coursen, C-5th Cavalry earns the Medal of Honor. (12 October)

1st Cavalry Division elements enter Pyongyang. (19 October)

Chinese Communist Forces surround 8th Cavalry Troopers at Unsan, Korea. (01 November)

1951
Seoul-Inchon, Korea area evacuated. (03 January)

1st LT Robert M. McGovern, A-5th Cavalry earns the Medal of Honor. (30 January)

Task Force Crombez breaks through to the encircled 23rd Regimental Combat Team and a French Army Battalion. (15 February)

1st Cavalry Division elements re-cross 38th parallel, Korea. (03 April)

1st Cavalry Division organized the defense of Seoul, Korea. (28 April)

1st Cavalry Division attacks the Iron Triangle. (30 May)

Operation COMMANDO begins. (03 October)

Operation COMMANDO ends. (19 October)

1st LT Lloyd L. Burke, G-5th Cavalry earns the Medal of Honor. (28 October)

1st LT James L. Stone, F-8th Cavalry earns the Medal of Honor. (21/22 November)

1st Cavalry Division begins rotation from Korea to Japan. 5th Cavalry leaves. (07 December)

7th Cavalry Regiment leaves Korea. (18 December)

77th Field Artillery fires the last 1st Cavalry Division round in the Korean War. (30 December)

8th Cavalry Regiment leaves Korea. (30 December)

1952
Korean GI Bill enacted. (16 July)

8th Cavalry Regiment returns to Korea. (16 October)

7th Cavalry sails for Pusan. (12 December)

1953
Korean Armistice signed. (27 July)

1957
24th Infantry Division redesignated as 1st Cavalry Division in Korea. (15 October)

1962
Howze Board findings submitted. (20 August)

1963
11th Air Assault Division (Test) activated. (15 February)

1965
1st Cavalry Division (AirMobile) activated. (01 July)

First 1st Cavalry Division combat troops sent to Vietnam. (14 August)

1st Cavalry Division advance party lands at An Khe, Vietnam. (27 August)

First combat of Vietnam War for 1st Cavalry Troopers in Operation GIBRALTAR. (18 September)

Operation GIBRALTAR ends. (20 September)

Operation HAPPY VALLEY begins. (03 October)

Operation SHINY BAYONET begins. (10 October)

Operation SILVER BAYONET begins. (23 October)

1-9th Cavalry fights Hospital Battle. (01 November)

Operation HAPPY VALLEY ends. (01 November)

1st Squadron, 9th Cavalry ambushes 66th NVA Regiment at LZ Betty. (03 November)

7th Cavalry air assaults into LZ X-Ray. (14 November)

CPT Edward W. Freeman, A-229th Assault Helicopter Battalion, earns the Medal of Honor. (14 November)

2nd LT Walter J. Marm, A-1/7th Cavalry earns the Medal of Honor. (14 November)

2nd Battalion,7th Cavalry, Company A, 1st Battalion, 5th Cavalry battle at LZ Albany. (17 November)

Operation SILVER BAYONET ends. (25 November)

Operation CLEAN HOUSE begins. (17 December)

Operation CLEAN HOUSE ends. (31 December)

Operation MATADOR begins. (31 December)

1966
Operation MATADOR ends. (13 January)

Operation MASHER/WHITE WING begins. (25 January)

Operation MASHER/WHITE WING ends. (06 March)

Operation JIM BOWIE begins. (13 March)

Operation LINCOLN HOUSE begins. (25 March)

Operation JIM BOWIE ends. (28 March)

Operation LINCOLN HOUSE ends. (08 April)

Operation MOSBY begins. (11 April)

Operation MOSBY ends. (03 May)

Operation LEWIS AND CLARK begins. (03 May)

Operation DAVY CROCKETT begins. (04 May)

Operation DAVY CROCKETT ends. (10 May)

Operation LEWIS AND CLARK ends. (16 May)

Operation CRAZY HORSE begins. (16 May)

SSG Jimmy G. Stewart, B-2/12th Cavalry earns the Medal of Honor. (18 May)

Sgt David C. Dolby, B-1/8th Cavalry earns the Medal of Honor. (21 May)

Operation CRAZY HORSE ends. (05 June)

Operation NATHAN HALE begins. (19 June)

Operation NATHAN HALE ends. (01 July)

Operation HENRY CLAY begins. (02 July)

Operation HENRY CLAY ends. (30 July)

Operation PAUL REVERE II begins. (02 August)

Operation PAUL REVERE II ends battle of Hill 534. (15 Aug)

Operation BYRD begins. (25 July)

Operation THAYER I begins. (13 September)

PFC Billy L. Lauffer, C-2/5th Cavalry earns the Medal of Honor. (21 September)

Operation THAYER I ends. (01 October)

Operation IRVING begins, over 2,000 enemy prisoners of war captured. (02 October)

Operation IRVING ends. (24 October)

Operation THAYER II begins. (26 October)

Operation PAUL REVERE IV begins. (31 October)

PFC Lewis Albanese, B-5/7th Cavalry earns the Medal of Honor. (01 December)

Operation ROVER begins. (09 December)

Operation ROVER ends. (14 December)

Operation PAUL REVERE IV ends. (27 December)

SSG Delbert O. Jennings, C-1/12th Cavalry earns The Medal of Honor. (27 December)

1967
Operation BYRD ends. (30 January)

Operation THAYER II ends. (12 February)

Operation PERSHING begins. (12 February)

PFC James H. Monroe, C-1/8th Cavalry earns the Medal of Honor. (16 February)

SP5 Charles C. Hagemeister, A-1/5th earns the Medal of Honor. (20 March)

Operation LEJEUNE begins. (07 April)

SP4 George A. Ingalls, A-2/5th Cavalry earns the Medal of Honor. (16 April)

Operation LEJEUNE ends. (22 April)

SP5 Edgar L. McWethy Jr., B-1/5th Cavalry earns the Medal of Honor. (21 June)

SP4 Carmel B. Harvey Jr., B-1/5th Cavalry earns the Medal of Honor. (21 June)

Operation SONG RE begins. (01 August)

Operation SONG RE ends. (20 August)

Operation WHELLER/WALLOWA begins. (02 October)

SGT Allen James Lynch, D-1/12th Cavalry earns the Medal of Honor. (15 December)

1968
SGT William D. Port, C-5/7th earns the Medal of Honor. (12 January)

Operation PERSHING ends. (21 January)

Operation PERSHING II begins. (22 January)

Operation JEB STUART begins. (22 January)

Operation WHELLER/WALLOWA ends. (25 January)

CW2 Fredrick Ferguson, C-227th Aviation Battalion earns the Medal of Honor. (31 January)

North Vietnamese and Viet Cong launch "Tet Offensive" at Hue. (31 January)

Operation PERSHING II ends. (18 February)

North Vietnamese and Viet Cong driven from the city of Hue. (22 February)

Operation JEB STUART ends. (31 March)

Operation PEGASUS/LAMSON 207 begins. (01 April)

Operation PEGASUS/LAMSON 207 ends. (15 April)

Operation DELAWARE/LAMSON 216 begins. (19 April)

CPT James M. Sprayberry, D-5/7th Cavalry earns the Medal of Honor. (25 April)

1st LT Douglas B. Fournet, B-1/7th Cavalry earns the Medal of Honor. (04 May)

Operation CONCORDIA SQUARE begins. (08 May)

Operation DELAWARE/LAMSON 217 ends. (17 May)

Operation JEB STUART III begins. (17 May)

Operation CONCORDIA SQUARE ends. (17 May)

SP4 Hector Santiago-Colon, B-5/7th Cavalry earns Medal of Honor. (28 June)

Operation COMANCHE FALLS begins. (11 September)

Operation LIBERTY CANYON begins. (28 October)

Operation JEB STUART III ends. (03 November)

Operation COMANCHE FALLS ends. (07 November)

Operation TOAN THANG begins. (12 November)

Operation LIBERTY CANYON ends. (15 November)

SGT John N. Holcomb, D-2/7th Cavalry earns the Medal of Honor. (03 December)

1969
Operation MONTANA SCOUT/RAIDER begins. (23 January)

Operation TOAN THANG ends. (16 February)

SP4 Donald R. Johnston, D-1/8th Cavalry earns the Medal of Honor. (21 March)

1LT Robert L. Poxon, B-1/9th Cavalry earns the Medal of Honor. (02 June)

SGT Rodney J. Evans, D-1/12th earns the Medal of Honor. (18 July)

SGT Donald S. Skidgel, D-1/9th Cavalry earns the Medal of Honor. (14 September)

2nd LT Robert R. Leisy, B-1/8th Cavalry earns the Medal of Honor. (02 December)

1970
SP4 John P. Baca, D-1/12th Cavalry earns the Medal of Honor. (10 February)

SGT Peter C. Lemon, E-2/8th Cavalry earns the Medal of Honor. (01 April)

Operation TOAN THANG (Fish Hook) begins. (01 May)

First Team enters Cambodia. (01 May)

Operation TOAN THANG (Fish Hook) ends. (29 June)

1971
CPT Jon E. Swanson, B-1/9th Cavalry earns the Medal of Honor. (26 February)

1971
1st Team stands down in Vietnam. (26 March)

1st Cavalry Division assigned to Ft. Hood, Texas. (05 May)

1972
Operation KEYSTONE PHEASANT begins. (26 June)

3rd Brigade departs Vietnam. (29 June)

1973
Cease-fire in Vietnam begins. (28 January)

US ground troops leave Vietnam. (28 March)

1974
1st Cavalry Division reorganized from TRICAP to Armor configuration. (06 December)

1982
"First" National Training Center rotation at Ft. Irwin, California. (September)

1990
Operation DESERT SHIELD begins. (07 August)

1st Cavalry Division Deployment to Southwest Asia begins. (16 September)

1st Cavalry Division closes in Saudi Arabia. (22 October)

1991
Operation DESERT STORM begins. (17 January)

First Iraqi enemy prisoners of war taken by the 1st Cavalry Division. (23 January)

1st Battalion, 5th Cavalry Recon Force initiates Operation Knight Strike into Iraq. (20 February)

3rd Brigade Recon Force initiates Operation Quick Strike into Iraq. (24 February)

Operation DESERT STORM cease-fire. (28 February)

First airplane with First Team Troopers return from Southwest Asia. (09 March)

Headquarters and Headquarters Company, 3rd Brigade, 1st Cavalry Division activated at Ft. Hood, Texas. (21 May)

Last airplane with First Team Troopers returns from Southwest Asia. (12 June)

1992
Operation INTRINSIC ACTION 92-03 - Kuwait initiated. (22 October)

1993
Operation IRIS GOLD 93-01 - Kuwait initiated. (15 January)

1994
Operation VIGILENT WARRIOR - Kuwait initiated. (October)

1995
Korean War Memorial dedicated. (27 July)
Operation VIGILENT SENTINEL - Kuwait initiated. (Aug)

Operation INTRINSIC ACTION 95-03 - Kuwait initiated. (22 August)

1996
Operation INTRINSIC ACTION 96-03 - Kuwait initiated. (26 August)

1997
Operation INTRINSIC ACTION 97-01 - Kuwait initiated. (February)

Operation INTRINSIC ACTION 97-03 - Kuwait initiated. (03 June)

1998
Deployment of the 1st Cavalry Division to Bosnia announced. (17 April)

Operation INTRINSIC ACTION 98-03 - Kuwait initiated. (May)

Advance party of 1st Cavalry Division departs Robert Gray Airfield for Bosnia. (20 August)

Main body of 1st Brigade Combat Team, 1st Cavalry Division deploys to Bosnia. (24 August)
The 1st Cavalry Division accepts authority for command of MultiNational Division - North. (07 October)

1999
The 1st Brigade Combat Team, 1st Cavalry Division begins redeployment to CONUS. (20 March)

The 1st Cavalry Division turns over command of Multi-National Division - North to the 10th Mountain Division. (04 August)

The 2nd Brigade Combat Team, 1st Cavalry Division turns over command of MultiNational Division - North to 2nd Brigade Combat Team, 10th Mountain Division. (01 October)

2000
Operation DESERT SPRING 00-01 - Kuwait initiated. (16 April)

2001
Operation INTRINSIC ACTION 01-01 - Kuwait initiated. (December)

"C" Troop, 10th Cavalry activated and assigned to the 1st Brigade, 1st Cavalry Division (18 September)

Appendix E: Decorations

Headquarters and Headquarters Company "The First Team"

The following decorations have been earned by the Headquarters and Headquarters Company, 1st Cavalry Division in the service for their country.

Presidential Unit Citation (Army), Streamer embroidered PLEIKU PROVINCE (Headquarters and Headquarters Company, 1st Cavalry Division, cited for the period 2 Oct to 26 Nov 1965; DA GO 40, 1967)

Presidential Unit Citation (Army), Streamer embroidered BINH THUAN PROVINCE (Detachment, Headquarters and Headquarters Company, 1st Cavalry Division, cited for the period 25 Aug 1966 to 04 Apr 1967; DA GO 2, 1973)

Philippine Presidential Unit Citation, Streamer embroidered 17 OCTOBER 1944 - 4 JULY 1945 (Headquarters, 1st Cavalry Division, cited; DA GO 47, 1950)

Republic of Korea Presidential Unit Citation, Streamer embroidered WAEGWAN-TAEGU (Headquarters, 1st Cavalry Division, cited; DA GO 35, 1951)

Valorous Unit Award, Streamer embroidered FISH HOOK (Headquarters and Headquarters Company, 1st Cavalry Division, cited for the period 01 May to 29 Jun 1970; DA GO 43, 1972)

Meritorious Unit Commendation (Army), Streamer embroidered SOUTHWEST ASIA (Headquarters and Headquarters Company, 1st Cavalry Division, cited for the period 16 Sep 1990 - 15 Mar 1991; DA GO 27, 1994)

Chryssoun Aristion Andrias (Bravery Gold Medal of Greece), Streamer embroidered KOREA (Headquarters, 1st Cavalry Division, cited; DA GO 2, 1956)

Republic of Vietnam Cross of Gallantry with Palm, Streamer embroidered VIETNAM 1965 - 1969 (Headquarters and Headquarters Company, 1st Cavalry Division, cited for the period 09 Aug 1965 to 19 May 1969; DA GO 59, 1969)

Republic of Vietnam Cross of Gallantry with Palm, Streamer embroidered VIETNAM 1969 - 1970 (Headquarters and Headquarters Company, 1st Cavalry Division, cited for the period May 1969 to Feb 1970; DA GO 11, 1973, amended DA GO 42, 1972)

Republic of Vietnam Cross of Gallantry with Palm, Streamer embroidered VIETNAM 1970 - 1971 (Headquarters and Headquarters Company, Rear (Prov) 1st Cavalry Division, cited 21 Feb 1970 to 29 Feb 1971; DA GO 42, 1972)

Republic of Vietnam Civil Action Honor Medal, First Class, Streamer embroidered VIETNAM 1969-1970 (Headquarters and Headquarters Company, 1st Cavalry Division, cited for the period 01 Jan 1969 to 01 Feb 1970; DA GO 42, 1972)

Headquarters and Headquarters Company 1st Brigade, 1st Cavalry Division "Ironhorse"

The following decorations have been earned by the Headquarters and Headquarters Company, 1st Brigade, 1st Cavalry Division in the service for their country.

Presidential Unit Citation (Army), Streamer embroidered LOS NEGROS ISLAND (Detachment, Headquarters and Headquarters Troop, 1st Brigade, 1st Cavalry Division, cited for the period 29 Feb to 04 Mar 1944; DA GO 16, 1949)

Presidential Unit Citation (Army), Streamer embroidered LUZON (Headquarters and Headquarters Troop, 1st Brigade, cited; DA GO 33, 1948)

Presidential Unit Citation (Army), Streamer embroidered PLEIKU PROVINCE (Headquarters and Headquarters Company, 1st Brigade, 1st Cavalry Division, cited for the period 23 Oct to 26 Nov 1965; DA GO 40, 1967)

Philippine Presidential Unit Citation, Streamer embroidered 17 OCTOBER 1944 - 4 JULY 1945 (Headquarters and Headquarters Troop, 1st Brigade, 1st Cavalry Division cited; DA GO 47, 1950)

Valorous Unit Award, Streamer embroidered FISH HOOK (Headquarters and Headquarters Company, 1st

Brigade, 1st Cavalry Division, cited for the period 01 May to 29 Jun 1970; DA GO 43, 1972)

Meritorious Unit Commendation (Army), Streamer embroidered SOUTHWEST ASIA (Headquarters and Headquarters Company, 1st Brigade, 1st Cavalry Division, cited for the period 16 Sep 1990 - 15 Mar 1991; DA GO 27, 1994)

Republic of Vietnam Cross of Gallantry with Palm, Streamer embroidered VIETNAM 1965 - 1969 (Headquarters and Headquarters Company, 1st Brigade, 1st Cavalry Division, cited for the period 09 Aug 1965 to 19 May 1969; DA GO 59, 1969)

Republic of Vietnam Cross of Gallantry with Palm, Streamer embroidered VIETNAM 1969 - 1970 (Headquarters and Headquarters Company, 1st Brigade, 1st Cavalry Division, cited for the period May 1969 to Feb 1970; DA GO 11, 1973, amended DA GO 42, 1972)

Republic of Vietnam Cross of Gallantry with Palm, Streamer embroidered VIETNAM 1970 - 1971 (Headquarters and Headquarters Company, 1st Brigade, 1st Cavalry Division, cited for the period 21 Feb 1970 to 28 Feb 1971; DA GO 42, 1972)

Republic of Vietnam Civil Action Honor Medal, First Class, Streamer embroidered VIETNAM 1969 - 1970 (Headquarters and Headquarters Company, 1st Brigade, 1st Cavalry Division, cited for the period 01 Jan 1969 to 01 Feb 1970; DA GO 42, 1972)

Headquarters and Headquarters Company 2nd Brigade, 1st Cavalry Division "Blackjack"

The following decorations have been earned by the Headquarters and Headquarters Company, 2nd Brigade, 1st Cavalry Division in the service for their country.

Presidential Unit Citation (Army), Streamer embroidered PLEIKU PROVINCE (Headquarters and Headquarters Company, 2nd Brigade, 1st Cavalry Division, cited for the period 23 Oct to 26 Nov 1965; DA GO 40, 1967)

Philippine Presidential Unit Citation, Streamer embroidered 17 October 1944 - 4 JULY 1945 (Headquarters and Headquarters Troop, 2nd Cavalry Brigade, cited; DA GO 47, 1950)

Valorous Unit Award, Streamer embroidered FISH HOOK (Headquarters and Headquarters Company, 2nd Brigade, 1st Cavalry Division, cited for the period 01 May to 29 Jun 1970; DA GO 43, 1972)

Valorous Unit Award, Streamer embroidered IRAQ (Headquarters and Headquarters Company, 2nd Brigade, 1st Cavalry Division, cited for the period 15 - 25 Feb 1991; DA GO 27, 1994)

Republic of Vietnam Cross of Gallantry with Palm, Streamer embroidered VIETNAM 1965 - 1969 (Headquarters and Headquarters Company, 2nd Brigade, 1st Cavalry Division, cited for the period 09 Aug 1965 to 19 May 1969; DA GO 59, 1969)

Republic of Vietnam Cross of Gallantry with Palm, Streamer embroidered VIETNAM 1969 - 1970 (Headquarters and Headquarters Company, 2nd Brigade, 1st Cavalry Division, cited for the period May 1969 to Feb 1970; DA GO 11, 1973, amended DA GO 42, 1972)

Republic of Vietnam Cross of Gallantry with Palm, Streamer embroidered VIETNAM 1970 - 1971 (Headquarters and Headquarters Company, 2nd Brigade, 1st Cavalry Division, cited for the period 21 Feb 1970 to 28 Feb 1971; DA GO 42, 1972)

Republic of Vietnam Civil Action Medal, First Class, Streamer embroidered VIETNAM 1969 - 1970 (Headquarters and Headquarters Company, 2nd Brigade, 1st Cavalry Division, cited for the period 01 Jan 1969 to 01 Feb 1970; DA GO 42, 1972)

Headquarters and Headquarters Company 3rd Brigade, 1st Cavalry Division "Greywolf"

The following decorations have been earned by the Headquarters and Headquarters Company, 3rd Brigade, 1st Cavalry Division in the service for their country.

Presidential Unit Citation (Army), Streamer embroidered PLEIKU PROVINCE (Headquarters and Headquarters Company, 3rd Brigade, 1st Cavalry Division, cited for the period 23 Oct to 26 Nov 1965; DA GO 40, 1967)

Valorous Unit Award, Streamer embroidered QUANG TIN PROVINCE (Headquarters and Headquarters Company, 3rd Brigade, 1st Cavalry Division, cited for the period 01 to 31 Oct 1967; DA GO 39, 1970)

Valorous Unit Award, Streamer embroidered FISH HOOK (Headquarters and Headquarters Company, 3rd Brigade, 1st Cavalry Division, cited for the period 01 May to 29 Jun 1970; DA GO 43, 1972)

Meritorious Unit Commendation (Army), Streamer embroidered EUROPE 1944 (Headquarters and Headquarters Company, 9th Armored Division Trains, cited for the period 1 Nov 1944 to 1 Jan 1945; DA GO 82, 1945)

Meritorious Unit Commendation (Army), Streamer embroidered EUROPE 1945 (Headquarters and

Headquarters Company, 9th Armored Division Trains, cited for the period 1 Jan to 1 Jul 1945; DA GO 143, 1945)

Republic of Vietnam Cross of Gallantry with Palm, Streamer embroidered VIETNAM 1965 - 1969 (Headquarters and Headquarters Company, 3rd Brigade, 1st Cavalry Division, cited for the period 09 Aug 1965 to 19 May 1969; DA GO 59, 1969)

Republic of Vietnam Cross of Gallantry with Palm, Streamer embroidered VIETNAM 1969 - 1970 (Headquarters and Headquarters Company, 3rd Brigade, 1st Cavalry Division, cited for the period May 1969 to Feb 1970; DA GO 11, 1973, amended DA GO 42, 1972)

Republic of Vietnam Cross of Gallantry with Palm, Streamer embroidered VIETNAM 1970 - 1971 (Headquarters and Headquarters Company, 3rd Brigade, 1st Cavalry Division, cited for the period 21 Feb 1970 to 28 Feb 1971; DA GO 11, 1972, amended DA GO 42, 1972)

Republic of Vietnam Cross of Gallantry with Palm, Streamer embroidered VIETNAM 1965 - 1972 (Headquarters and Headquarters Company, 3rd Brigade, 1st Cavalry Division, cited for the period 17 Sep 1965 to 29 Jun 1972; DA GO 54, 1974)

Republic of Vietnam Civil Action Honor Medal, First Class, Streamer embroidered VIETNAM 1969 - 1970 (Headquarters and Headquarters Company, 3rd Brigade, 1st Cavalry Division, cited for the period 01 Jan 1969 to 01 Feb 1970; DA GO 42, 1972)

Headquarters and Headquarters Company Aviation Brigade, 1st Cavalry Division "Warriors"

The following decorations have been earned by the Headquarters and Headquarters Company, Aviation Brigade, 1st Cavalry Division in the service for their country.

Meritorious Unit Commendation (Army), Streamer embroidered SOUTHWEST ASIA (Headquarters and Headquarters Company, Aviation Brigade, 1st Cavalry Division, cited for the period 25 Sep 1990 - 13 Apr 1991; DA GO 27, 1994)

Headquarters and Headquarters Battery 1st Cavalry Division Artillery "The Red Team"

The following decorations have been earned by the Headquarters and Headquarters Battery, Division Artillery, 1st Cavalry Division in the service for their country.

Presidential Unit Citation (Army), Streamer embroidered PLEIKU PROVINCE (Headquarters and Headquarters Battery, 1st Cavalry Division Artillery, cited for the period 23 Oct to 26 Nov 1965; DA GO 40, 1967)

Philippine Presidential Unit Citation, Streamer embroidered 17 OCTOBER 1944 - 4 JULY 1945 (Headquarters and Headquarters Battery, 1st Cavalry Division Artillery, cited; 1950)

Republic of Korea Presidential Unit Citation, Streamer embroidered WAEGWAN-TAEGU (Headquarters and Headquarters Battery, 1st Cavalry Division Artillery, cited; 1951)

Valorous Unit Award, Streamer embroidered FISH HOOK (Headquarters and Headquarters Battery, 1st Cavalry Division Artillery, cited for the period 01 May to 29 Jun 1970; DA GO 43, 1972)

Meritorious Unit Commendation, Streamer embroidered VIETNAM 1966 - 1967 (Headquarters and Headquarters Battery, 1st Cavalry Division Artillery, cited for the period 13 Sep 1966 to 31 Jul 1967; DA GO 17, 1968)

Meritorious Unit Commendation, Streamer embroidered VIETNAM 1967 - 1969 (Headquarters and Headquarters Battery, 1st Cavalry Division Artillery, cited for the period 01 Aug 1967 to 11 Feb 1969; DA GO 36, 1970)

Meritorious Unit Commendation (Army), Streamer embroidered SOUTHWEST ASIA (Headquarters and Headquarters Battery, 1st Cavalry Division Artillery, cited for the period 29 Sep 1990 - 06 Apr 1991; DA GO 27, 1994)

Chryssoun Aristion Andrias (Bravery Gold Medal of Greece), Streamer embroidered KOREA (Headquarters and Headquarters Battery, 1st Cavalry Division Artillery, cited; 1956)

Republic of Vietnam Cross of Gallantry with Palm, Streamer embroidered VIETNAM 1969 - 1970 (Headquarters and Headquarters Battery, 1st Cavalry

Division Artillery, cited for the period May 1969 to Feb 1970; DA GO 11, 1973, amended DA GO 42, 1972)

Republic of Vietnam Cross of Gallantry with Palm, Streamer embroidered VIETNAM 1970 - 1971 (Headquarters Battery, 1st Cavalry Division Artillery, cited for the period 21 Feb 1970 to 28 Feb 1971; DA GO 42, 1972)

Republic of Vietnam Civil Action Honor Medal, First Class, Streamer embroidered VIETNAM 1969 - 1970 (Headquarters and Headquarters Battery, 1st Cavalry Division Artillery, cited for the period 01 Jan 1969 to 28 Feb 1971; DA GO 42, 1972)

Defense of Saudi-Arabia, Streamer embroidered SAUDI ARABIA 1990 - 1991 (Headquarters and Headquarters Battery, 1st Cavalry Division Artillery, cited;)

Liberation and Defense of Kuwait, Streamer embroidered KUWAIT 1990 - 1991 (Headquarters and Headquarters Battery, 1st Cavalry Division Artillery, cited;)

Headquarters and Headquarters Detachment Engineer Brigade, 1st Cavalry Division "Sapper Forward"

None

Headquarters and Headquarters Company 1st Cavalry Division Support Command "Wagon Masters"

The following decorations have been earned by the Headquarters and Headquarters Company, Support Command, 1st Cavalry Division in the service for their country.

Presidential Unit Citation (Army), Streamer embroidered PLEIKU PROVINCE (Headquarters, Headquarters Company and Band, 1st Cavalry Division Support Command, cited for the period 23 Oct to 26 Nov 1965; DA GO 40, 1967)

Valorous Unit Award, Streamer embroidered FISH HOOK (Headquarters, Headquarters Company and Band, 1st Cavalry Division Support Command, cited for the period 01 May to 29 Jun 1970; DA GO 43, 1972)

Meritorious Unit Commendation, Streamer embroidered VIETNAM 1967 - 1968 (Headquarters, Headquarters Company and Band, 1st Cavalry Division Support Command, cited for the period 01 Jan 1967 to 31 May 1968; DA GO 28, 1969)

Meritorious Unit Commendation, Streamer embroidered VIETNAM 1968 - 1969 (Headquarters, Headquarters Company and Band, 1st Cavalry Division Support Command, cited for the period 01 Jun 1968 to 30 Nov 1969; DA GO 2, 1971)

Meritorious Unit Commendation, Streamer embroidered VIETNAM 1970 - 1971 (Headquarters, Headquarters Company and Band, 1st Cavalry Division Support Command, cited for the period 01 Jan 1970 to 21 Feb 1971; DA GO 18, 1979)

Meritorious Unit Commendation (Army), Streamer embroidered SOUTHWEST ASIA (Headquarters and Headquarters Company, 1st Cavalry Division Support Command, cited for the period 29 Sep 1990 - 06 Apr 1991; DA GO 27, 1994)

Republic of Vietnam Cross of Gallantry with Palm, Streamer embroidered VIETNAM 1965 - 1969 (Headquarters, Headquarters Company and Band, 1st Cavalry Division Support Command, cited for the period 09 Aug 1965 to 19 May 1969; DA GO 59, 1969)

Republic of Vietnam Cross of Gallantry with Palm, Streamer embroidered VIETNAM 1969 - 1970 (Headquarters, Headquarters Company and Band, 1st Cavalry Division Support Command, cited for the period May 1969 to Feb 1970; DA GO 11, 1973, amended DA GO 42, 1972)

Republic of Vietnam Cross of Gallantry with Palm, Streamer embroidered VIETNAM 1970 - 1971 (Headquarters, Headquarters Company and Band, 1st Cavalry Division Support Command, cited for the period 21 Feb 1970 to 28 Feb 1971; DA GO 42, 1972)

Republic of Vietnam Civil Action Honor Medal, First Class, Streamer embroidered VIETNAM 1969 - 1970 (Headquarters, Headquarters Company and Band, 1st Cavalry Division Support Command, cited for the period 01 Jan 1969 to 01 Feb 1970; DA GO 42, 1972)

Appendix F: Campaign Streamers

The concept of campaign streamers came to prominence in the Civil War when Army organizations embroidered the names of battles on their organizational colors. This was discontinued in 1890, when units were authorized to place silver bands, engraved with the names of battles, around the staffs of their organizational colors. When American Expedition Forces Units in World War I were unable to obtain silver bands, General Pershing authorized the use of small ribbons bearing the names of the World War I operations. In 1921 all color-bearing Army organizations were authorized to use the large campaign streamers currently displayed.

Only campaign streamers (2 3/4 inches wide and 4 feet long), embroidered with the designation of a campaign and the year(s) in which it occurred, are displayed on the Army flag. The 174 streamers attached to the Army Flag staff denote campaigns fought by the Army throughout our nation's history. The identification colors of the streamer is identical to the color design of the campaign ribbon authorized for service in that particular theater of war.

Streamers (2 3/4 inches wide and 3 feet in length) are currently authorized for display by organizations authorized a distinguishing flag to indicate the following:

- **Campaign Streamers. - The streamer reflects campaign participation credit and is the campaign or service ribbon design authorized for the soldiers for that specific conflict or operation during the period. Organizations display the streamers on the colors for campaigns or service for which they have received campaign participation credit as shown on the unit's lineage and honors. The inscription is as shown on the unit's lineage and honors/statement of service. Campaign streamers displayed by organizations do not have the date of the campaign embroidered on the streamer unless it is part of the name (e.g. Mexico 1916-1917). The 174 campaign streamers displayed on the Army flag include the date of the action.**

- **War Service Streamers/Battle Streamers. - War service streamers without inscriptions were awarded during or prior to World War II to units that were located in the theater but did not participate in designated campaigns nor specified battles/locations. Prior to 1945, units may also have been awarded streamers for participation in battles other than designated campaigns. The lineage and honors and the inscription on the streamer indicates location and year rather than the designated campaign (e.g. Civil War Service may reflect "GEORGIA 1863" or WWI reflects "LORRAINE 1918" rather than the name of a designated campaign).**

- **Unit Award Streamers. - Unit award streamers indicate the unit has been awarded a unit decoration. Both foreign and US decorations are reflected through the use of unit award streamers. In addition to the regular size unit award streamers for display by organizations authorized a distinguishing flag, unit award streamers (1 3/8 inches wide and 2 feet long) are displayed by separate units authorized a guidon. The inscription, normally location and/or date(s), is as indicated in the unit's lineage and honor.**

Streamers are affixed to the attaching device in a counterclockwise manner. LEXINGTON 1775 is the first streamer and KOSOVO AIR CAMPAIGN will be the 174th streamer (with an additional streamer to be added at a later date). When the United States flag is not being carried, the streamers will be arranged in such a manner that LEXINGTON 1775 is displayed in the preeminent position, i.e., in the center facing forward and completely identifiable. The streamer awarded for the most recent campaign will also be visible.

Separate Table of Organization and Equipment (TOE) units authorized a guidon will indicate campaign participation credit by use of a silver band in lieu of a campaign streamer. The silver band is placed on the guidon staff immediately below the guidon. Campaign participation credit for companies of battalions are reflected by the battalion streamer since honors for those companies accrue to the battalion if the battalion received credit. If the battalion did not receive credit but individual companies received credit, the silver band may be displayed by the company.

Headquarters and Headquarters Company
"The First Team"

The following campaign streamers have been earned by the Headquarters and Headquarters Company, 1st Cavalry Division in service for their country.

World War II, Pacific Theater

New Guinea
Bismarck Archipelago
Leyte (with arrowhead)
Luzon

Korean War

UN Defensive
UN Offensive
CCF Intervention
First UN Counteroffensive
CCF Spring Offensive
UN Summer-Fall Offensive
Second Korean Winter

Vietnam

Defense
Counteroffensive
Counteroffensive, Phase II
Counteroffensive, Phase III
Tet Counteroffensive
Counteroffensive, Phase IV
Counteroffensive, Phase V
Counteroffensive, Phase VI
Tet 69/Counteroffensive
Summer-Fall 1969
Winter-Spring 1970
Sanctuary Counteroffensive
Counteroffensive, Phase VII

SouthWest Asia

Defense of Saudi Arabia
Liberation and Defense of Kuwait

Headquarters and Headquarters Company
1st Brigade, 1st Cavalry Division
"Iron Horse"

The following campaign streamers have been earned by the Headquarters and Headquarters Company, 1st Brigade, 1st Cavalry Division in service for their country.

World War II, Pacific Theater

New Guinea
Bismarck Archipelago (with arrowhead)
Leyte (with arrowhead)
Luzon

Vietnam

Defense
Counteroffensive
Counteroffensive, Phase II
Counteroffensive, Phase III
Tet Counteroffensive
Counteroffensive, Phase IV
Counteroffensive, Phase V
Counteroffensive, Phase VI
Tet 69\Counteroffensive
Summer-Fall 1969
Winter-Spring 1970
Sanctuary Counteroffensive
Counteroffensive, Phase VII

SouthWest Asia

Defense of Saudi Arabia
Liberation and Defense of Kuwait

Headquarters and Headquarters Company 2nd Brigade, 1st Cavalry Division "Blackjack"

The following campaign streamers have been earned by the Headquarters and Headquarters Company, 2nd Brigade, 1st Cavalry Division in service for their country.

World War II, Pacific Theater

New Guinea
Bismarck Archipelago
Leyte W/Arrowhead
Luzon

Vietnam

Defense
CounterOffensive
CounterOffensive, Phase II
CounterOffensive, Phase III
Tet CounterOffensive
CounterOffensive, Phase IV
CounterOffensive, Phase V
CounterOffensive, Phase VI
Tet/69 CounterOffensive
Summer-Fall 1969
Winter-Spring 1970
Sanctuary CounterOffensive
CounterOffensive VII

SouthWest Asia

Defense of Saudi Arabia
Liberation of Kuwait

Headquarters and Headquarters Company 3rd Brigade, 1st Cavalry Division "Graywolf"

The following campaign streamers have been earned by the Headquarters and Headquarters Company, 3rd Brigade, 1st Cavalry Division in service for their country.

World War II, European Theater

Rineland
Ardennes-Alsace
Central Europe

Vietnam

Defense
Counteroffensive
Counteroffensive, Phase II
Counteroffensive, Phase III
Tet Counteroffensive
Counteroffensive, Phase IV
Counteroffensive, Phase V
Counteroffensive, Phase VI
Tet 69\Counteroffensive
Summer-Fall 1969
Winter-Spring 1970
Sanctuary Counteroffensive
Counteroffensive, Phase VII
Consolidation I
Consolidation II
Cease Fire

Headquarters and Headquarters Company Aviation Brigade, 1st Cavalry Division "Warriors"

The following campaign streamers have been earned by the Headquarters and Headquarters Company, Aviation Brigade, 1st Cavalry Division in service for their country.

SouthWest Asia

Defense of Saudi Arabia
Liberation of Kuwait
Cease Fire

Headquarters and Headquarters Battery 1st Cavalry Division Artillery "The Red Team"

The following campaign streamers have been earned by the Headquarters and Headquarters Battery, Division Artillery, 1st Cavalry Division in service for their country.

World War II, Pacific Theater

New Guinea
Bismarck Archipelago
Leyte (W/Arrowhead)
Luzon

Korean War

UN Defensive
UN Offensive
CCF Intervention
First UN CounterOffensive
CCF Spring Offensive
UN Summer-Fall Offensive
Second Korean Winter

Vietnam

Defense
CounterOffensive
CounterOffensive, Phase II
CounterOffensive, Phase III
Tet CounterOffensive
CounterOffensive, Phase IV
CounterOffensive, Phase V
CounterOffensive, Phase VI
Tet/69 CounterOffensive
Summer-Fall 1969
Winter-Spring 1970
Sanctuary CounterOffensive
CounterOffensive VII

SouthWest Asia

Defense of Saudi Arabia
Liberation and Defense of Kuwait

Headquarters and Headquarters Detachment Engineer Brigade, 1st Cavalry Division "Sapper Forward"

The following campaign streamers have been earned by the Headquarters and Headquarters Detachment, Engineer Brigade, 1st Cavalry Division in service for their country.

None

Headquarters and Headquarters Company 1st Cavalry Division Support Command "Wagon Masters"

The following campaign streamers have been earned by the Headquarters and Headquarters Company, Support Command, 1st Cavalry Division in service for their country.

Vietnam

Defense
Counteroffensive
Counteroffensive, Phase II
Counteroffensive, Phase III
Tet Counteroffensive
Counteroffensive, Phase IV
Counteroffensive, Phase V
Counteroffensive, Phase VI
Tet 69/Counteroffensive
Summer-Fall 1969
Winter-Spring 1970
Sanctuary Counteroffensive
Counteroffensive, Phase VII

SouthWest Asia

Defense of Saudi Arabia
Liberation and Defense of Kuwait

Appendix G: Heraldic Items

"The First Team"

The insignia selected for the First Team patch was designed by Colonel and Mrs. Ben Dorcy. The Colonel was then commander of the 7th Cavalry Regiment at Fort Bliss, Texas. Mrs. Dorcy related that the combination of the golden sunset at Fort Bliss and the traditional colors of the Cavalry; blue and yellow, were a great influence on the background color and the insignia. The choice of the horse's head for the insignia was made by the family after they observed a mounted trooper ride by their home on a beautiful blue-black thoroughbred. Later, to improve visibility, the color scheme was modified replacing the blue with black, the symbolic color of iron and armor.

On a "sunset" yellow triangular Norman Shield with rounded corners 5 1/4 inches in height, a black diagonal stripe extends over the shield from upper left to the lower right. In the upper right, a head of a black horse cut off diagonally at the neck, appears within one-eighth inch of the Army Green border of piping. The traditional Cavalry color of yellow and the head of a horse is symbolic of the original organizational structure of the Cavalry. The color black is symbolic of iron, alluding to the organizational transition from mounted horses to tanks and heavy armor. The black stripe, in heraldry, termed a Sable Bend, represents a Baldric (a standard Army issue belt worn over the right shoulder to the opposite hip - sometimes referred to as a Sam Browne belt) which retains either a scabbard which sheaths the trooper's saber or revolver holster.

During the Vietnam engagements, the sunset yellow field of the patch was changed to a subdued Olive Drab (OD) green for the Battle Dress Uniform (BDU) in order to minimize targeting of personnel. For operations in desert environments, the field of the patch was again changed to a tan (Khaki) color and the emblazoned, Sable black elements were changed to saddle brown (Spice) so that the contrast against the Desert Battle Dress Uniform (DBDU) was minimized. The sunset yellow/black patch is retained for Class "A" dress uniform. Otherwise the patch has not changed from the original design and shape.

HQ & HQ Company

The design of the insignia of Headquarters and Headquarters Company, 1st Cavalry Division, as well as all the non color bearing direct subordinate units in the division, centers on a gold metal and enamel device one inch in height consisting of a gold Norman shield emblazoned with a sable (black) horse's head couped in sinister chief and a sable bend (diagonal stripe) charged with two gold five-pointed stars.

1st Cavalry Division Patch

The patch of the 1st Cavalry Division has a history as colorful as its design, reflecting the proud heritage of the United States Cavalry in a timeless manner.

Appendix H: Order of Battle

"As one of the Army's two on-call heavy contingency force divisions, the First Team has an on-order mission to deploy to a designated contingency area of operations by sea, air or land, conduct reception, staging, onward movement and integration; and on order, conduct combat operations, and redeployment."

The roots of the 1st Cavalry Division go deep and wide, protecting the basic rights established by the founders of this great nation. They span the United States and extend to theaters around the world where the Division fought in the Second World War, the Korean War, the Vietnam War and the Persian Gulf War. In these conflicts the Division protected the interests of the United States and defended other countries from aggression. The recorded exploits of the 1st Cavalry Division are rich with individual and unit heroics of attack, defense, and daring rescue. The Division, with superb leadership, employed devastating firepower to overcome the aggressors. In newer roles, the First Team has served as a model of an international force in maintaining peace between waring factions in Southwest Asian countries and Bosnia. The theaters of operations for the numerous worldwide deployments of the 1st Cavalry Division are highlighted below:

The evolution of the capabilities and units comprising the 1st Cavalry Division has been brought about by an orderly transition of continuous training and education of the intelligent, physically fit, highly motivated trooper who is the backbone of the division. This training takes advantage of the full power of the emerging, advanced state-of-the-art simulations and training devices.

Backing up the human element, which provides the spirit and dedicated enthusiasm for carrying out a mission, ever changing equipment technologies and services capabilities have been tailored, tested and provided to match the threats of our nation.

The following data identifies and summarizes the ever changing command structure and subordinate units that have been assigned or attached to the 1st Cavalry Division and have participated, beginning with its organization and activation in 1921 to the present day, in their many missions of defending our nation, achieving victory in war and in peace and providing humanitarian support anywhere in the world. In order to present a pictorial of the time representation showing continuity of the command of individual units, along with their designations (and redesignations), the data is listed, matching the periods of actions described in the prior history chapters.

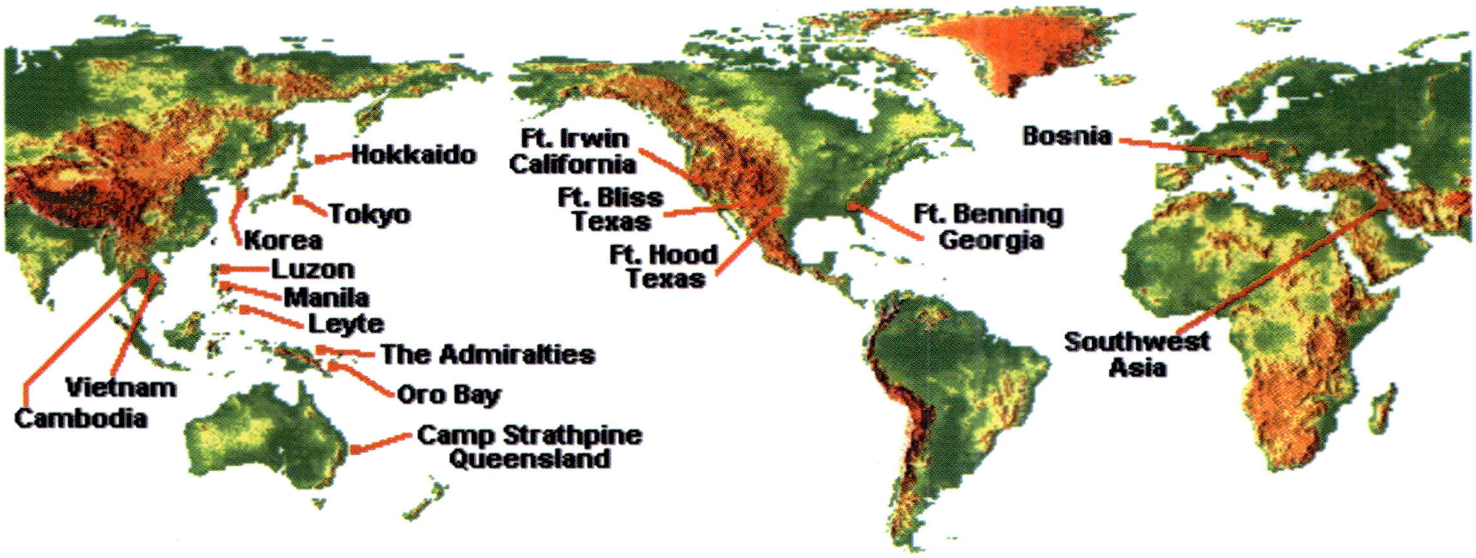

THE EARLY YEARS
13 September 1921 - 06 December 1941

EARLY YEARS TIME LINE	13 Sep 1921 '22 '23 '24 '25 '26 '27 '28 '29 '30 '31 '32 '33 '34 '35 '36 '37 '38 '39 '40 '41
Headquarters, 1st Cavalry Division	13 Sep 1921 — — — '41~ (Con't next era)
1st BRIGADE, 1st Cavalry Division	
"HHT"	20 Aug 1921 — — — '41~ (Con't next era)
1st Machine Gun Squadron	'21 — — — '28; Dsbnd'd 01 Feb
1st Cavalry Regiment	20 Aug 1921 — — — '33; Relv'd 03 Jan from 1st Cav Div
"HHT"	'21 — — — '33; Relv'd 03 Jan from 1st Cav Div
Machine Gun Squadron	01 Dec '28 — — — '33; Relv'd 03 Jan from 1st Cav Div
1st Squadron, 1st Cavalry Regt	'21 — — — '33; Relv'd 03 Jan from 1st Cav Div
"A" Troop	'21 — — — '33; Relv'd 03 Jan from 1st Cav Div
"B" Troop	'21 — — — '33; Relv'd 03 Jan from 1st Cav Div
"C" Troop	'21 — — — '28; Dsbnd'd 01 Feb
2nd Squadron, 1st Cavalry Regt	'21 — — — '33; Relv'd 03 Jan from 1st Cav Div
"D" Troop	'21 — — — '28; Dsbnd'd 01 Feb
"E" Troop	'21 — — — '33; Relv'd 03 Jan from 1st Cav Div
"F" Troop	'21 — — — '33; Relv'd 03 Jan from 1st Cav Div
5th Cavalry Regiment	18 Dec '22 — — — '41~ (Con't next era)
"HHT"	'22 — — — '41~ (Con't next era)
Machine Gun Squadron	01 Dec '28 — — — '41~ (Con't next era)
1st Squadron, 5th Cavalry Regt	'22 — — — '41~ (Con't next era)
"A" Troop	'22 — — — '41~ (Con't next era)
"B" Troop	'22 — — — '41~ (Con't next era)
"C" Troop	'22 — — — '28; Dsbnd'd 01 Feb Aug '40/'41~ (Con't next era)
2nd Squadron, 5th Cavalry Regt	'22 — — — '41~ (Con't next era)
"E" Troop	'22 — — — '41~ (Con't next era)
"F" Troop	'22 — — — '41~ (Con't next era)
"G" Troop	01 Aug '40/'41~ (Con't next era)

EARLY YEARS TIME LINE	13 Sep 1921 '22 '23 '24 '25 '26 '27 '28 '29 '30 '31 '32 '33 '34 '35 '36 '37 '38 '39 '40 '41
10th Cavalry Regiment	13 Sep '21/'22; Relv'd 18 Dec from 1st Cav Div
"HHT"	'21/'22; Relv'd 18 Dec from 1st Cav Div
1st Squadron, 10th Cavalry Regt	'21/'22; Relv'd 18 Dec from 1st Cav Div
"A" Troop	'21/'22; Relv'd 18 Dec from 1st Cav Div
"B" Troop	'21/'22; Relv'd 18 Dec from 1st Cav Div
"C" Troop	'21/'22; Relv'd 18 Dec from 1st Cav Div
2nd Squadron, 10th Cavalry Regt	'21/'22; Relv'd 18 Dec from 1st Cav Div
"D" Troop	'21/'22; Relv'd 18 Dec from 1st Cav Div
"E" Troop	'21/'22; Relv'd 18 Dec from 1st Cav Div
"F" Troop	'21/'22; Relv'd 18 Dec from 1st Cav Div
12th Cavalry Regiment	03 Jan '33 — '41~ (Con't next era)
"HHT"	'33 — '41~ (Con't next era)
Machine Gun Troop	'33 — '41~ (Con't next era)
1st Squadron, 12th Cavalry Regt	'33 — '41~ (Con't next era)
"A" Troop	'33 — '41~ (Con't next era)
"B" Troop	'33 — '41~ (Con't next era)
"C" Troop	01 Aug '40/'41~ (Con't next era)
2nd Squadron, 12th Cavalry Regt	'33 — '41~ (Con't next era)
"D" Troop	01 Aug '40/'41~ (Con't next era)
"E" Troop	'33 — '41~ (Con't next era)
"F" Troop	'33 — '41~ (Con't next era)
"G" Troop	01 Aug '40/'41~ (Con't next era)
2nd BRIGADE, 1st Cavalry Division	
"HHT"	20 Aug '21 — '41~ (Con't next era)
2nd Machine Gun Squadron	'21 — '28; Dsbnd'd 01 Feb
7th Cavalry Regiment	13 Sep '21 — '41~ (Con't next era)
"HHT"	'21 — '41~ (Con't next era)
Machine Gun Squadron	01 Dec '28 — '41~ (Con't next era)
1st Squadron, 7th Cavalry Regt	'21 — '41~ (Con't next era)
"A" Troop	'21 — '41 (Con't next era)
"B" Troop	'21 — '41~ (Con't next era)
"C" Troop	'21 — '28; Dsbnd'd 01 Feb; Aug '40/'41~ (Con't next era)
2nd Squadron, 7th Cavalry Regt	'21 — '41~ (Con't next era)
"E" Troop	'21 — '41~ (Con't next era)
"F" Troop	'21 — '41~ (Con't next era)
"G" Troop	'21 — '28; Dsbnd'd 31 Jan; Aug '40/'41~ (Con't next era)

EARLY YEARS TIME LINE	13 Sep 1921 '22 '23 '24 '25 '26 '27 '28 '29 '30 '31 '32 '33 '34 '35 '36 '37 '38 '39 '40 '41
8th Cavalry Regiment	13 Sep '21 — — — '41~ (Con't next era)
"HHT"	'21 — — — '41~ (Con't next era)
Machine Gun Squadron	01 Dec '28 — — — '41~ (Con't next era)
1st Squadron, 8th Cavalry Regt	'21 — — — '41~ (Con't next era)
"A" Troop	'21 — — — '41 (Con't next era)
"B" Troop	'21 — — — '41~ (Con't next era)
"C" Troop	'21 — — — '28; Dsbnd'd 01 Feb Aug '40/'41~ (Con't next era)
2nd Squadron, 8th Cavalry Regt	'21 — — — '41~ (Con't next era)
"E" Troop	'21 — — — '41~ (Con't next era)
"F" Troop	'21 — — — '41~ (Con't next era)
"G" Troop	01 Aug '40/'41~ (Con't next era)
SPECIAL TROOPS, 1st Cavalry Division	
"HHT"	13 Sep '21 — — — '41~ (Con't next era)
10th Light Tank Company	'21 — — — '28; Inactv'd
13th Signal Troop	13 Sep '21 —— '23~ Reorg'd & Dsgn'd 13 Dec as 1st Signal Trp, below
1st Signal Troop	13 Dec '23 '41~ (Con't next era)
15th Veterinary Company	'21 — — — '26;
27th Ordnance Company	20 Sep '21 — — — '41~ (Con't next era)
1st Cavalry Division Air Service	
1st Observation Squadron, U.S. Air Corps	Oct '27 —— '29; Feb
1st Armored Car Squadron	Nov '28 — — — '41~ Redsgn'd 91st Recon Sqdn, below
91st Reconnaissance Squadron	03 Jan '41~ (Con't next era)
1st Antitank Troop	03 Jan '41~ (Con't next era)
43rd Ambulance Company	'21 — — — '26~
1st Medical Squadron	01 Jun '26 — — — '41~ (Con't next era)
Military Police Plt, Hdqtrs Troop	15 Feb '39 —— '41~ (Con't next era)

EARLY YEARS TIME LINE	13 Sep 1921 '22 '23 '24 '25 '26 '27 '28 '29 '30 '31 '32 '33 '34 '35 '36 '37 '38 '39 '40 '41
82nd Field Artillery Battalion	09 Sep '21 — '30~ Dsgn'd 17 Mar as 1st Bn, below
"A" Btry, 82nd Field Arty Bn	'21 — '30~ Dsgn'd 17 Mar as "A" Co, 1st Bn, below
"B" Btry, 82nd Field Arty Bn	'21 — '30~ Dsgn'd 17 Mar as "B" Co, 1st Bn, below
"C" Btry, 82nd Field Arty Bn	'21 — '30~ Dsgn'd 17 Mar as "C" Co, 1st Bn, below
1st Battalion, 82nd Field Artillery	17 Mar '30 — '41~
"A" Btry, 82nd Field Arty Bn	17 Mar '30 — '41~
"B" Btry, 82nd Field Arty Bn	17 Mar '30 — '41~
"C" Btry, 82nd Field Arty Bn	17 Mar '30 — '34~ Inactv'd 01 Dec at Ft. Bliss, TX
"D" Btry, 82nd Field Arty Bn	17 Mar '30 — '41~
"E" Btry, 82nd Field Arty Bn	17 Mar '30 — '41~
"F" Btry, 82nd Field Arty Bn	17 Mar '30 — '41~
2nd Battalion, 82nd Field Artillery	01 Dec '34 — '41~
DIVISION ARTILLERY, 1st Cavalry Division	
"HHB"	03 Jan '41~ (Con't next era)
1st Battalion, 82nd Field Artillery	03 Jan '41~ (Con't next era)
"A" Btry, 1st Bn, 82nd Field Arty	03 Jan '41~ (Con't next era)
"B" Btry, 1st Bn, 82nd Field Arty	03 Jan '41~ (Con't next era)
"C" Btry, 1st Bn, 82nd Field Arty	03 Jan '41~ (Con't next era)
"D" Btry, 1st Bn, 82nd Field Arty	03 Jan '41~ (Absrb'd by "A" Btry)
"E" Btry, 1st Bn, 82nd Field Arty	03 Jan '41~ (Absrb'd by "B" Btry)
"F" Btry, 1st Bn, 82nd Field Arty	03 Jan '41~ (Con't next era)
3rd Battalion, 82nd Field Artillery	03 Jan '41~ (Con't next era)
61st Field Artillery Battalion	03 Jan '41~ (Con't next era)
62nd Field Artillery Battalion	10 Feb '41~ (Con't next era)
8th Engineer Battalion (Mounted)	27 Jul '21 — '30~ Reorg'd & Dsgn'd 01 Jan as 8th Engr Sqdn, below
8th Engineer Squadron	01 Jun '30 — '41~ (Con't next era)
TRAINS, 1st Cavalry Division	20 Sep '21 — '36~ Reorg'd & Dsgn'd 01 May as 16th QtrMstr Sqdn, below
"HHT"	20 Sep '21 — '36~ Asgn'd 01 May to 16th QtrMstr Sqdn, below
1st Pack Train	20 Sep '21 — '36;
2nd Pack Train	20 Sep '21 — '36;
3rd Pack Train	20 Scp '21 — '36;
4th Pack Train	20 Sep '21 — '36;
25th Wagon Company	20 Sep '21 — '28;
26th Wagon Company	20 Sep '21 — '36;
49th Motor Transport Company	'28 — '36;
81st Motor Repair Section	'28 — '36;
250th Motor Transport Company	(Inactive)

EARLY YEARS TIME LINE	13 Sep 1921 '22 '23 '24 '25 '26 '27 '28 '29 '30 '31 '32 '33 '34 '35 '36 '37 '38 '39 '40 '41
16th Quartermaster Squadron	01 May '36 — — — — — '41~ (Con't next era)
"HHT"	01 May '36 — — — — — '41~ (Con't next era)
ATTACHED UNITS	
Cavalry	
56th Cavalry Brigade (TXNG)	
"HHT"	18 Nov '40/'41~ (Con't next era)
111th Cavalry Regiment (New Mexico)	18 Nov '40/'41;
112th Cavalry Regiment (North Texas)	18 Nov '40/'41~ (Con't next era)
124th Cavalry Regiment (South Texas)	18 Nov '40/'41~ (Con't next era)

WORLD WAR II
07 December 1941 - 02 September 1945

WORLD WAR II TIME LINE	07 Dec 1941 '42 '43 '44 '45
"HQ"s, 1st Cavalry Division	07 Dec '41 ——'43~ Reorg'd & Dsgn'd 04 Dec "HHT", below
"HHT", 1st Cavalry Division, (Special)	04 Dec '43 ——'45~ (Con't next era)
1st BRIGADE, 1st Cavalry Division	
"HHT"	07 Dec '41 — — — —'45~ (Con't next era)
Weapons Trp, 1st Cavalry Brigade	07 Dec '41 ——'43;
5th Cavalry Regiment	07 Dec '41 — — — —'45~ (Con't next era)
1st Squadron, 5th Cavalry Regt	
"A" Troop	07 Dec '41 — — — —'45~ (Con't next era)
"B" Troop	07 Dec '41 — — — —'45~ (Con't next era)
"C" Troop	07 Dec '41 — — — —'45~ (Con't next era)
"D" Troop	04 Dec '43——'45~ (Con't next era)
2nd Squadron, 5th Cavalry Regt	
"E" Troop	07 Dec '41 — — — —'45~ (Con't next era)
"F" Troop	07 Dec '41 — — — —'45~ (Con't next era)
"G" Troop	07 Dec '41 — — — —'45~ (Con't next era)
"H" Troop	04 Dec '43 ——'45~ (Con't next era)
12th Cavalry Regiment	07 Dec '41 — — — —'45~ (Con't next era)
1st Squadron, 12th Cavalry Regt	
"A" Troop	07 Dec '41 — — — —'45~ (Con't next era)
"B" Troop	07 Dec '41 — — — —'45~ (Con't next era)
"C" Troop	07 Dec '41 — — — —'45~ (Con't next era)
"D" Troop	04 Dec '43——'45~ (Con't next era)

WORLD WAR II TIME LINE	07 Dec 1941 '42 '43 '44 '45
2nd Squadron, 12th Cavalry Regt	
"E" Troop	07 Dec '41 — — — —'45~ (Con't next era)
"F" Troop	07 Dec '41 — — — —'45~ (Con't next era)
"G" Troop	07 Dec '41 — — — —'45~ (Con't next era)
"H" Troop	04 Dec '43 ——'45~ (Con't next era)
2nd BRIGADE, 1st Cavalry Division	
"HHT"	07 Dec '41 — — — —'45~ (Con't next era)
Weapons Trp, 2nd Cavalry Brigade	07 Dec '41 ——'43;
7th Cavalry Regiment	07 Dec '41 — — — —'45~ (Con't next era)
1st Squadron, 7th Cavalry Regt	
"A" Troop	07 Dec '41 — — — —'45~ (Con't next era)
"B" Troop	07 Dec '41 — — — —'45~ (Con't next era)
"C" Troop	07 Dec '41 — — — —'45~ (Con't next era)
"D" Troop	04 Dec '43 ——'45~ (Con't next era)
2nd Squadron, 7th Cavalry Regt	
"E" Troop	07 Dec '41 — — — —'45~ (Con't next era)
"F" Troop	07 Dec '41 — — — —'45~ (Con't next era)
"G" Troop	07 Dec '41 — — — —'45~ (Con't next era)
"H" Troop	04 Dec '43 ——'45~ (Con't next era)
8th Cavalry Regiment	07 Dec '41 — — — —'45~ (Con't next era)
1st Squadron, 8th Cavalry Regt	
"A" Troop	07 Dec '41 — — — —'45~ (Con't next era)
"B" Troop	07 Dec '41 — — — —'45~ (Con't next era)
"C" Troop	07 Dec '41 — — — —'45~ (Con't next era)
"D" Troop	04 Dec '43 ——'45~ (Con't next era)
2nd Squadron, 8th Cavalry Regt	
"E" Troop	07 Dec '41 — — — —'45~ (Con't next era)
"F" Troop	07 Dec '41 — — — —'45~ (Con't next era)
"G" Troop	07 Dec '41 — — — —'45~ (Con't next era)
"H" Troop	04 Dec '43 ——'45~ (Con't next era)
SPECIAL TROOPS, 1st Cavalry Division	
"HHT"	07 Dec '41 — — — —'45~ (Con't next era)
1st Signal Troop	07 Dec '41 — — — —'45~ (Con't next era)
27th Ordnance Company	07 Dec '41 ——'43~ Reorg'd & Dsgn'd 02 Mar as 27th Ord Med Maint Co, below
27th Ordnance Medium Maintenance Co	02 Mar '43 ——'45~ (Con't next era)
1st Armored Car Squadron	07 Dec '41~ Dsgn'd 91st Recon Sqdn, below
91st Reconnaissance Squadron	'41 ——'43; Transferred to North Africa Theatre
7th Reconnaissance Squadron	16 Jan '42/'43~ Reorgn'd & Dsgn'd 04 Dec as 302nd Recon Trp & 603rd Med Tank Bn, below
302nd Reconnaissance Troop	04 Dec '43/'44~ Dsgn'd 30 Oct as 302nd Recon Troop, Mechanized, below
302nd Reconnaissance Troop, Mechanized	30 Oct '44/'45~ (Con't next era)
3rd Reconnaissance Squadron	'42/'43

WORLD WAR II TIME LINE	07 Dec 1941 '42 '43 '44 '45
1st Antitank Troop	07 Dec '41 —— '43
603rd Light Tank Company	04 Dec '43 —— '45~ Reorgn'd 09 Mar as 603rd Med Tank Co, below
603rd Medium Tank Company	09 Mar '45~ (Con't next era)
1st Medical Squadron	07 Dec '41 — — — — '45~ (Con't next era)
168th Language Detachment	31 Jan '45~ (Con't next era)
Mltry Police Plt, Hdqtrs Troop	07 Dec '41 /'42~ Dsgn'd 04 Jul as Mltry Police Plt, 1st Cav Div, below
Mltry Police Plt, 1st Cav Div'	04 Jul '42/'43~ Dsgn'd Mltry Police Plt, 1st Cav Div (Spc'l), below
Mltry Police Plt, 1st Cav Div (Spc'l)	04 Dec '43 —— '45~ (Con't next era)
801st Counter Intelligence Corps Detachment	'44
DIVISION ARTILLERY, 1st Cavalry Division	
"HHB"	07 Dec '41 — — — — '45~ (Con't next era)
1st Battalion, 82nd Field Artillery	07 Dec '41 — — — — '45~ (Con't next era)
"A" Btry, 82nd Field Arty	07 Dec '41 — — — — '45~ (Con't next era)
"B" Btry, 82nd Field Arty	07 Dec '41 — — — — '45~ (Con't next era)
"C" Btry, 82nd Field Arty	07 Dec '41 — — — — '45~ (Con't next era)
61st Field Artillery Battalion	07 Dec '41 — — — — '45~ (Con't next era)
62nd Field Artillery Battalion	07 Dec '41 /'42~
99th Field Artillery Battalion	02 Mar '43 —— '45~ (Con't next era)
153rd Field Artillery Battalion	15 Nov '42/'43; Relv'd 15 Mar from 1st Cav Div
271st Field Artillery Battalion	11 Oct '43 —— '45~ (Con't next era)
8th Engineer Squadron	07 Dec '41 —— '43~ Reorg'd & Dsgn'd 15 Mar as 8th Engr Combat Sqdn, below
8th Engineer Combat Squadron	15 Mar '43 —— '45 (Con't next era)
161st Engineer Squadron	15 Nov '42/'43; Inactv'd 21 Apr, Assets & Personnel absrb'd by 8th CES (above)
Band, 1st Cavalry Division	03 Jun '45~ (Con't next era)
16th Quartermaster Squadron	07 Dec '41 —— '43~ Dsgn'd 15 Mar as 16th Cav QtrMstr Sqdn, below
"HHT"	07 Dec '41 —— '43~ Reorgn'd 15 Mar, below

WORLD WAR II TIME LINE	07 Dec 1941 ’42 ’43 ’44 ’45
16th Cavalry Quartermaster Squadron	15 Mar ’43 ——’45~ (Con’t next era)
“HHT”	15 Mar ’43 ——’45~ (Con’t next era)
Trp “A”, 16th Quartermaster Sqdn	15 Mar ’43 ——’45~ (Con’t next era)
Trp “B”, 16th Quartermaster Sqdn	15 Mar ’43 ——’45~ (Con’t next era)
Trp “C”, 16th Quartermaster Sqdn	15 Mar ’43 ——’45~ (Con’t next era)
ATTACHED UNITS	
Cavalry	
56th Cavalry Brigade (TXNG)	
“HHT”	07 Dec ’41— — —’44~ Dsgn’d as 56th Recon Trp Mech, below
112th Cavalry Regt (North Texas)	07 Dec ’41 /’42~ Relv’d 08 Jul & Asgn’d to the Americal Div
112th Regimental Combat Team	14 Nov ’44/’45~ Relv’d 31 Jan, See below
112th Cavalry Regiment (Special)	14 Nov ’44/’45~ Relv’d 31 Jan, See below
148th FA Battalion (105mm Howitzer)	14 Nov ’44/’45~ Relv’d 31 Jan, See below
112th Regimental Combat Team	09 Feb ’45; Relv’d 19 Feb
112th Cavalry Regiment (Special)	09 Feb ’45; Relv’d 19 Feb
148th FA Battalion (105mm Howitzer)	09 Feb ’45; Relv’d 19 Feb
124th Cavalry Regt (South Texas)	07 Dec ’41— — —’44~ Relv’d 16 May & Asgn’d to Ft. Riley, KS
56th Reconnaissance Troop, Mechanized	’44
Infantry	
1st Battalion, 19th Infantry Regiment	09 Nov ’44; Relv’d 15 Nov
103rd Regimental Combat Team (43rd Div)	10 Mar ’45; Relv’d 12 Mar
Armor	
44th Tank Battalion	
Co “B”, 44th Tank Bn	Oct ’44/’45~ (Con’t next era)
Co “C”, 44th Tank Bn	Oct ’44/’45~ (Con’t next era)
Co “D”, 44th Tank Bn	Oct ’44/’45~ (Con’t next era)
754th Tank Battalion	Feb ’45; Relv’d Mar
Artillery	
“C” Btry, 168th A/A Arty Battalion (Gun)	29 Feb ’44; Relv’d 18 May
“A” Btry, 211th A/A Arty Battalion (AW)	29 Feb ’44; Relv’d 18 May
974th Field Arty Bn (155 Howitzer)	
Support Units	
85th Chemical Mortar Bn	
30th Portable Surgical Hospital	29 Feb ’44; Relv’d 18 May

WORLD WAR II TIME LINE	07 Dec 1941 ’42 ’43 ’44 ’45
Navy	
40th Naval Construction Battalion (SeaBees)	
Marine Air Group (MAG) #24, 1st Marine Air Wing	
Mobile Explosive Investigation Unit #1, US Navy	
Co “E”, Shore Bn, 592nd Engineer Boat and Shore Regt	
Naval Gunfire Support Party and Air Liaison Party	
Air Force	
Air Force Supervisory Detachment	
“C” Co, 583rd Signal Aircraft Warning Bn	29 Feb ’44; Relv’d 4 Mar
Allied Units	
Australian New Guinea Administration Group	29 Feb ’44; Relv’d 18 May

OCCUPATION OF JAPAN
03 September 1945 - 18 July 1950

OCCUPATION OF JAPAN TIME LINE	03 Sep 1945 ’46 ’47 ’48 ’49 ’50
“HHT”, 1st Cavalry Division, (Spc’l)	03 Sep ’45 — — — —’49~ Reorg’d & Dsgn’d 25 Mar as “HHC”, below
“HHC”, 1st Cavalry Division	25 Mar ’49/’50~ (Con’t next era)
1st BRIGADE, 1st Cavalry Division	
“HHT”	03 Sep ’45 — — — —’49~ Inactv’d 25 Mar in Japan
5th Cavalry Regiment	03 Sep ’45 — — — —’49~ Reorgn’d 25 Mar with Trps Dsgn’d as Cos, below
1st Squadron, 5th Cavalry Regt	
“A” Troop	03 Sep ’45 — — — —’49 Reorgn’d 25 Mar as Company, below
“B” Troop	03 Sep ’45 — — — —’49 Reorgn’d 25 Mar as Company, below
“C” Troop	03 Sep ’45 — — — —’49 Reorgn’d 25 Mar as Company, below
“D” Troop	03 Sep ’45 — — — —’49 Reorgn’d 25 Mar as Company, below
2nd Squadron, 5th Cavalry Regt	
“E” Troop	03 Sep ’45 — — — —’49 Reorgn’d 25 Mar as Company, below
“F” Troop	03 Sep ’45 — — — —’49 Reorgn’d 25 Mar as Company, below
“G” Troop	03 Sep ’45 — — — —’49 Reorgn’d 25 Mar as Company, below
“H” Troop	03 Sep ’45 — — — —’49 Reorgn’d 25 Mar as Company, below
5th Cavalry Regiment (Inf)	25 Mar ’49/’50~ (Con’t next era)
1st Battalion, 5th Cavalry Regt	
“A” Company	25 Mar ’49/’50~ (Con’t next era)
“B” Company	25 Mar ’49/’50~ (Con’t next era)
“C” Company	25 Mar ’49/’50~ (Con’t next era)
“D” Company	25 Mar ’49/’50~ (Con’t next era)

OCCUPATION OF JAPAN TIME LINE	03 Sep 1945 ’46 ’47 ’48 ’49 ’50
2nd Battalion, 5th Cavalry Regt	
“E” Company	25 Mar ’49/’50~ (Con’t next era)
“F” Company	25 Mar ’49/’50~ (Con’t next era)
“G” Company	25 Mar ’49/’50~ (Con’t next era)
“H” Company	25 Mar ’49/’50~ (Con’t next era)
12th Cavalry Regiment	03 Sep ’45 — — — —’49; Relv’d 25 Mar in Japan
1st Squadron, 12th Cavalry Regt	
“A” Troop	03 Sep ’45 — — — —’49; Relv’d 25 Mar in Japan
“B” Troop	03 Sep ’45 — — — —’49; Relv’d 25 Mar in Japan
“C” Troop	03 Sep ’45 — — — —’49; Relv’d 25 Mar in Japan
“D” Troop	03 Sep ’45 — — — —’49; Relv’d 25 Mar in Japan
2nd Squadron, 12th Cavalry Regt	
“E” Troop	03 Sep ’45 — — — —’49; Relv’d 25 Mar in Japan
“F” Troop	03 Sep ’45 — — — —’49; Relv’d 25 Mar in Japan
“G” Troop	03 Sep ’45 — — — —’49; Relv’d 25 Mar in Japan
“H” Troop	03 Sep ’45 — — — —’49; Relv’d 25 Mar in Japan
2nd BRIGADE, 1st Cavalry Division	
“HHT”	03 Sep ’45 — — — —’49~ Inactv’d 25 Mar in Japan
7th Cavalry Regiment	03 Sep ’45 — — — —’49~ Reorgn’d 25 Mar with Trps Dsgn’d as Cos, below
1st Squadron, 7th Cavalry Regt	
“A” Troop	03 Sep ’45 — — — —’49 Reorgn’d 25 Mar as Company, below
“B” Troop	03 Sep ’45 — — — —’49 Reorgn’d 25 Mar as Company, below
“C” Troop	03 Sep ’45 — — — —’49 Reorgn’d 25 Mar as Company, below
“D” Troop	03 Sep ’45 — — — —’49 Reorgn’d 25 Mar as Company, below
2nd Squadron, 7th Cavalry Regt	
“E” Troop	03 Sep ’45 — — — —’49 Reorgn’d 25 Mar as Company, below
“F” Troop	03 Sep ’45 — — — —’49 Reorgn’d 25 Mar as Company, below
“G” Troop	03 Sep ’45 — — — —’49 Reorgn’d 25 Mar as Company, below
“H” Troop	03 Sep ’45 — — — —’49 Reorgn’d 25 Mar as Company, below
7th Cavalry Regiment (Inf)	25 Mar ’49/’50~ (Con’t next era)
1st Battalion, 7th Cavalry Regt	
“A” Company	25 Mar ’49/’50~ (Con’t next era)
“B” Company	25 Mar ’49/’50~ (Con’t next era)
“C” Company	25 Mar ’49/’50~ (Con’t next era)
“D” Company	25 Mar ’49/’50~ (Con’t next era)
2nd Battalion, 7th Cavalry Regt	
“E” Company	25 Mar ’49/’50~ (Con’t next era)
“F” Company	25 Mar ’49/’50~ (Con’t next era)
“G” Company	25 Mar ’49/’50~ (Con’t next era)
“H” Company	25 Mar ’49/’50~ (Con’t next era)
8th Cavalry Regiment	03 Sep ’45 — — — —’49~ Reorgn’d 25 Mar with Trps Dsgn’d as Cos, below
1st Squadron, 8th Cavalry Regt	
“A” Troop	03 Sep ’45 — — — —’49 Reorgn’d 25 Mar as Company, below
“B” Troop	03 Sep ’45 — — — —’49 Reorgn’d 25 Mar as Company, below
“C” Troop	03 Sep ’45 — — — —’49 Reorgn’d 25 Mar as Company, below
“D” Troop	03 Sep ’45 — — — —’49 Reorgn’d 25 Mar as Company, below

OCCUPATION OF JAPAN TIME LINE	03 Sep 1945 '46 '47 '48 '49 '50
2nd Squadron, 8th Cavalry Regt	
"E" Troop	03 Sep '45 — — — —'49 Reorgn'd 25 Mar as Company, below
"F" Troop	03 Sep '45 — — — —'49 Reorgn'd 25 Mar as Company, below
"G" Troop	03 Sep '45 — — — —'49 Reorgn'd 25 Mar as Company, below
"H" Troop	03 Sep '45 — — — —'49 Reorgn'd 25 Mar as Company, below
8th Cavalry Regiment (Inf)	25 Mar '49/'50~ (Con't next era)
1st Battalion, 8th Cavalry Regt	
"A" Company	25 Mar '49/'50~ (Con't next era)
"B" Company	25 Mar '49/'50~ (Con't next era)
"C" Company	25 Mar '49/'50~ (Con't next era)
"D" Company	25 Mar '49/'50~ (Con't next era)
2nd Battalion, 8th Cavalry Regt	
"E" Company	25 Mar '49/'50~ (Con't next era)
"F" Company	25 Mar '49/'50~ (Con't next era)
"G" Company	25 Mar '49/'50~ (Con't next era)
"H" Company	25 Mar '49/'50~ (Con't next era)
1st Cavalry Division Special Troops	
"HHT"	03 Sep '45 — — — —'49;
1st Signal Troop	03 Sep '45 — — — —'49~ Reorg'd & Dsgn'd 25 Mar as 13th Signal Co,, below
13th Signal Company	25 Mar '49/'50~ (Con't next era)
27th Ordnance Medium Maint Company	03 Sep '45 — — — —'49~ Dsgn'd as 27th Ord Maint Bn, below
27th Ordnance Maintenance Company	25 Mar '49/'50~ (Con't next era)
168th Language Detachment	03 Sep '45 — — —'48
302nd Mechanized Cavalry Recon Troop	03 Sep '45 — — — —'49~ Reorg'd & Dsgn'd 25 Mar as 16th Recon Co,, below
16th Reconnaissance Company	25 Mar '49/'50~ (Con't next era)
603rd Medium Tank Company	03 Sep '45 — — —'48; Relv'd 18 Mar from 1st Cav Div at Camp Drake
71st Heavy Tank Battalion	25 Mar '49/'50~ (Con't next era)
95th Light Tank Company	30 Sep '47~
1st Medical Squadron	03 Sep '45 — — — —'49~ Reorg'd & Dsgn'd 25 Mar as 15th Medical Bn, below
Medical Det, 1st Cavalry Division	'49/'50~ (Con't next era)
15th Medical Battalion	25 Mar '49/'50~ (Con't next era)
DIVISION ARTILLERY, 1st Cavalry Division	
"HHB", Division Artillery	03 Sep '45 — — — —'49~ Reorg'd 25 Mar, below
"HHB", Division Artillery	25 Mar '49/'50~ (Con't next era)
Medical Detachment	'49/'50~ (Con't next era)

OCCUPATION OF JAPAN TIME LINE	03 Sep 1945 '46 '47 '48 '49 '50
82nd Field Artillery Battalion	03 Sep '45 — — — — —'50~ (Con't next era)
"A" Btry, 82nd Field Arty	03 Sep '45 — — — — —'50~ (Con't next era)
"B" Btry, 82nd Field Arty	03 Sep '45 — — — — —'50~ (Con't next era)
"C" Btry, 82nd Field Arty	03 Sep '45 — — — — —'50~ (Con't next era)
3rd Battalion, 82nd Field Arty	03 Sep '45 — — — — —'50~ (Con't next era)
61st Field Artillery Battalion	03 Sep '45 — — — — —'50~ (Con't next era)
99th Field Artillery Battalion	03 Sep '45 — — — — —'50~ (Con't next era)
271st Field Artillery Battalion	03 Sep '45 — — — —'49~ Dsgn'd 25 Mar as 77th Field Arty Bn, below
77th Field Artillery Battalion	25 Mar '49/'50~ (Con't next era)
92nd A/A Arty Automatic Wpns Bn	14 Mar '49/'50~ (Con't next era)
8th Engineer Combat Squadron	03 Sep '45 — — — —'49~ Reorg'd & Dsgn'd 25 Mar as 8th Combat Bn, below
8th Engineer Combat Battalion	25 Mar '49/'50~ (Con't next era)
Band, 1st Cavalry Division	03 Sep '45 — — — — —'50~ (Con't next era)
16th Cavalry Quartermaster Squadron	03 Sep '45/'46~ Reorg'd, Convrt'd & Dsgn'd as follows:
"HHT"	03 Sep '45/'46~ Reorg'd & Dsgn'd 15 Nov as "HHT" 16th QtrMstr Sqdn, below
Trp "A", 16th Cav QtrMstr Sqdn	03 Sep '45/'46~ Reorg'd & Dsgn'd 15 Nov as 18th Transport'n Truck Trp, below
Trp "B", 16th Cav QtrMstr Sqdn	03 Sep '45/'46~ Reorg'd & Dsgn'd 15 Nov as 19th Transport'n Truck Trp, below
Trp "C", 16th Cav QtrMstr Sqdn	03 Sep '45/'46~ Reorg'd & Dsgn'd 15 Nov as 15th QtrMster Pack Trp, below
16th Quartermaster Squadron	15 Nov '46 — —'49~ Convert'd & Dsgn'd 15 Mar as 15th Replc'mt Co, below
"HHT"	15 Nov '46 — —'49; Inactv'd 31 May at Osaka Japan
"HHC"	25 Mar '49/'50~ (Con't next era)
15th Quartermaster Pack Troop	15 Nov '46/'47~ Dsgn'd 31 Mar as 15th QtrMster Trp, below
15th Quartermaster Troop	31 Mar '47 ——'49~ Dsgn'd 25 Mar as 15th QtrMstr Co, below
15th Quartermaster Company	25 Mar '49/'50~ (Con't next era)
18th Transportation Truck Trp	15 Nov '46~
19th Transportation Truck Trp	15 Nov '46~
15th Replacement Company	25 Mar '49/'50 (Con't next era)
Mltry Police Plt, 1st Cav Div (Spc'l)	03 Sep '45 — — — —'49~ Dsgn'd 25 Mar as 545th Mltry Police Co, below
545th Mltry Police Company	25 Mar '49/'50~ (Con't next era)
OPCON ROUND UP UNITS	
3rd Battalion, 141st Inf Regt (TXARNG)	'46 — — — —'50~ (Con't next era)

KOREAN WAR
18 July 1950 - 30 December 1951

KOREAN WAR TIME LINE	18 Jul 1950 '51
"HHC", 1st Cavalry Division	18 Jul '50/'51~ (Con't next era)
Medical Det, 1st Cavalry Division	18 Jul '50/'51~ (Con't next era)
5th Cavalry Regiment (Inf)	18 Jul '50/'51~ (Con't next era)
1st Battalion, 5th Cavalry Regt	18 Jul '50/'51~ (Con't next era)
"A" Company	18 Jul '50/'51~ (Con't next era)
"B" Company	18 Jul '50/'51~ (Con't next era)
"C" Company	18 Jul '50/'51~ (Con't next era)
"D" Company	18 Jul '50/'51~ (Con't next era)
2nd Battalion, 5th Cavalry Regt	18 Jul '50/'51~ (Con't next era)
"E" Company	18 Jul '50/'51~ (Con't next era)
"F" Company	18 Jul '50/'51~ (Con't next era)
"G" Company	18 Jul '50/'51~ (Con't next era)
"H" Company	18 Jul '50/'51~ (Con't next era)
3rd Battalion, 5th Cavalry Regt	29 Aug '50/'51~ (Con't next era)
"I" Company	29 Aug '50/'51~ (Con't next era)
"K" Company	29 Aug '50/'51~ (Con't next era)
"L" Company	29 Aug '50/'51~ (Con't next era)
"M" Company	29 Aug '50/'51~ (Con't next era)
7th Cavalry Regiment (Inf)	18 Jul '50/'51~ (Con't next era)
1st Battalion, 7th Cavalry Regt	18 Jul '50/'51~ (Con't next era)
"A" Company	18 Jul '50/'51~ (Con't next era)
"B" Company	18 Jul '50/'51~ (Con't next era)
"C" Company	18 Jul '50/'51~ (Con't next era)
"D" Company	18 Jul '50/'51~ (Con't next era)
2nd Battalion, 7th Cavalry Regt	18 Jul '50/'51~ (Con't next era)
"E" Company	18 Jul '50/'51~ (Con't next era)
"F" Company	18 Jul '50/'51~ (Con't next era)
"G" Company	18 Jul '50/'51~ (Con't next era)
"H" Company	18 Jul '50/'51~ (Con't next era)
3rd Battalion, 7th Cavalry Regt	29 Aug '50/'51~ (Con't next era)
"I" Company	29 Aug '50/'51~ (Con't next era)
"K" Company	29 Aug '50/'51~ (Con't next era)
"L" Company	29 Aug '50/'51~ (Con't next era)
"M" Company	29 Aug '50/'51~ (Con't next era)

KOREAN WAR TIME LINE	18 Jul 1950 '51
8th Cavalry Regiment (Inf)	18 Jul '50/'51~ (Con't next era)
1st Battalion, 8th Cavalry Regt	18 Jul '50/'51~ (Con't next era)
"A" Company	18 Jul '50/'51~ (Con't next era)
"B" Company	18 Jul '50/'51~ (Con't next era)
"C" Company	18 Jul '50/'51~ (Con't next era)
"D" Company	18 Jul '50/'51~ (Con't next era)
2nd Battalion, 8th Cavalry Regt	18 Jul '50/'51~ (Con't next era)
"E" Company	18 Jul '50/'51~ (Con't next era)
"F" Company	18 Jul '50/'51~ (Con't next era)
"G" Company	18 Jul '50/'51~ (Con't next era)
"H" Company	18 Jul '50/'51~ (Con't next era)
3rd Battalion, 8th Cavalry Regt	29 Aug '50/'51~ (Con't next era)
"I" Company	29 Aug '50/'51~ (Con't next era)
"K" Company	29 Aug '50/'51~ (Con't next era)
"L" Company	29 Aug '50/'51~ (Con't next era)
"M" Company	29 Aug '50/'51~ (Con't next era)
13th Signal Company	18 Jul '50/'51~ (Con't next era)
27th Ordnance Maintenance Company	18 Jul '50/'51~ (Con't next era)
16th Reconnaissance Company	18 Jul '50/'51~ (Con't next era)
70th Tank Battalion	10 Nov '50/'51~ (Con't next era)
71st Heavy Tank Battalion	18 Jul '50~ Reorg'd & Dsgn'd 05 Aug as 71st Tank Bn, below
71st Tank Battalion	05 Aug '50/'51~ (Con't next era)
15th Medical Battalion	18 Jul '50/'51~ (Con't next era)
DIVISION ARTILLERY, 1st Cavalry Division	
"HHB"	18 Jul '50/'51~ (Con't next era)
Medical Detachment	18 Jul '50/'51~ (Con't next era)
82nd Field Artillery Battalion	18 Jul '50/'51~ (Con't next era)
"A" Btry, 82nd Field Arty	18 Jul '50/'51~ (Con't next era)
"B" Btry, 82nd Field Arty	18 Jul '50/'51~ (Con't next era)
"C" Btry, 82nd Field Arty	18 Jul '50/'51~ (Con't next era)
3rd Battalion, 82nd Field Arty	18 Jul '50/'51~ (Con't next era)
50th A/A Arty AW Bn (SP)	10 Nov '51~ (Con't next era)
61st Field Artillery Battalion	18 Jul '50/'51~ (Con't next era)
77th Field Artillery Battalion	18 Jul '50/'51~ (Con't next era)

KOREAN WAR TIME LINE	18 Jul 1950 '51
92nd A/A Arty Automatic Wpns Bn	18 Jul '50/'51; Inactv'd 28 Dec in Korea
99th Field Artillery Battalion	18 Jul '50/'51~ (Con't next era)
8th Engineer Combat Battalion	18 Jul '50/'51~ (Con't next era)
Band, 1st Cavalry Division	18 Jul '50/'51~ (Con't next era)
15th Quartermaster Company	18 Jul '50/'51~ (Con't next era)
15th Replacement Company	18 Jul '50/'51~ (Con't next era)
191st Counter Intelligence Corps Det	06 Oct '50/'51~ (Con't next era)
545th Mltry Police Company	18 Jul '50/'51~ (Con't next era)
OPCON ROUND UP UNITS	
3rd Bn, 141st Inf Regt (TXARNG)	18 Jul '50/'51~ (Con't next era)
ATTACHED UNITS	
Infantry	
5th Regimental Combat Team	14 Sep '50/'51; Relv'd 30 Jan
4th Ranger Infantry Company	31 Dec '50/'51; Relv'd 01 Aug
3rd Battalion, 19th Inf Regt	29 Jul '50; Relv'd 01 Aug
21st Infantry Regiment	29 Jul '50/'51; Relv'd 18 Feb
23rd Infantry Regiment	18 Aug '50; Relv'd 05 Sep
24th Infantry Regiment	14 Feb '51; Relv'd 01 Aug
27th Infantry Regiment	18 Aug '50; Relv'd 01 Sep
1st Battalion, 35th Inf Regiment	26 Jul '50; Relv'd 29 Jul
52nd Inf Counterfire Platoon	30 Jul '51; Relv'd 28 Oct
58th Inf Counterfire Platoon	13 Aug '51; Relv'd 28 Oct
64th Inf Counterfire Platoon	31 Aug '51; Relv'd 28 Oct
65th Infantry Regiment	18 Nov '51; Relv'd 20 Nov
187th Inf Counterfire Platoon	03 Nov '51; Relv'd 28 Nov
Armor	
Co "B", 6th Tank Bn, 24th Inf Div	12 Oct '50; Relv'd 13 Oct
Co "D", 6th Tank Bn, 24th Inf Div	14 Feb '51; Relv'd 16 Feb during Task Force "Crombez"
64th Tank Battalion	15 May '51; Relv'd 19 Oct
70th Tank Battalion	12 Aug '50/'51; Relv'd Asign'd 10 Nov to 1st Cav Div, above
Co "C", 73rd Tank Battalion	03 Sep '50; Relv'd 04 Sep
Co "A", 79th Tank Battalion	31 Jul '51; Relv'd 03 Aug
89th Tank Battalion	10 Oct '50; Relv'd 20 Oct

Artillery	
1st Field Arty Observation Battalion	29 Aug '50; Relv'd 24 Nov
2nd Chemical Mortar Battalion	10 Nov '50; Relv'd 22 Nov
9th Field Arty Battalion	28 Aug '50; Relv'd 20 Nov
10th Anti Aircraft Arty Group	15 Sep '50; Relv'd 10 Nov
13th Field Arty Battalion	11 Oct '50; Relv'd 15 Oct
15th Field Arty Battalion	06 Nov '50; Relv'd 10 Nov
17th Field Arty Battalion	13 Sep '50/'51; Relv'd 30 Oct
34th Countermortar Radar Detachment	11 Sep '51; Relv'd 28 Oct
36th Countermortar Radar Detachment	18 Oct '51; Relv'd 28 Oct
Btry "C", 37th Field Arty Battalion	29 Aug '50; Relv'd 05 Sep
39th Field Arty Battalion	19 Nov '51; Relv'd 20 Nov
Btry "B", 49th Field Arty Battalion	06 Oct '51; Relv'd 19 Oct
58th Armored Field Arty Battalion	18 Nov '51; Relv'd 20 Nov
64th Field Arty Battalion	08 Jul '50; Relv'd 20 Jul
68th Anti Aircraft Arty Gun Bn	15 Sep '50; Relv'd 21 Sep
78th Anti Aircraft Arty Gun Bn	15 Sep '50; Relv'd 10 Nov
90th Field Arty Battalion	11 Oct '50/'51; Relv'd 12 May
92nd Armored Field Arty Battalion	14 Feb '51; Relv'd 24 Mar
96th Field Arty Battalion	26 Jan '51; Relv'd 06 Feb
204th Field Arty Battalion	22 Jun '51; Relv'd 18 Oct
436th Field Arty Battalion	17 Jul '51; Relv'd 30 Jul
555th Field Arty Battalion	14 Sep '50/'51; Relv'd 30 Jan
936th Field Arty Battalion	29 May '50/'51; Relv'd 21 Oct
955th Field Arty Battalion	08 Oct '51; Relv'd 30 Oct
999th Armored Field Arty Battalion	22 Feb '51; Relv'd 27 Apr
Support Units	
37th Preventative Medicine Company	18 Oct '51; Relv'd 20 Dec
38th Medical Malaria Control Det	07 Jul '51; Relv'd 23 Jul
1st Plt, 126th Signal Service Company	22 Oct '51; Relv'd 25 Oct
501st Army Postal Unit	01 Oct '51~ (Con't next era)
8063rd Mltry Army Surgeon Hospital	07 Jul '50; Relv'd 28 Jul

KOREAN WAR TIME LINE	18 Jul 1950 '51
Allied Units	
16th New Zealand Arty Regiment	20 Mar '51; Relv'd 22 Mar
21st Thailand Infantry Regiment	19 Jan '51; Relv'd 13 May
21st Thailand Infantry Battalion	19 Jan '51; Relv'd 22 Dec
25th Canadian Infantry Brigade	16 Jun '51; Relv'd 16 Jul
27th British Commonwealth Inf Bde	01 Sep '50; Relv'd 08 Dec
28th British Commonwealth Inf Bde	
The King's Shropshire Light Inf	
The King's Own Scottish Borderers	
Royal Australian Regiment	
29th British Commonwealth Inf Bde	29 May '51; Relv'd 03 Jun
Belgian Infantry Battalion	15 Oct '51; Relv'd 12 Nov
Greek Infantry Brigade	13 Dec '50/'51; Relv'd 28 Nov
Netherlands Battalion	Oct '50;
Philippine 10th Bn Combat Team	13 Dec '50/'51; Relv'd 10 Jan
Turkish Brigade	Oct '50;

RETURN TO JAPAN
30 December 1951 - 15 October 1957

RETURN TO JAPAN TIME LINE	30 Dec 1951 '52 '53 '54 '55 '56 '57
"HHC", 1st Cavalry Division	30 Dec '51 — — — — — — — '57~ Con't next era
Medical Det, 1st Cavalry Division	30 Dec '51 — — — — — — — '57~ Inactv'd 15 Oct from 1st Cav Div
5th Cavalry Regiment (Inf)	30 Dec '51 — — — — — — — '57~ Relv'd 15 Oct from 1st Cav Div
1st Battalion, 5th Cavalry Regt	30 Dec '51 — — — — — — — '57~ Relv'd 15 Oct from 1st Cav Div
"A" Co, 5th Cavalry Regt	30 Dec '51 — — — — — — — '57~ Dsgn'd 15 Oct as "HHC", 1st Btl Grp, 5th Cav (Con't next era)
2nd Battalion, 5th Cavalry Regt	30 Dec '51 — — — — — — — '57~ Relv'd 15 Oct from 1st Cav Div
3rd Battalion, 5th Cavalry Regt	30 Dec '51 — — — — — — — '57~ Relv'd 15 Oct from 1st Cav Div
Tank Co, 5th Cavalry Regt	30 Dec '51 — — — — — — — '57; Inactv'd & Relv'd 15 Oct from 1st Cav Div
7th Cavalry Regiment (Inf)	30 Dec '51 — — — — — — — '57~ Relv'd 15 Oct from 1st Cav Div
1st Battalion, 7th Cavalry Regt	30 Dec '51 — — — — — — — '57~ Relv'd 15 Oct from 1st Cav Div
"A" Co, 7th Cavalry Regt	30 Dec '51 — — — — — — — '57~ Dsgn'd 15 Oct as "HHC", 1st Btl Grp, 7th Cav (Con't next era)
2nd Battalion, 7th Cavalry Regt	30 Dec '51 — — — — — — — '57~ Relv'd 15 Oct from 1st Cav Div
3rd Battalion, 7th Cavalry Regt	30 Dec '51 — — — — — — — '57~ Relv'd 15 Oct from 1st Cav Div
Tank Co, 7th Cavalry Regt	30 Dec '51 — — — — — — — '57; Inactv'd & Relv'd 15 Oct from 1st Cav Div

RETURN TO JAPAN TIME LINE	30 Dec 1951 ’52 ’53 ’54 ’55 ’56 ’57
8th Cavalry Regiment (Inf)	30 Dec ’51 — — — — — — — ’57~ Relv’d 15 Oct from 1st Cav Div
1st Battalion, 8th Cavalry Regt	30 Dec ’51 — — — — — — — ’57~ Relv’d 15 Oct from 1st Cav Div
“A” Co, 8th Cavalry Regt	30 Dec ’51 — — — — — — — ’57~ Dsgn’d 15 Oct as “HHC”, 1st Btl Grp, 8th Cav (Con’t next era)
Tank Co, 8th Cavalry Regt	30 Dec ’51 — — — — — — — ’57~ Relv’d 15 Oct from 1st Cav Div
2nd Battalion, 8th Cavalry Regt	30 Dec ’51 — — — — — — — ’57~ Relv’d 15 Oct from 1st Cav Div
3rd Battalion, 8th Cavalry Regt	30 Dec ’51 — — — — — — — ’57~ Inactv’d & Relv’d 15 Oct
16th Reconnaissance Company	30 Dec ’51 — — — — — — — ’57~ (Con’t next era)
70th Tank Battalion	30 Dec ’51 — — — — — — — ’57; Inactv’d & Relv’d 15 Oct from 1st Cav Div
41st Infantry Scout Dog Platoon	21 Feb ’54~
DIVISION ARTILLERY, 1st Cavalry Division	
“HHB”, 1st Cavalry Division Arty	30 Dec ’51 — — — — — — — ’57~ (Con’t next era)
Medical Det, Division Arty	30 Dec ’51 — — — — — — — ’57~ Dsgn’d 15 Oct as “HHB” Med Sectn (Con’t next era)
29th A/A Arty Automatic Wpns Bn	27 Dec ’51 —— ’53~ Dsgn’d 01 Apr as 29th A/A Arty Bn, below
“A” Btry, 29th A/A Arty AW Bn	27 Dec ’51 —— ’53~
“B” Btry, 29th A/A Arty AW Bn	27 Dec ’51 —— ’53~
“C” Btry, 29th A/A Arty AW Bn	27 Dec ’51 —— ’53~
29th A/A Artillery Battalion	01 Apr ’53 — — — — ’57; Inactv’d & Relv’d 15 Oct from 1st Cav Div
50th A/A Arty AW Bn (SP)	30 Dec ’51
61st FA Battalion (105mm How)	30 Dec ’51 — — — — — — — ’57; Inactv’d 15 Oct
77th FA Battalion (105mm How)	30 Dec ’51 — — — — — — — ’57~ Inactv’d 15 Oct
82nd FA Battalion (155mm How)	30 Dec ’51 — — — — — — — ’57~ Inactv’d 15 Oct
“A” Btry, 82nd FA Battalion	30 Dec ’51 — — — — — — — ’57; Inactv’d 15 Oct
“B” Btry, 82nd FA Battalion	30 Dec ’51 — — — — — — — ’57; Inactv’d 15 Oct
“C” Btry, 82nd FA Battalion	30 Dec ’51 — — — — — — — ’57; Inactv’d 15 Oct
3rd Battalion, 82nd Field Arty	30 Dec ’51 — — — — — — — ’57~ Inactv’d 15 Oct
99th FA Battalion (105mm How)	30 Dec ’51 — — — — — — — ’57; Inactv’d 15 Jul in Japan
268th Field Artillery Battalion (Gyroscope Unit)	’56/’57; Inactv’d in Japan
Band, 1st Cavalry Division	30 Dec ’51 — — — — — — — ’57~ (Con’t next era)

RETURN TO JAPAN TIME LINE	30 Dec 1951 '52 '53 '54 '55 '56 '57
8th Engineer Combat Battalion	30 Dec '51 — — —'54~ Reorgn'd & Desgn'd 10 Mar as 8th Engr Bn, below
8th Engineer Battalion	10 Mar '54 — —'57~ (Con't next era)
Medical Detachment	10 Mar '54 — —'57; Inactv'd 15 Oct
"A" Co, 8th Engineer Bn	10 Mar '54 — —'57~ (Con't next era)
"B" Co, 8th Engineer Bn	10 Mar '54 — —'57~ (Con't next era)
"C" Co, 8th Engineer Bn	10 Mar '54 — —'57~ (Con't next era)
"D" Co, 8th Engineer Bn	10 Mar '54 — —'57~ (Con't next era)
"E" Co, 8th Engineer Bn	10 Mar '54 — —'57~ (Con't next era)
15th Medical Battalion	30 Dec '51 — — — — — — — —'57~ (Con't next era)
15th Quartermaster Company	30 Dec '51 — — — — — — — —'57~ (Con't next era)
15th Replacement Company	30 Dec '51 — — — — — — — —'57~ Dsgn'd 15 Oct as 15th Admin Co (Con't next era)
27th Ordnance Maintenance Company	30 Dec '51 ——'53~ Desgn'd 01 Jan as 27th Ord Bn, below
27th Ordnance Battalion	01 Jan '53 — — — —'57~ (Con't next era)
SEPARATE BATTALIONS/COMPANIES	
13th Signal Company	30 Dec '51 — — — — — — — —'57~ (Con't next era as HHC 13th Signal Bn)
191st Counter Intelligence Det	30 Dec '51 — — — — — — —'56
545th Military Police Company	30 Dec '51 — — — — — — — —'57; Inactv'd 15 Oct & Relv'd from 1st Cav Div
1st Traffic Plt	30 Dec '51 — — — — — — — —'57; Inactv'd 15 Oct & Relv'd from 1st Cav Div
2nd Traffic Plt	30 Dec '51 — — — — — — — —'57; Inactv'd 15 Oct & Relv'd from 1st Cav Div
3rd Traffic Plt	30 Dec '51 — — — — — — — —'57; Inactv'd 15 Oct & Relv'd from 1st Cav Div
ATTACHED UNITS	
Support Units	
36th Ordnance Company	01 Oct '52/'53; 27 Jul
49th Ordnance Company	01 Oct '52/'53; 27 Jul
501st Army Postal Unit	30 Dec '51 ——'53; 27 Jul
510th Military Police Platoon	01 Oct '52/'53; 27 Jul
8016th Army Unit	01 Sep '52/'53; 27 Jul
8029th Army Unit	01 Oct '52/'53; 27 Jul
8145th Army Unit	01 Oct '52/'53; 27 Jul
8196th Army Unit	01 Oct '52/'53; 27 Jul

DEMILITARIZED ZONE
15 October 1957 - 01 July 1965

DEMILTARIZED ZONE TIME LINE	15 Oct 1957 '58 '59 '60 '61 '62 '63 '64 '65
"HHC", 1st Cavalry Division	15 Oct '57 — — — — — — — — — — '65~ (Con't next era)
2nd Battle Group, 4th Cavalry Regiment	15 Oct '57 — — — — — — — '63~ Relv'd 10 Aug Asgn'd to 4th Armor Div
"HHC", 2nd Battle Group	
Mortar Btry, 2nd Battle Group	
"A" Co, 4th Cavalry Regt	
"B" Co, 4th Cavalry Regt	
"C" Co, 4th Cavalry Regt	
"D" Co, 4th Cavalry Regt	
1st Battle Group, 5th Cavalry Regiment	15 Oct '57 — — — — — — — '63~ Relv'd XX XXX from 1st Cav Div
"HHC", 1st Battle Group	
Mortar Btry, 1st Battle Group	
"A" Co, 5th Cavalry Regt	
"B" Co, 5th Cavalry Regt	
"C" Co, 5th Cavalry Regt	
"D" Co, 5th Cavalry Regt	
1st Battle Group, 7th Cavalry Regiment	15 Oct '57 — — — — — — — '63~ Relv'd XX XXX from 1st Cav Div
"HHC", 1st Battle Group	
Mortar Btry, 1st Battle Group	
"A" Co, 7th Cavalry Regt	
"B" Co, 7th Cavalry Regt	
"C" Co, 7th Cavalry Regt	
"D" Co, 7th Cavalry Regt	
1st Battle Group, 8th Cavalry Regiment	15 Oct '57 — — — — — — — '63~ Relv'd XX XXX from 1st Cav Div
"HHC", 1st Battle Group	
Mortar Btry, 1st Battle Group	
"A" Co, 8th Cavalry Regt	
"B" Co, 8th Cavalry Regt	
"C" Co, 8th Cavalry Regt	
"D" Co, 8th Cavalry Regt	

DEMILTARIZED ZONE TIME LINE	15 Oct 1957 '58 '59 '60 '61 '62 '63 '64 '65
2d Battle Group, 12th Cavalry Regiment	15 Oct '57— — — — — — — —'63~ Relv'd XX XXX from 1st Cav Div
"HHC", 2nd Battle Group	
Mortar Btry, 2nd Battle Group	
"A" Co, 12th Cavalry Regt	
"B" Co, 12th Cavalry Regt	
"C" Co, 12th Cavalry Regt	
"D" Co, 12th Cavalry Regt	
1st BRIGADE, 1st Cavalry Division	
"HHC", "All The Way"	15 Jul '63—'65~ (Con't next era)
1st Battalion, 8th Cavalry Regiment	01 Sep '63—'65~ (Con't next era)
2nd Battalion, 8th Cavalry Regiment	01 Sep '63—'65~ (Con't next era)
1st Battalion, 12th Cavalry Regiment	01 Sep '63—'65~ (Con't next era)
2nd BRIGADE, 1st Cavalry Division	
"HHC", "Black Horse"	15 Jul '63—'65~ (Con't next era)
1st Battalion, 5th Cavalry Regiment	01 Sep '63—'65~ (Con't next era)
2nd Battalion, 5th Cavalry Regiment	01 Aug '63—'65~ (Con't next era)
2nd Battalion, 12th Cavalry Regiment	01 Sep '63; Relv'd 01 Sep from 1st Cav Div
3rd BRIGADE, 1st Cavalry Division	
"HHC", "GarryOwen"	15 Jul '63—'65~ (Con't next era)
1st Battalion, 7th Cavalry Regiment	01 Sep '63—'65~ (Con't next era)
2nd Battalion, 7th Cavalry Regiment	01 Sep '63—'65~ (Con't next era)
DIVISION ARTILLERY, 1st Cavalry Division	
"HHB", 1st Cavalry Division Arty	15 Oct '57— — — — — — — — — — —'65~ (Con't next era)
2nd Howitzer Bn, 19th Arty (105 mm)	15 Oct '57— — — — — — — —'63~ Reorg'd & Dsgn'd 01 Sept as 2nd Bn, 19th Arty, below
"HHB", 2nd Bn, 19th Arty	
"A" Btry, 2nd Bn, 19th Arty	
"B" Btry, 2nd Bn, 19th Arty	
"C" Btry, 2nd Bn, 19th Arty	
"D" Btry, 2nd Bn, 19th Arty	
"E" Btry, 2nd Bn, 19th Arty	
Service Btry, 2nd Bn, 19th Arty	
2nd Battalion, 19th Artillery	01 Sep '63—'65~ (Con't next era)
2nd Battalion, 20th FA (Rkt/How)	15 Oct '57— — —'60~ Reorg'd & Dsgn'd 01 Jul as 2nd Rkt/How Bn, below
"HHB", 2nd Bn, 20th Arty	
"A" Btry, 2nd Bn, 20th Arty	
"B" Btry, 2nd Bn, 20th Arty	
"C" Btry, 2nd Bn, 20th Arty	
"D" Btry, 2nd Bn, 20th Arty	
Service Btry, 2nd Bn, 20th Arty	

DEMILTARIZED ZONE TIME LINE	15 Oct 1957 ’58 ’59 ’60 ’61 ’62 ’63 ’64 ’65
2nd Rocket Howitzer Bn, 20th Artillery	01 Jul ’60 — — —’63~ Reorg’d & Dsgn’d as 2nd Bn, 20th Arty, below
2nd Battalion, 20th Artillery (Blue Max)	01 Sep ’63 —’65~ (Con’t next era)
1st Howitzer Battalion, 21st Artillery	Jul ’60 — —’63~ Reorgn’d & Redsgn’d 01 Sep, below
1st Battalion, 21st Artillery	01 Sep ’63 —’65~ (Con’t next era)
5th Howitzer Bn, 31st Artillery	01 Sep ’60 — —’63~ Inactv’d 01 Sep Korea
5th Howitzer Bn, 82nd Artillery	01 Jul ’60 — —’63~ Reorgn’d & Dsgn’d 01 Sep as 5th Bn, 82nd Arty, below
5th Battalion, 82nd Artillery	01 Sep ’63 —’65~ Reorg’d & Dsgn’d 01 Jul “E” Btry, 82nd Arty (Con’t next era)
DIVISION TRAINS	
“HHD”, Trains	01 Nov ’57— — — — — — — —’63~ SUPT CMD 1st Cav Div, below
Band, 1st Cavalry Division	01 Nov ’57— — — — — — — —’63~ SUPT CMD 1st Cav Div, below
15th Administrative Service Co	01 Nov ’57— — — — — — — —’63~ SUPT CMD 1st Cav Div, below
15th Aviation Company	01 Nov ’57— — — — — — — —’63~ Reorg’d & Desgn’d 01 Sep as 15th Avn Bn, below
15th Medical Battalion (Inf)	01 Nov ’57— — — — — — — —’63~ SUPT CMD 1st Cav Div, below
“HHD”,	01 Nov ’57— — — — — — — —’63~ SUPT CMD 1st Cav Div, below
“A” Co (Ambulance)	01 Nov ’57— — — — — — — —’63~ SUPT CMD 1st Cav Div, below
“B” Co (Clearing)	01 Nov ’57— — — — — — — —’63~ SUPT CMD 1st Cav Div, below
15th Quartermaster Company (Inf)	01 Nov ’57— — — — — — — —’63~ Desgn’d 01 Sep as 15th Sply & Trans, below
23rd Transportation Battalion (Inf)	01 Nov ’57— — — — — — — —’63;
“HHC”,	01 Nov ’57— — — — — — — —’63;
“A” Co (Truck)	01 Nov ’57— — — — — — — —’63;
“B” Co (Armed Carrier)	01 Nov ’57— — — — — — — —’63;
“B” Co (Armed Carrier)	01 Nov ’57— — — — — — — —’63;
27th Ordnance Battalion (Inf)	01 Nov ’57— — — — — — — —’63~ Desgn’d 01 Sep as 27th Maint Bn, below
“HHC” Co (Main Spt)	01 Nov ’57— — — — — — — —’63~
“A” Co (Fwd Spt)	01 Nov ’57— — — — — — — —’63~
“B” Co (Main Spt)	01 Nov ’57— — — — — — — —’63~
DIVISION SUPPORT, 1st Cavalry Division	
“HHC” and Band, 1st Cavalry	01 Sep ’63 —’65~ (Con’t next era)
15th Administrative Company	01 Sep ’63 —’65~ (Con’t next era)
15th Aviation Battalion	01 Sep ’63 —’65~ Redesignated 01 Jul as 15th Trans Corps Bn
15th Medical Battalion	01 Sep ’63 —’65~ (Con’t next era)

DEMILTARIZED ZONE TIME LINE	15 Oct 1957 '58 '59 '60 '61 '62 '63 '64 '65
15th Supply and Transport Battalion	01 Sep '63 —'65~ (Con't next era)
27th Maintenance Battalion	01 Sep '63 —'65~ (Con't next era)
SEPARATE BNs/COs	
"HHC", 1st Cavalry Division	
"A" Trp, 9th Cavalry Regiment	15 Oct '57~ Absrb'd W/16th Recon Co & Dsgn'd "HHT", 1st Recon Sqdn, 9th Cav, below
1st Reconnaissance Sqdn, 9th Cavalry	01 Nov '57 — — — — —'63~ Desgn'd 01 Sep as 1st Sqdn, 9th Cav, below
"HHT"	01 Nov '57 — — — — — — — —'63~
"A" Troop	01 Nov '57 — — — — — — — —'63~
"B" Troop	01 Nov '57 — — — — — — — —'63~
"C" Troop	01 Nov '57 — — — — — — — —'63~
1st Sqdn, 9th Cavalry Regiment	01 Sep '63 —'65 (Con't Next Era)
1st Battalion, 15th Armored	01 Jul '63 —'65; Inactv'd 01 Jul at Ft. Benning, GA
2nd Battalion, 15th Armored	01 Aug '63 —'65; Inactv'd 01 Jul at Ft. Benning, GA
3rd Medium Tank Bn, 40th Armor	15 Oct '57 — — — — — — — —'63; Relv'd 15 Jul from 1st Cav Div
"HHC", 3rd Med Tank Battalion	15 Oct '57 — — — — — — — —'63; Relv'd 15 Jul from 1st Cav Div
"A" Co, 3rd Bn	15 Oct '57 — — — — — — — —'63; Relv'd 15 Jul from 1st Cav Div
"B" Co, 3rd Bn	15 Oct '57 — — — — — — — —'63; Relv'd 15 Jul from 1st Cav Div
"C" Co, 3rd Bn	15 Oct '57 — — — — — — — —'63; Relv'd 15 Jul from 1st Cav Div
"D" Co, 3rd Bn	15 Oct '57 — — — — — — — —'63; Relv'd 15 Jul from 1st Cav Div
"E" Co, 3rd Bn	15 Oct '57 — — — — — — — —'63; Relv'd 15 Jul from 1st Cav Div
8th Engineer Battalion (Inf)	15 Oct '57 — — — — — — — — — — —'65~ (Con't next era)
"HHC", 8th Engineer Battalion	15 Oct '57 — — — — — — — — — — —'65~ (Con't next era)
"A" Co, 8th Engr Bn	15 Oct '57 — — — — — — — — — — —'65~ (Con't next era)
"B" Co, 8th Engr Bn	15 Oct '57 — — — — — — — — — — —'65~ (Con't next era)
"C" Co, 8th Engr Bn	15 Oct '57 — — — — — — — — — — —'65~ (Con't next era)
"D" Co, 8th Engr Bn	15 Oct '57 — — — — — — — — — — —'65~ (Con't next era)
"E" Co, 8th Engr Bn	15 Oct '57 — — — — — — — — — — —'65~ (Con't next era)
13th Signal Battalion (Inf)	15 Oct '57 — — — — — — — — — — —'65~ (Con't next era)
"HHC",	15 Oct '57 — — — — — — — — — — —'65~ (Con't next era)
"A" Co, Command Operations	15 Oct '57 — — — — — — — — — — —'65~ (Con't next era)
"B" Co, Forward Communications	15 Oct '57 — — — — — — — — — — —'65~ (Con't next era)
191st Military Intelligence Detachment	15 May '59 — — — — — — — —'65~ (Con't next era)
545th Military Police Company	15 Jul '63 —'65~ (Con't next era)

DEMILTARIZED ZONE TIME LINE	15 Oct 1957 ’58 ’59 ’60 ’61 ’62 ’63 ’64 ’65
DMZ Police Company (Prov)	15 Oct ’57— — —’60; Absrb’d 01 Apr W/1st Recon Sqdn, 9th Cav, above

FORT BENNING, AIRMOBILE
01 July 1965 - 15 August 1965

FORT BENNING, AIRMOBILE TIME LINE	01 Jul 1965
“HHC”, 1st Cavalry Division (Airmobile)	
1st Brigade, 1st Cavalry Division	
“HHC”, “All The Way”	01 Jul ’65~ (Con’t next era)
1st Bn (Airborne), 8th Cavalry	01 Jul ’65~ (Con’t next era)
2nd Bn (Airborne), 8th Cavalry	01 Jul ’65~ (Con’t next era)
1st Bn (Airborne), 12th Cavalry	01 Jul ’65~ (Con’t next era)
2nd Brigade, 1st Cavalry Division	
“HHC”, “Blackhorse”	01 Jul ’65~ (Con’t next era)
1st Bn, 5th Cavalry Regiment	01 Jul ’65~ (Con’t next era)
2nd Bn, 5th Cavalry Regiment	01 Jul ’65~ (Con’t next era)
2nd Bn, 12th Cavalry Regiment	01 Jul ’65~ (Con’t next era)
3rd Brigade, 1st Cavalry Division	
“HHC, “GarryOwen”	01 Jul ’65~ (Con’t next era)
1st Bn, 7th Cavalry Regiment	01 Jul ’65~ (Con’t next era)
2nd Bn, 7th Cavalry Regiment	01 Jul ’65~ (Con’t next era)
DIVISION ARTILLERY	
“HHB”, Division Arty	01 Jul ’65~ (Con’t next era)
2nd Bn, 19th Arty (105mm)	01 Jul ’65~ (Con’t next era)
2nd Bn, 20th Arty (ARA)	01 Jul ’65~ (Con’t next era)
1st Bn, 21st Arty (105mm)	01 Jul ’65~ (Con’t next era)
1st Bn, 77th Arty (105mm)	01 Jul ’65~ (Con’t next era)
DIVISION SUPPORT COMMAND	
“HHC” & Band, Support Command	01 Jul ’65~ (Con’t next era)
8th Engineer Battalion	01 Jul ’65~ (Con’t next era)
13th Signal Battalion	01 Jul ’65~ (Con’t next era)
15th Medical Battalion	01 Jul ’65~ (Con’t next era)
15th Supply & Service Battalion	01 Jul ’65~ (Con’t next era)
15th Transportation Battalion	01 Jul ’65~ (Con’t next era)
27th Maintenance Battalion	01 Jul ’65~ (Con’t next era)
15th Administrative Company	01 Jul ’65~ (Con’t next era)

FORT BENNING, AIRMOBILE TIME LINE	01 Jul 1965
15th Supply & Service Battalion	01 Jul '65~ (Con't next era)
545th Mltry Police Company	01 Jul '65~ (Con't next era)
191st Mltry Intelligence Detachment	01 Jul '65~ (Con't next era)
"C" Co, 313th ASA Bn (AKA 10th RRU)	01 Jul '65~ (Con't next era)
11th AVIATION GROUP (CAG)	
"HHC, 11th Avn Group (Airborne)	01 Jul '65~ (Con't next era)
11th Aviation Company	01 Jul '65~ (Con't next era)
1st Sqdn, 9th Cavalry Regt	01 Jul '65~ (Con't next era)
227th Avn Battalion	
"HHC, 227th Avn Battalion	01 Jul '65~ (Con't next era)
"A" Co, 227th Avn Battalion	01 Jul '65~ (Con't next era)
"B" Co, 227th Avn Battalion	01 Jul '65~ (Con't next era)
"C" Co, 227th Avn Battalion	01 Jul '65~ (Con't next era)
"D" Co, 227th Avn Battalion	01 Jul '65~ (Con't next era)
228th Avn Battalion	
" HHC", 228th Avn Battalion	01 Jul '65~ (Con't next era)
"A" Co, 228th Avn Battalion	01 Jul '65~ (Con't next era)
"B" Co, 228th Avn Battalion	01 Jul '65~ (Con't next era)
"C" Co, 228th Avn Battalion	01 Jul '65~ (Con't next era)
229th Avn Battalion	
"HHC", 229th Avn Battalion	01 Jul '65~ (Con't next era)
"A" Co, 229th Avn Battalion	01 Jul '65~ (Con't next era)
"B" Co, 229th Avn Battalion	01 Jul '65~ (Con't next era)
"C" Co, 229th Avn Battalion	01 Jul '65~ (Con't next era)
"D" Co, 229th Avn Battalion	01 Jul '65~ (Con't next era)

VIETNAM WAR
16 August 1965 - 28 June 1972

VIETNAM WAR TIME LINE	16 Aug 1965 '66 '67 '68 '69 '70 '71 '72
"HHC", 1st Cavalry Division, (Airmobile) Headquarters and Headquarters Det Personnel Services Company (Rear)	

VIETNAM WAR TIME LINE	16 Aug 1965 '66 '67 '68 '69 '70 '71 '72
1st Brigade, 1st Cavalry Division	
"HHC", "All The Way"	20 Sep '65 — — — — — — — —'71~ (Con't next era)
1st Bn (Airborne), 8th Cavalry	20 Sep '65 — — — — — — — —'71~ (Con't next era)
2nd Bn (Airborne), 8th Cavalry	20 Sep '65 — — — — — — — —'71~ (Asgn'd 30 Apr to 3rd Bde (SEP Below)
1st Bn (Airborne), 12th Cavalry	20 Sep '65 — — — — — — — —'71~ (Asgn'd 30 Apr to 3rd Bde (SEP Below)
2nd Brigade, 1st Cavalry Division	
"HHC", "Black Horse"	14 Sep '65 — — — — — — — —'71~ (Con't next era)
1st Bn, 5th Cavalry Regt	14 Sep '65 — — — — — — — —'71~ (Con't next era)
2nd Bn, 5th Cavalry Regt	14 Sep '65 — — — — — — — —'71~ (Asgn'd 30 Apr to 3rd Bde (SEP), below
2nd Bn, 12th Cavalry Regt	14 Sep '65 — — — — — — — —'71~ (Con't next era)
3rd Brigade, 1st Cavalry Division	
"HHC", "GarryOwen"	20 Sep '65 — — — — — — — —'71~ Reorg'd & Dsgn'd 30 Apr as "Separate", below
"HHC", "Separate"	30 Apr '71/'72~ Reorg'd & Dsgn'd 28 June as "Greywolf" (Con't next era)
Detachment 10, (Ranger)	23 Apr '71~ Reasgn'd 08 Aug to "H" Co (Ranger), 75th Infantry, below
1st Bn, 7th Cavalry Regt	20 Sep '65 — — — — — — — — —'72~ (Con't next era)
2nd Bn, 7th Cavalry Regt	20 Sep '65 — — — — — — — —'71~ (Con't next era)
5th Bn, 7th Cavalry Regt	19 Aug '66 — — — — —'71; Inactv'd 29 Mar at Ft. Lewis, WA
2nd Bn, 5th Cavalry Regt	30 Apr '71/'72~ (Con't next era)
2nd Bn (Airborne), 8th Cavalry	30 Apr '71/'72~ (Con't next era)
1st Bn (Airborne), 12th Cavalry	30 Apr '71/'72~ (Con't next era)
Aviation Brigade	
"HHC" - (Special)	05 May '71/'72~ Inactv'd 30 Jun at Ft. Hood, TX
1st Aviation Platoon	'71/'72;
"F" Troop, 9th Cavalry	30 Apr '71/'72;
229th Aviation Battalion (AHB)	30 Apr '71/'72; Inactv'd 22 Aug & Relv'd from the 1st Cav Div.
"HHC", 229th Avn Battalion (AHB)	30 Apr '71/'72; Inactv'd 22 Aug & Relv'd from the 1st Cav Div.
"A" Company, 229th Avn Battalion (AHB)	30 Apr '71/'72; Inactv'd 22 Aug & Relv'd from the 1st Cav Div.
"B" Company, 229th Avn Battalion (AHB)	30 Apr '71/'72; Inactv'd 22 Aug & Relv'd from the 1st Cav Div.
"C" Company, 229th Avn Battalion (AHB)	30 Apr '71/'72; Inactv'd 22 Aug & Relv'd from the 1st Cav Div.
"D" Company, 229th Avn Battalion (AHB)	30 Apr '71/'72; Inactv'd 22 Aug & Relv'd from the 1st Cav Div.
362nd Aviation Company (ASH) (Prov)	30 Jun '71/'72; Dsbnd'd Aug in Vietnam, assets Xfer'd to Vietnamese Air Force
Artillery Brigade	
1st Battalion, 21st Artillery	20 Apr '71~ Dsgn'd 01 Sep as 1st Bn. 21st FA, below
1st Battalion, 21st Field Artillery	01 Sep '71/'72; Reliev'd 13 Sep from asgnm't to 1st Cav Div
"F" Battery, 26th Artillery (Prov)	'71/'72
"F" Battery, 77th Artillery, Aviation	01 Sep '71/'72; Inactv'd 22 Aug in Vietnam
"F" Battery, 79th Artillery (ARA) "Blue Max"	30 Jun '71~ Dsgn'd 01 Sep as "F" Battery, 79th FA, below
"F" Battery, 79th Field Artillery	01 Sep '71/'72; Inactv'd 22 Aug in Vietnam

VIETNAM WAR TIME LINE	16 Aug 1965 '66 '67 '68 '69 '70 '71 '72
Support	
14th Military History Detachment	'71/'72;
26th Chemical Detachment	'71/'72;
215th Support Battalion	30 Jun '71/'72; Inactv'd 30 Jun
501st Engineer Company	'71/'72; Inactv'd 21 Aug at Ft Belvoir, VA
525th Signal Company	30 Jun '71/'72; Inactv'd 20 Aug in Vietnam
Temporary Duty Unit	
2nd Squadron, 11th Armored Cavalry	'71/'72;
Attached Units	
34th Inf Platoon, (Scout Dog Patrol)	'71/'72;
62nd Inf Platoon, (Combat Trackers)	16 Feb '68 — — — —'72; 15 Aug
75th Infantry Rangers, "H" Company	'71/'72;
Brigade Radio Research Detachment (Prov)	03 Apr '71; Inactv'd 30 Sep. Mission asgn'd to 405th RRD, below
191st Military Intelligence Company	30 Apr '71/'72: Inactv'd 15 Aug Vietnam
405th Radio Receiver Detachment	'71/'72:
483rd Military Police Platoon	'71/'72:
11th AVIATION GROUP (CAG)	
"HHC, 11th Avn Group (Airborne)	13 Sep '65 — — — — — — — —'71; Xfer'd 05 May to 1st Aviation Brigade
11th Pathfinder Platoon	
227th Pathfinder Team	
228th Pathfinder Team	
229th Pathfinder Team	
Air Traffic Control Platoon (Prov)	
11th Aviation Company (GS)	13 Sep '65 — — — — — — — —'71; Asgn'd 19th Apr, 4th US Army, inactv'd 11 Jun at Ft Hood, TX
1st Cavalry ASTA Platoon	13 Sep '65 — — — —'69; Asgn'd Nov to 73rd Surveillance Airplane Division
"Pathfinders", 11th Avn Co.	
227th Aviation Battalion (AB)	13 Sep '65 — — — — — — — —'71~ (Con't next era)
"HHC, 227th Avn Battalion	13 Sep '65 — — — — — — — —'71~ (Con't next era)
"A" Co, 227th Avn Battalion	13 Sep '65 — — — — — — — —'71~ (Con't next era)
"B" Co, 227th Avn Battalion	13 Sep '65 — — — — — — — —'71~ (Con't next era)
"C" Co, 227th Avn Battalion	13 Sep '65 — — — — — — — —'71~ (Con't next era)
"D" Co, 227th Avn Battalion	13 Sep '65 — — — — — — — —'71~ Inactv'd 30 Aug Vietnam
228th Aviation Battalion (AB)	13 Sep '65 — — — — — — — —'71~ Inactv'd 14 Jun at Ft Lewis, WA
"HHC", 228th Avn Battalion	13 Sep '65 — — — — — — — —'71~ Inactv'd 14 Jun at Ft Lewis, WA
"A" Co, 228th Avn Battalion	13 Sep '65 — — — — — — — —'71~ Inactv'd 14 Jun at Ft Lewis, WA
"B" Co, 228th Avn Battalion	13 Sep '65 — — — — — — — —'71~ Dsgn'd 30 Jun as 362nd ASH, 3rd Bde above
"C" Co, 228th Avn Battalion	13 Sep '65 — — — — — — — —'71~ Inactv'd 14 Jun at Ft Lewis, WA

VIETNAM WAR TIME LINE	16 Aug 1965 '66 '67 '68 '69 '70 '71 '72
229th Aviation Battalion (AB)	13 Sep '65 — — — — — — — —'71~ Asgn'd 30 Apr to 3rd Bde (SEP), above
"HHC", 229th Avn Battalion	13 Sep '65 — — — — — — — —'71~ Asgn'd 30 Apr to 3rd Bde (SEP), above
"A" Co, 229th Avn Battalion	13 Sep '65 — — — — — — — —'71~ Asgn'd 30 Apr to 3rd Bde (SEP), above
"B" Co, 229th Avn Battalion	13 Sep '65 — — — — — — — —'71~ Asgn'd 30 Apr to 3rd Bde (SEP), above
"C" Co, 229th Avn Battalion	13 Sep '65 — — — — — — — —'71~ Asgn'd 30 Apr to 3rd Bde (SEP), above
"D" Co, 229th Avn Battalion	13 Sep '65 — — — — — — — —'71~ Asgn'd 30 Apr to 3rd Bde (SEP), above
1st Sqdn, 9th Cavalry Regiment	13 Sep '65 — — — — — — — —'71~ Asgn'd 05 Dec to 9th Air Cavalry Bde, below
9th (Air) Cavalry Brigade (Provisional)	05 Dec '70/'71; Asgn'd 10 Apr to 12th Avn Grp
1st Sqdn, 9th Cavalry Regiment	05 Dec '70/'71; Asgn'd 10 Apr to
3rd Sqdn, 17th Cavalry Regiment	02 Nov '70/'71; Asgn'd 10 Apr to 1st Avn Bde
2nd Battalion, 12th Infantry	
2nd Battalion, 20th Arty (Blue Max)	
62nd Infantry Platoon, Combat Trackers	
"H" Co, 75th Infantry (Ranger)	
DIVISION ARTILLERY	
"HHB", 1st Cav Div Artillery	05 Sep '65 — — — — — — — —'71~ (Con't next era)
2nd Battalion, 19th Arty (105mm)	05 Sep '65 — — — — — — — —'71; Inactv'd 02 Apr at Ft Lewis, WA
2nd Battalion, 20th Arty (AR)	05 Sep '65 — — — — — — — —'71; Inactv'd 10 Apr at Ft Lewis, WA
1st Battalion, 21st Arty (105mm)	05 Sep '65 — — — — — — — —'71; Asgn'd 01 Sep to 3rd Bde above
1st Battalion, 30th Field Arty	01 Jan '68~ — —'71; Inactv'd 06 Apr
1st Battalion, 77th Arty (105mm)	05 Sep '65 — — — — — — — —'71~ (Con't next era)
"E" Battery, 82nd Artillery	'65 — — — — — — — —'71; Inactv'd 10 Apr in Vietnam
An Khe Airfield Command (Prov)	15 Dec '65 —'67; Oct
Rplcm't Trng Schl, 1st Team Academy (Prov)	01 Oct '66 — — — — —'71;
First Team Academy (AKA Rplcm't Trng Schl)	01 Oct '66 — — — — —'71;
1st Personnel Services Battalion (Prov)	26 Jan '68/'69; Inactv'd 01 Jul
1st Personnel Services Company	01 Jul '69
US Army Special Security Det	
DIVISION SUPPORT COMMAND	
"HHC" & Band, Support Command	12 Sep '65 — — — — — — — —'71~ (Con't next era)
1st Forward Service Support Element	Vietnam Era
2nd Forward Service Support Element	Vietnam Era
3rd Forward Service Support Element	Vietnam Era
4th Forward Service Support Element	Vietnam Era
5th Forward Service Support Element	Vietnam Era
8th Engineer Battalion	16 Aug '65 — — — — — — — —'71~ (Con't next era)
13th Signal Battalion	05 Sep '65 — — — — — — — —'71~ (Con't next era)
15th Administrative Company	05 Sep '65 — — — — — — — —'71~ Dsgn'd 15 Mar as 15th Finance Bn, below

VIETNAM WAR TIME LINE	16 Aug 1965 '66 '67 '68 '69 '70 '71 '72
15th Finance Battalion	15 Mar '71~ (Con't next era)
15th Medical Battalion	05 Sep '65 — — — — — — — —'71~ (Con't next era)
15th Supply & Service Battalion	05 Sep '65 — — — — — — — —'71~ (Con't next era)
Aerial Equipment Support Company (Prov)	'65/ '66;
15th Transportation Battalion (AM&S)	05 Sep '65 — — — — — — — —'71~ Inactv'd 28 Apr Ft Lewis, WA
27th Maintenance Battalion	05 Sep '65 — — — — — — — —'71~ (Con't next era)
191st Mltry Intelligence Detachment	05 Sep '65 — — — —'69~ Dsgn'd 26 Dec as Co, below
LRRP Det (Prov)	07 Jan '66/'67~ Asgn'd 02 Feb, below
LRRP Det	02 Feb '67~ Dsgn'd 14 Jun as Company, below
LRRP Company, 1st Cavalry Division	14 Jun '67~ Dsgn'd 20 Dec as "E" Co, 52nd Inf, below
191st Mltry Intelligence Company	26 Dec '69 ——'71~ Asgn'd 30 Apr to 3rd Bde (SEP), above
"C" Co, 313th ASA Bn (AKA 10th RRU)	05 Sep '65/ '66~ Dsgn'd 15 Oct as 371st ASA Co, below
371st Army Security Agency Company	15 Dec '66 — — — — —'71~ (Con't next era)
545th Mltry Police Company	05 Sep '65 — — — — — — — —'71~ (Con't next era)
51st Aircraft Maint/Trans Det	'69 ——'71;
80th Aircraft Maint/Trans Det	'69 ——'71;
98th Aircraft Maint/Trans Det	'69 ——'71;
150th Aircraft Maint/Trans Det	'69 ——'71;
151st Aircraft Maint/Trans Det	'69 ——'71;
165th Aircraft Maint/Trans Det	'69 ——'71;
166th Aircraft Maint/Trans Det	'69 ——'71;
171st Aircraft Maint/Trans Det	'69 ——'71;
225th Aircraft Maint/Trans Det	'69 ——'71;
255th Aircraft Maint/Trans Det	'69 ——'71;
329th Aircraft Maint/Trans Det	'69 ——'71;
369th Aircraft Maint/Trans Det	'69 ——'71;
382nd Aircraft Maint/Trans Det	'69 ——'71;
390th Aircraft Maint/Trans Det	'69 ——'71;
391st Aircraft Maint/Trans Det	'69 ——'71;
392nd Aircraft Maint/Trans Det	'69 ——'71;
393rd Aircraft Maint/Trans Det	'69 ——'71;
394th Aircraft Maint/Trans Det	'69 ——'71;
400th Aircraft Maint/Trans Det	'69 ——'71;
545th Aircraft Maint/Trans Det	'69 ——'71;
564th Aircraft Maint/Trans Det	'69 ——'71;
571st Aircraft Maint/Trans Det	'69 ——'71;
575th Aircraft Maint/Trans Det	'69 ——'71;
576th Aircraft Maint/Trans Det	'69 ——'71;
652nd Aircraft Maint/Trans Det	'69 ——'71;

VIETNAM WAR TIME LINE	16 Aug 1965 '66 '67 '68 '69 '70 '71 '72
ATTACHED UNITS	
Cavalry	
3rd Sqdn, 4th Cavalry, 25th Inf Div	01 May '70; 29 Jun
3rd Sqdn, 5th Cavalry Regiment	15 Feb '68 Relv'd 15 Apr
919th Engr Co, 11th Armored Cavalry Rgm't	01 May '70; 29 Jun
Infantry	
7th Combat Trackers	'68
8th Combat Trackers	'68
5th Bn, 12th Infantry, 199th Inf Bde	01 May '70; 29 Jun
AVLB Plt, "E" Co, 15th Engr Bn, 9th Inf	15 Feb '68 Relv'd from Atchm't 15 Apr
25th Infantry Plt (Scout Dog Patrol)	20 Jun '66 — — — — — '71; 20 Mar
34th Infantry Plt (Scout Dog Patrol)	21 Nov '66 — — — — — — — — '72; 20 Aug
2nd Bn, 34th Armor, 25 Inf Div	01 May '70; 29 Jun
37th Infantry Plt (Scout Dog Patrol)	15 Nov '69 —— '71; 07 Mar
2nd Bn, 47th Infantry, 3rd Bde, 9th Inf Div	01 May '70; 29 Jun
1st Battalion, 50th Mechanized Infantry	22 Sep '67/'68; 15 Mar
"E" Co, 52nd Infantry Regt (LRP)	20 Dec '67 —— '69; Inactv'd 01 Feb. Assets allocated to "H" Co, 75th Inf Rangers, below.
"H" Co, 75th Infantry Rangers	01 Feb '69 — — '72; 15 Aug
54th Infantry Det (Ground Radar)	Sep '65 — — — '68; 17 Jan
5th Bn, 60th Infantry, 3rd Bde, 9th Inf Div	01 May '70; 29 Jun
3rd Platoon, 571 Engr Co, 3rd Bde, 9th Inf Div	01 May '70; 29 Jun
62nd Infantry Plt, (Combat Trackers)	Aug '68 — — — — '72; 15 Aug
Aviation	
17th Aviation Company (CV-2)	15 Sep '65 — — — — — — — '70; Relv'd 01 Jan
53rd Aviation Detachment (Prov)	Dec '66 — '68; Dsbnd'd Feb
125th Aviation Company	01 May '70; 29 Jun
273rd Aviation Company (HH)	Nov '68 — — '71; Relv'd Mar
344th Aviation Company (AW)	05 Dec '70/'71; Relv'd 17 Jan
478th Aviation Company (HH)	Mar '66 — — '69; Relv'd 28 Feb
Armor	
1st Sqdn, 11th Armored Cavalry Regt	12 Apr '69 —— '71; Apr
2nd Sqdn, 11th Armored Cavalry Regt	12 Apr '69 —— '71; Apr
3rd Sqdn, 11th Armored Cavalry Regt	12 Apr '69 —— '71; Apr
"C" Co, 2nd Bn, 34th Armored, 25th Div	19 Feb '68 Relv'd 15 Apr
1st Battalion, 50th Inf (Mechanized)	22 Sep '67/'68; Relv'd 08 Mar

VIETNAM WAR TIME LINE	16 Aug 1965 '66 '67 '68 '69 '70 '71 '72
Artillery	
3rd Battalion, 2nd Field Arty	Dec '69
"A" Btry, 5th Bn, 2nd Arty	Jan '69 ——'71; Mar
7th Bn, 9th Artillery	01 May '70; 29 Jun
2nd Bn, 12th Artillery, 23rd Arty Group	01 May '70; 29 Jun
2nd Bn, 13th Artillery	01 May '70; 29 Jun
6th Battalion, 14th Field Arty	29 Oct '65/'66; 01 Feb
"F" Trp, 16th Artillery	01 May '70; 29 Jun
"C" Btry, 6th Bn, 16th Field Arty	Dec '66 / '67
2nd Battalion, 17th Field Arty	25 Oct '65/'66; 01 Feb
3rd Battalion, 18th Field Arty	26 Oct '65/'66; 01 Oct
6th Bn, 27th Artillery	01 May '70; 29 Jun
2nd Plt, "B" Btry, 29th Field Arty	23 Oct '65 — — —'68; Oct
1st Battalion, 30th Field Arty	'66 ——'68~ Asgn'd 01 Jan '68 to 1st Cav Div, above
"A" Btry, 2nd Bn, 32nd Arty	01 May '70; 29 Jun
"B" Btry, 2nd Bn, 32nd Arty	01 May '70; 29 Jun
"C" Btry, 6th Bn, 32nd Arty	25 Aug '66/'67; 04 Apr
"A" Btry, 2nd Bn, 35th Arty, 23rd Arty Group	01 May '70; 29 Jun
"A" Btry, 2nd Bn, 40th Arty, 199th Inf Group	01 May '70; 29 Jun
"D" Btry, 2nd Bn, 40th Arty, 199th Inf Group	01 May '70; 29 Jun
"E" Btry, 41st Artillery	25 Aug '66/'67; 04 Apr
5th Bn, 42nd Artillery	01 May '70; 29 Jun
"A" Btry, 2nd Bn, 44th Arty	01 May '70; 29 Jun
"B" Btry, 2nd Bn, 44th Arty	01 May '70; 29 Jun
"C" Btry, 4th Bn, 60th Field Arty	'68/'69
"C" Btry, 6th Bn, 60th Field Arty	'68/'69
268th Arty Detachment (Radar)	14 Apr '69 ——'71; 02 Jan
273rd Arty Detachment (Radar)	05 May '69 ——'71; 15 Apr
Support Units	
"HHQ" Det, 1st Logistical Cmd	26 Aug '66/'67; 04 Apr
Luong Son Det, 5th SFG, 1st SF	26 Aug '66/'67; 04 Apr
5th Weather Sqdn, 24th Det	Dec '66 — — — — — — —'71; Mar
5th Weather Sqdn, 31st Det	Dec '66 — — — — —'71; Mar
10th Radio Research Unit	Sep '65 ——'67; Jan Mission Asgn'd to 371st ASA, above
14th Engineer Battalion	22 Mar '68; 31 Oct
14th Mltry History Det	15 May '68 — — — —'72; 20 Mar
26th Chemical Detachment	16 Oct '65 — — — — — — — — —'72; 26 Mar
30th Weather Sqdn, (elements)	Oct '65/'66; Nov
34th Quartermaster Battalion	'68
34th Supply and Service Battalion	01 May '70; 29 Jun
41st Civil Affairs	03 Jun '67/'68;
41st Public Affairs Office	01 May '70; 29 Jun
41st Public Information Det	21 Mar '67 — — — —'71; 30 Jun
42nd Public Affairs Office	01 May '70; 29 Jun
42nd Public Information Det	21 Mar '67 — — — —'71; 29 Apr

VIETNAM WAR TIME LINE	16 Aug 1965 '66 '67 '68 '69 '70 '71 '72
54th Signal Battalion	Oct '65 —— '67; 04 Apr
CIPOW Team 55th MI Company	26 Aug '66/'67 04 Apr
65th Transportation Company	'65
70th Engineer Battalion	Aug '65
"B" Co, 84th Construction Battalion	
131st Engineer Company (LE)	
444th Engineer Detachment	
511th Engineer Company (Panel Bridge)	
630th Engineer Company (LE)	
31st Engr Bn, 79th Engr Group	01 May '70; 29 Jun
60th Engr Co, 79th Engr Group	01 May '70; 29 Jun
588th Engr Bn, 79th Engr Group	01 May '70; 29 Jun
2nd Pltn, "B" Co, 87th Engineer	03 Sep '66; 22 Dec
184th Chemical Plt (Direct Support)	16 Oct '65 — — — — — — — — '71; 01 Mar
241st Signal Detachment	'66 —— '68
34th Pltn, 272nd MP Company	14 Nov '66/'67; 04 Apr
Team 8, 215th Psychological Opns Co.	25 Aug '66/'67 04 Apr
330 Radio Research, 8th Det	20 Apr '68 — — '71; Inactv'd 30 Aug
391st Quartermaster Detachment	01 May '70; 29 Jun
583rd Mltry Intelligence Det	20 Jul '67 — — — — '71; 29 Mar
586th Signal Company	Sep '65 —— '67; Jan
American Red Cross, Female Contingent	Vietnam Era
Navy	
Det 102 Helicopter Support Sqdn 5 (US Navy)	01 May '70; 29 Jun
Air Force	
19th TASS Traffic Air Control Party (USAF)	01 May '70; 29 Jun

TRICAP TO ARMOR
05 May 1971 - 16 September 1990

TRICAP TO ARMOR TIME LINE	05 May 1971 '72 '73 '74 '75 '76\'77 '78 '79 '80 '81 '82 '83 '84 '85 '86 '87 '88 '89 '90
"HHC", 1st Cavalry Division (TRICAP)	05 May '71 — — — — '75 Reorgn'd 21 Feb, below
"HHC", 1st Cavalry Division (ARMOR)	21 Feb '75 — — — — — — — — — — — — — — — — — — — '90~ (Con't next era)

TRICAP TO ARMOR TIME LINE	05 May 1971 '72 '73 '74 '75 '76 '77 '78 '79 '80 '81 '82 '83 '84 '85 '86 '87 '88 '89 '90
1st Armored Brigade	
"HHC", "All The Way"	05 May '71~ Replc'd by "Ironhorse", below
"HHC", 1st Brigade, "Iron Horse"	Jun '72 — '90~ (Con't next era)
1st Bn (Armored), 13th Armor	05 May '71 — —'74; Inactv'd 21 Jun. Rplac'd by 1st Bn, 7th Cav, below
1st Bn (Armor), 7th Cavalry Regt	20 Jun '74 — — — — — — — — — — — — — — —'86~ Reorgn'd & Dsgn'd 16 Oct as 1st Sqdn, 7th Cav, below
2nd Bn, 33rd Armored Regt	16 Oct '86; Reflg'd 16 Oct as 3rd Bn, 32nd Armor, below
3rd Bn, 32nd Armor Regt	16 Oct '86 — — — —'90~ (Con't next era)
2nd Bn (Mech/Inf), 12th Cavalry	05 May '71/'72: Replc'd 29 Jun by 1st Bn, 12th Cav, below
1st Bn (Mech/Inf), 12th Cavalry	29 Jun '72 '83; Relv'd 15 Jun from 1st Cav Div
1st Bn (Armored), 81st Armor	05 May '71 — — —'74; Inactv'd 20 Jul; Replc'd by 2nd Bn, 5th Cav, below
2nd Bn, 5th Cavalry Regt	20 Jun '74 '86; Inactv'd 16 Sep & Asgn'd to 3rd Armor Div
2nd Bn, 5th Cavalry Regt	16 Jan '87 — —'90~ (Con't next era)
2nd Bn, 8th Cavalry Regt	20 Apr '74 — — — — — — — — — — — — — — — — — — —'90~ (Con't next era)
3rd Bn, 141st Inf Regt (TXARNG)	05 May '71 —'90;
2nd Air Cavalry Combat Brigade	
"HHC", 2nd Air Cavalry Combat Bde	05 May '71 — — — —'75 ~ Replc'd by "Blackjack", below
"HHC", 2nd Brigade, "Blackjack"	14 Feb '75 — — — — — — — — — — — — — — — — — —'90~ (Con't next era)
3rd Sqdn, 1st Cavalry Regt	05 May '71 — — — — — — — —'78; Inactv'd 21 Jun at Ft Hood, TX
2nd Bn (Armored), 13th Armored	05 May '71 — —'74; Inactv'd 23 Apr. Replc'd by 2nd Bn. 8th Cav, 1st Bde, above
1st Bn (Am/Inf), 5th Cavalry Regt	30 Jun '72 —'90~ (Con't next era)
1st Bn (Am/Inf), 8th Cavalry Regt	30 Jun '72 —'90~ (Con't next era)
3rd Battalion, 10th Cavalry Regt	01 Jul '77 ——'79; Inactv'd & Relv'd July from 1st Cav Div
1st Battalion, 32nd Armored Regt	16 Oct '86 — — — —'90~ (Con't next era)
4th Sqdn (Air), 9th Cavalry Regt	05 May '71~ Relv'd 30 Jun & Reactv'd 28 Feb '72, below
4th Sqdn (Air), 9th Cavalry Regt	28 Feb '72; Inactv'd 13 Sep & Relv'd from 1st Cav Div
4th Sqdn (Air), 9th Cavalry Regt	19 Nov '74/'75; Relv'd 21 Jul from 1st Cav Div & asgn'd to 6th Cav Bde
7th Sqdn (Air), 17th Cavalry Regt	13 Sep '72 — —'75; Relv'd 21 Jul from 1st Cav Div & asgn'd to 6th Cav Bde
170th Aviation Company	05 May '71/'72; Inactv'd 28 Feb at Ft Hood, TX
155th Armored Brigade (MSARNG)	'86 — — — —'90~
3rd Brigade, (Separate) NOTE: The 3rd Brigade (Separate) remained in Vietnam until 28 Jun 1972	
"HHC", 3rd Brigade, (Separate)	29 Jun '72~ Replc'd "HHC" 4th Bde on 29 Jun, below
2nd Battalion, 5th Cavalry Regiment	29 Jun '72; Inactv'd 31 Jul at Ft Hood, TX
1st Battalion, 7th Cavalry Regiment	29 Jun '72; Inactv'd 22 Aug at Ft Hood, TX
2nd Battalion, 8th Cavalry Regiment	29 Jun '72; Inactv'd 28 Jun at Ft Hood, TX
1st Battalion, 12th Cavalry Regiment	29 Jun '72~ Replc'd 2nd Bn, 12th Cav Regt on 29 Jun, 1st Bde, above
3rd Brigade, 1st Cavalry Division	
"HHC" 3rd Brigade, Greywolf"	29 Jun '72 — — — — — — — — — —'80; Inactv'd 15 Sep at Ft Hood, TX
2nd Bn, (Am/Inf) 7th Cavalry Regt	30 Jun '72 — — — — — — — — — — — — — — — — — — —'87~ Relv'd 16 Jan from 1st Cav Div
5th Bn, 7th Cavalry Regt	21 Sep '75 — —'78; Inactv'd 30 Sep at Ft Hood, TX
1st Bn, 67th Armor Regt, 2nd AD	'81 — — — — — — — — — — — —'90~ (Con't next era)

TRICAP TO ARMOR TIME LINE	05 May 1971 '72 '73 '74 '75 '76 '77 '78 '79 '80 '81 '82 '83 '84 '85 '86 '87 '88 '89 '90
4th Brigade, AirMobile Infantry	
"HHC", 4th AirMobile Inf Brigade	05 May '71/'72; Inactv'd 30 Jun at Ft Hood, TX
"HHC", 3rd Brigade, "GarryOwen"	
1st Bn, (Am/Inf) 5th Cavalry Regt	05 May '71/'72~ Xfer'd 30 Jun to 2nd Bde, above
2nd Bn, (Am/Inf) 7th Cavalry Regt	05 May '71/'72~ Xfer'd 30 Jun to 3rd Bde, above
1st Bn, (Am/Inf) 8th Cavalry Regt	05 May '71/'72~ Xfer'd 30 Jun to 2nd Bde, above
AVIATION BRIGADE, 1st Cav Div - Provisional	05 May '74 — — — — '84~ COMBAT AVIATION BRIGADE, below
"HHC", Aviation Brigade	05 May '74 — — — — '84~ COMBAT AVIATION BRIGADE, below
1st Sqdn, 6th Cavalry Regiment	22 Jun '73 — '75; Reliev'd 21 Feb from 1st Cav Div & asgn'd to 6th Cav Bde
1st Squadron, 9th Cavalry Regiment	08 Aug '77 — — — — '84~ COMBAT AVIATION BRIGADE, below
19th Combat Support Aviation Company	08 Aug '77 — — — — '84~ COMBAT AVIATION BRIGADE, below
General Support Aviation Company	08 Aug '77 — — — — '84~ COMBAT AVIATION BRIGADE, below
227th Aviation Battalion	30 Jun '72 — '74; Reorgn'd & Dsgn'd 19 Nov as "HHC", 227 Avn Co, below
"HHC", 227th Avn Battalion	30 Jun '72 — '74; Inactv'd 19 Nov at Ft Hood, TX
"A" Co, 227th Avn Battalion	30 Jun '72 — '74; Inactv'd 19 Nov at Ft Hood, TX
"B" Co, 227th Avn Battalion	30 Jun '72 — '74; Inactv'd 19 Nov at Ft Hood, TX
"C" Co, 227th Avn Battalion	30 Jun '72 — '74; Inactv'd 19 Nov at Ft Hood, TX
227th Aviation Company	19 Nov '74 — — — — '78; Reorgn'd & Dsgn'd 21 May as "HHC", 227 Avn Bn, below
"HHC", 227th Aviation Company	19 Nov '74 — — — — '78; Inactv'd 21 May at Ft Hood, TX
227th Aviation Battalion (AB)	21 May '78 — — — — '84~ COMBAT AVIATION BRIGADE, below
"HHC", 227th Aviation (AB)	21 May '78 — — — — '84~ COMBAT AVIATION BRIGADE, below
"A" Company, 227th Aviation Bn (AB)	21 May '78 — — — — '83; Inactv'd 30 Sep at Ft Hood, TX
"A" Company, 227th Aviation BN (AB)	16 Jun '83~ Dsgn'd 01 Oct as 228th Helicopter Battalion, below
"B" Company, 227th Aviation Bn (AB)	21 May '78 — — — — '84~ COMBAT AVIATION BRIGADE, below
"C" Company, 227th Aviation Bn (AB)	21 May '78 — — — — '84~ COMBAT AVIATION BRIGADE, below
"D" Company, 227th Aviation (General Support)	15 Dec '80 — — — — '84~ COMBAT AVIATION BRIGADE, below
"E" Company, 227th Aviation Bn (AB)	21 May '78 — — — — '83; Inactv'd 30 Sep & Relv'd from 1st Cav Div at Ft Hood, TX
228th Aviation Battalion	01 Oct '83/'84~ COMBAT AVIATION BRIGADE, below
"HHC", 228th Aviation Battalion (AH)	01 Oct '83/'84~ COMBAT AVIATION BRIGADE, below
"A" Company, 228th Aviation Battalion (AH)	01 Oct '83/'84~ COMBAT AVIATION BRIGADE, below
"B" Company, 228th Aviation Battalion (AH)	01 Oct '83/'84~ COMBAT AVIATION BRIGADE, below
"C" Company, 228th Aviation Battalion (AH)	01 Oct '83/'84~ COMBAT AVIATION BRIGADE, below
"D" Company, 228th Aviation Battalion (AH)	01 Oct '83/'84~ COMBAT AVIATION BRIGADE, below
COMBAT AVIATION BRIGADE, 1st Cavalry Division (Provisional)	14 Apr '84; Actv'd 01 Sep at Ft Hood, TX

TRICAP TO ARMOR TIME LINE	05 May 1971 '72 '73 '74 '75 '76 '77 '78 '79 '80 '81 '82 '83 '84 '85 '86 '87 '88 '89 '90
COMBAT AVIATION BRIGADE, 1st Cavalry Division	
"HHC" - "Warriors"	01 Sep '84 ——————'90~ (Con't next era)
1st Sqdn, 9th Cavalry Regiment	01 Sep '84 —'86~ Inactv'd & Relv'd 16 Oct
1st Sqdn, 7th Cavalry Regiment	18 Oct '86 ————'90~ (Con't next era)
7th Sqdn, 6th Cavalry Regiment (USAR Roundout Unit)	'84 ——————'90~ (Con't next era)
19th Combat Support Aviation Company	01 Sep '85 —'87~ Dsgn'd 16 Jul as "E" Co, 227th Avn Regt, below
General Support Aviation Company	'84/'85~ Co "D" & "E", below
1st Battalion, 227th Aviation (AB)	'84/'85~ Deactv'd 20 Dec & Relv'd from the 1st Cav Div
1st Battalion, 227th Aviation (AB)	16 Jul '87 — —'90~ (Con't next era)
"B" Company, 227th Aviation (AB)	01 Sep '84/'85; Inactv'd 01 Sep at Ft Hood, TX
"E" Company, 227th Aviation (AB)	01 Sep '84/'85 Inactv'd 01 Sept at Ft. Hood, TX
"E" Company, 227th Aviation (AB)	16 Jul '87 — —'90~ (Con't next era)
"D" Company, 227th Aviation (AB)	16 Jul '87 — —'90~ (Con't next era)
"E" Company, 227th Aviation (AB)	16 Jul '87 — —'90~ (Con't next era)
"F" Company, 227th Aviation (AB)	16 Jul '87 — —'90~ (Con't next era)
228th Aviation Battalion	01 Sep '84 — —'87; Inactv'd 16 Jul & Relv'd from the 1st Cav Div
"HHC", 228th Aviation Battalion (AB)	01 Sep '84 — —'87; Inactv'd 16 Jul & Relv'd from the 1st Cav Div
"A" Company, 228th Aviation Battalion (AB)	01 Sep '84 — —'87; Inactv'd 16 Jul & Relv'd from the 1st Cav Div
"B" Company, 228th Aviation Battalion (AB)	01 Sep '84 — —'87; Inactv'd 16 Jul & Relv'd from the 1st Cav Div
"C" Company, 228th Aviation Battalion (AB)	01 Sep '84 — —'87; Inactv'd 16 Jul & Relv'd from the 1st Cav Div
"D" Company, 228th Aviation Battalion (AB)	01 Sep '84 — —'87; Inactv'd 16 Jul & Relv'd from the 1st Cav Div
DIVISION ARTILLERY	
"HHB", Division Arty	05 May '71 ————————————————'90~ (Con't next era)
1st Battalion, 6th Artillery	05 May '71~ Reorg'd & Dsgn'd 01 Sep as 1st Bn, 6th FA, below
1st Battalion, 6th Field Artillery	01 Sep '71 ————'75; Relv'd 21 Jun from 1st Cav Div
2nd Battalion, 19th Field Artillery	20 Apr '74—————'79; Inactv'd 30 Mar at Ft Hood, TX
2nd Battalion, 20th Field Artillery	01 Sep '71/'72; Relv'd 13 Sep from 1st Cav Div
1st Battalion, 21st Field Artillery	01 Sep '71/'72 '75 ——————————'86;
"A" Btry, 21st Field Artillery	02 Jul '86 ————'90~ (Con't next era)
"I" Btry, 29th Field Artillery	21 Jun '76/'77; Inactv'd 21 June at Ft Hood, TX
1st Battalion, 68th Air Dfnce Artry	13 Sep '72 ——————————————'88; Inactv'd & Relv'd 16 Nov at Ft Hood, TX
"E" Btry, 1st Bn, 68th Air Dfnce Arty	01 Jul '77 ——————————'88; Inactv'd & Relv'd 16 Nov at Ft Hood, TX
1st Battalion, 77th Artillery	05 May '71~ Reorg'd & desg'd 01 Sep as 1st Bn, 77th FA, below
1st Battalion, 77th Field Artillery	01 Sep '71 ——————————————'86; Inactv'd 16 Jun & Relv'd from 1st Cav Div
2nd Battalion, 77th Field Artillery	'75 —'77;
"F" Btry, 79th Field Artillery	21 Mar '74; Inactv'd 20 Dec at Ft Hood, TX

TRICAP TO ARMOR TIME LINE	05 May 1971 '72 '73 '74 '75 '76 '77 '78 '79 '80 '81 '82 '83 '84 '85 '86 '87 '88 '89 '90
1st Battalion, 82nd Field Artillery	21 Jun '75 — — '90~ (Con't next era)
"E' Btry, 1st Bn, 82nd Field Arty	21 Jan '77 — — '87
3rd Battalion, 82nd Field Artillery	16 Jun '86 — — '90~ (Con't next era)
"A" Btry, 333rd Field Artillery	16 Aug '87 — — '90~ (Con't next era)
29th Chemical Detachment	31 Dec '76 — — '85; Inactv'd 15 Sep
68th Chemical Company	01 Jul '77 — — '90~ (Con't next era)
DIVISION SUPPORT COMMAND	
"HHC", Support Command, 1st Cav Div	05 May '71 — — '90~ (Con't next era)
Material Mgm't Center Bn, 1st Cav Div	21 Sep '76 — — '85; Inactv'd 15 Sep at Ft Hood, TX
15th Medical Battalion	05 May '71 — — '85; Inactv'd 15 Sep at Ft Hood, TX
15th Supply & Transport Battalion	05 May '71 — — '85; Inactv'd 15 Sep at Ft Hood, TX
27th Maintenance Battalion	05 May '71 — — '85; Inactv'd 15 Sep at Ft Hood, TX
4th Main Support Battalion	20 Jun '85 — '87~ Dsgn'd 10 May as 27th SB, below
27th Support Battalion (Main)	10 May '87 — — '90~ (Con't next era)
34th Support Battalion	30 Nov '71 — — '75; Relv'd 21 Jul from 1st Cav Div & asgn'd to 6th Cav Bde
38th Support Battalion	'74/'75
1st Forward Supt Bn, 1st Cav Div	01 Oct '84 — — '87~ Dsgn'd 01 May as 115 SB, below
115th Support Battalion (FWD)	01 May '87 — — '90 (Con't next era)
2nd Forward Supt Bn, 1st Cav Div	01 Oct '84 — — '87~ Dsgn'd 01 May as 15th SB, below
15th Support Battalion, (FWD)	01 May '87 — — '90~ (Con't next era)
315th Composite Supt Battalion	05 May '71
493rd Transportation Maintenance Company	20 Jun '85 — '87~ Dsgn'd 01 May 227th TAM, below
227th Transportation Maintenance Company	01 May '87 — — '90~ (Con't next era)
SEPARATE BATTALIONS/COMPANIES	
"HHC", 1st Cavalry Division (TRICAP)	05 May '71 — — '75~ Dsgn'd ARMOR 21 Feb
"HHC", 1st Cavalry Division (ARMOR)	21 Feb '75 — — '90~ (Con't next era)
Band, 1st Cavalry Division	16 Feb '87 — — '90~ (Con't next era)
1st Cavalry Horse Detachment	Dec '76 — — '90~ (Con't next era)

TRICAP TO ARMOR TIME LINE	05 May 1971 '72 '73 '74 '75 '76 '77 '78 '79 '80 '81 '82 '83 '84 '85 '86 '87 '88 '89 '90
8th Engineer Battalion	05 May '71 — — — '90~ (Con't next era)
13th Signal Battalion	05 May '71 — — — '90~ (Con't next era)
55th Signal Company	21 Feb '75; Relv'd 21 Jul from 1st Cav Div & asgn'd to 6th Cav Bde
15th Adjutant General Company	05 May '71 — — — '87;
15th Data Processing Unit	05 May '71
15th Finance Company	05 May '71 — — — '82~ Relv'd 16 Mar Xfer'd DIRECT SUPPORT, below
191st Military Intelligence Company (Prov)	02 Jul '77 — — — '81~ Mission asgn'd to 312 MI, below
312th Military Intelligence Company	01 Oct '81 — — — '90~ (Con't next era)
371 Army Security Agency (ASA)	05 May '71 — — — '76; Relv'd Oct & Asgn'd to FORSCOM
545 Military Police Company	05 May '71 — — — '90~ (Con't next era)
230th Avn Battalion (Note 1)	05 May '71; Inactv'd 30 June at Ft Hood, TX

Note 1: The 230th Avn Battalion was an accounting organization setup to receive assets of the Vietnam Avn Organizations.

DIRECT SUPPORT UNITS	
15th Finance Company	07 Apr '83 — — — '87~ Dsgn'd 16 Apr as 15th Finance Supt Unit, below
15th Finance Support Unit	16 Apr '87 — — '90~ (Con't next era)
15th Personnel Service Company	01 Oct '85 — — — '90~ (Con't next era)

PERSIAN GULF WAR
16 September 1990 - 07 March 1991

PERSIAN GULF WAR TIME LINE	16 Sep 1990 '91
"HHC", 1st Cavalry Division	16 Sep '90/'91~ Ret'rd to CONUS 07 Mar (Con't next era)
1st BRIGADE, 1st Cavalry Division "HHC", 1st Brigade - "Iron Horse"	16 Sep '90/'91~ Ret'rd to CONUS 07 Mar (Con't next era)

PERSIAN GULF WAR TIME LINE	16 Sep 1990 '91
2nd Battalion, 5th Cavalry Regiment	16 Sep '90/'91~ Ret'rd to CONUS (Con't next era)
"HHC", 2nd Bn, 5th Cavalry Regt	
"A" Co, 2nd Bn, 5th Cavalry Regt	
"B" Co, 2nd Bn, 5th Cavalry Regt	
"C" Co, 2nd Bn, 5th Cavalry Regt	
"D" Co, 2nd Bn, 5th Cavalry Regt	
2nd Battalion, 8th Cavalry Regiment	16 Sep '90/'91~ Ret'rd to CONUS (Con't next era)
"HHC", 2nd Bn, 8th Cavalry Regt	
"A" Co, 2nd Bn, 8th Cavalry Regt	
"B" Co, 2nd Bn, 8th Cavalry Regt	
"C" Co, 2nd Bn, 8th Cavalry Regt	
"D" Co, 2nd Bn, 8th Cavalry Regt	
3rd Battalion, 32nd Armor Regiment	16 Sep '90/'91~ Ret'rd to CONUS (Con't next era)
"HHC", 3rd Bn, 32nd Armor Regt	
"A" Co, 3rd Bn, 32nd Armor Regt	
"B" Co, 3rd Bn, 32nd Armor Regt	
"C" Co, 3rd Bn, 32nd Armor Regt	
"D" Co, 3rd Bn, 32nd Armor Regt	
2nd BRIGADE, 1st Cavalry Division	
"HHC", 2nd Brigade - "Blackjack"	16 Sep '90/'91~ Ret'rd to CONUS (Con't next era)
1st Battalion, 5th Cavalry Regiment	16 Sep '90/'91~ Ret'rd to CONUS (Con't next era)
"HHC", 1st Bn, 5th Cavalry Regt	
"A" Co, 1st Bn, 5th Cavalry Regt	
"B" Co, 1st Bn, 5th Cavalry Regt	
"C" Co, 1st Bn, 5th Cavalry Regt	
"D" Co, 1st Bn, 5th Cavalry Regt	
1st Battalion, 8th Cavalry Regiment	16 Sep '90/'91~ Ret'rd to CONUS (Con't next era)
"HHC", 1st Bn, 8th Cavalry Regt	
"A" Co, 1st Bn, 8th Cavalry Regt	
"B" Co, 1st Bn, 8th Cavalry Regt	
"C" Co, 1st Bn, 8th Cavalry Regt	
"D" Co, 1st Bn, 8th Cavalry Regt	
1st Battalion, 32nd Armor Regiment	16 Sep '90/'91~ Ret'rd to CONUS (Con't next era)
"HHC", 1st Bn, 32nd Armor Regt	
"A" Co, 1st Bn, 32nd Armor Regt	
"B" Co, 1st Bn, 32nd Armor Regt	
"C" Co, 1st Bn, 32nd Armor Regt	
"D" Co, 1st Bn, 32nd Armor Regt	

PERSIAN GULF WAR TIME LINE	16 Sep 1990 '91
1st (TIGER) BRIGADE	
"HHC", 1st (Tiger) Brigade	04 Oct '90; Relv'd 30 Dec & Asgn'd to 2nd Marine Div
3rd Battalion, 41st Infantry Regt	04 Oct '90; Relv'd 30 Dec & Asgn'd to 2nd Marine Div
"HHC", 3rd Bn, 41st Inf Regt	
"A" Co, 3rd Bn,41st Inf Regt	
"B" Co, 3rd Bn,41st Inf Regt	
"C" Co, 3rd Bn,41st Inf Regt	
"D" Co, 3rd Bn,41st Inf Regt	
1st Battalion, 67th Armor Regt	04 Oct '90; Relv'd 30 Dec & Asgn'd to 2nd Marine Div
"HHC", 1st Bn, 67th Armor Regt	
"A" Co, 1st Bn, 67th Armor Regt	
"B" Co, 1st Bn, 67th Armor Regt	
"C" Co, 1st Bn, 67th Armor Regt	
"D" Co, 1st Bn, 67th Armor Regt	
3rd Battalion, 67th Armor Regt	04 Oct '90; Relv'd 30 Dec & Asgn'd to 2nd Marine Div
"HHC", 3rd Bn, 67th Armor Regt	
"A" Co, 3rd Bn, 67th Armor Regt	
"B" Co, 3rd Bn, 67th Armor Regt	
"C" Co, 3rd Bn, 67th Armor Regt	
"D" Co, 3rd Bn, 67th Armor Regt	
COMBAT AVIATION BRIGADE, 1st Cavalry Division	
"HHC", Combat Aviation, "Warriors"	16 Sep '90/'91~ Ret'rd to CONUS (Con't next era)
1st Battalion, 3rd Aviation Regt	16 Sep '90/'91~ Ret'rd to CONUS (2nd Armored Div)
"HHC", 1st Bn, 3rd Avn Regt	
"A" Co, 1st Bn, 3rd Avn Regt	
"B" Co, 1st Bn, 3rd Avn Regt	
"C" Co, 1st Bn, 3rd Avn Regt	
"D" Co, 1st Bn, 3rd Avn Regt	
1st Battalion, 227th Aviation Regt	16 Sep '90/'91~ Ret'rd to CONUS Apr (Con't next era)
"HHC", 1st Bn, 227th Avn Regt	
"A" Co, 1st Bn, 227th Avn Regt	
"B" Co, 1st Bn, 227th Avn Regt	
"C" Co, 1st Bn, 227th Avn Regt	
"D" Co, 1st Bn, 227th Avn Regt	
"E" Co, 1st Bn, 227th Avn Regt	
"F" Co, 1st Bn, 227th Avn Regt	

PERSIAN GULF WAR TIME LINE	16 Sep 1990 '91
ARTILLERY, 1st Cavalry Division	
"HHB", Division Arty - "Red Team"	16 Sep '90/'91~ Ret'rd to CONUS (Con't next era)
1st Battalion, 3rd Field Artillery	
"HHB", 1st Bn, 3rd Field Arty	
"A" Btry, 1st Bn, 3rd Field Arty	
"B" Btry, 1st Bn, 3rd Field Arty	
"C" Btry, 1st Bn, 3rd Field Arty	
Service Btry, 1st Bn, 3rd Field Arty	
1st Battalion, 82nd Field Arty	08 Oct '90/'91~ Ret'rd to CONUS 06 Apr (Con't next era)
"HHB", 1st Bn, 82nd Field Arty	
"A" Btry, 1st Bn, 82nd Field Arty	
"B" Btry, 1st Bn, 82nd Field Arty	
"C" Btry, 1st Bn, 82nd Field Arty	
Service Btry, 1st Bn, 82nd Field Arty	
3rd Battalion, 82nd Field Artillery	08 Oct '90/'91~ Ret'rd to CONUS 06 Apr (Con't next era)
"HHB", 3rd Bn, 82nd Field Arty	
"A" Btry, 1st Bn, 82nd Field Arty	
"B" Btry, 1st Bn, 82nd Field Arty	
"C" Btry, 1st Bn, 82nd Field Arty	
Service Btry, 3rd Bn, 82nd Field Arty	
"A" Btry, 21st Field Artillery	07 Oct '90/'91~ Ret'rd to CONUS 21 Apr (Con't next era)
"A" Btry, 92nd Field Artillery	
"A" Btry, 333rd Field Artillery	05 Oct '90/'91~ Ret'rd to CONUS 21 Apr (Con't next era)
DIVISION SUPPORT CMD, 1st Cavalry Division	
"HHC", Support Cmd - "Wagon Masters"	29 Sep '90/'91~ Ret'rd to CONUS 06 Apr (Con't next era)
15th Support Battalion (FWD)	12 Oct '90/'91~ Ret'rd to CONUS 21 Apr (Con't next era)
"HHC", 15th Support Bn (FWD)	
"A" Co, 15th Support Bn (FWD)	
"B" Co, 15th Support Bn (FWD)	
"C" Co, 15th Support Bn (FWD)	
27th Support Battalion (MAIN)	25 Sep '90/'91~ Ret'rd to CONUS 18 Apr (Con't next era)
"HHC", 27th Support Bn (MAIN)	
"A" Co, 27th Support Bn (MAIN)	
"B" Co, 27th Support Bn (MAIN)	
"C" Co, 27th Support Bn (MAIN)	
"D" Co, 27th Support Bn (MAIN)	
"E" Co, 27th Support Bn (MAIN)	

PERSIAN GULF WAR TIME LINE	16 Sep 1990 '91
43rd Corps Support Group	25 Oct '90/'91; 19 Apr
"HHC", 43rd Support Group	25 Oct '90/'91; 19 Apr
"HHC", 544th Command Support Bn	26 Oct '90/'91; 19 May
53rd Quartermaster Co	26 Oct '90/'91; 19 May
418th Transportation Co	26 Oct '90/'91; 19 May
"HHC", 169th Command Support Bn	25 Oct '90/'91; 19 Apr
96th Transportation Co	11 Oct '90/'91; 21 Apr
254th Maintenance Det	21 Oct '90/'91; 19 May
340th Service and Support Co	22 Dec '90/'91; 19 May
452nd Ordnance Company	25 Oct '90/'91; 19 Apr
530th Maintenance Company	22 Oct '90/'91; 22 Apr
597th Maintenance Company	15 Oct '90/'91; 19 May
644th Transportation Co (HVY)	25 Oct '90/'91; 19 Apr
915th Transportation Co (MED)	25 Oct '90/'91; 19 Apr
1001st Field Service Co	12 Nov '90/'91; 22 Apr
1544th Transportation Co	06 Nov '90/'91; 23 Apr
91st Chemical Company	05 Oct '90/'91; 26 Mar
115th Support Battalion (FWD)	
"HHC", 115th Support Bn (FWD)	08 Oct '90/'91~ Ret'rd to CONUS 21 Apr (Con't next era)
"A" Co, 115th Support Bn (FWD)	
"B" Co, 115th Support Bn (FWD)	
"C" Co, 115th Support Bn (FWD)	
"HHD", 169th Maintenance Battalion	04 Nov '90/'91; 19 May
183rd Maintenance Company	02 Nov '90/'91; 19 Apr
240th Supply and Service Company	11 Nov '90/'91; 01 May
254th Maintenance Company	21 Sep '90/'91; 19 May
340th Supply and Service Company	22 Dec '90/'91; 19 May
502nd Support Battalion (FWD)	
SEPARATE BNs/COs, 1st Cavalry Division - "Maverick"	
"HHC", 1st Cavalry Division	
Band, 1st Cavalry Division	
1st Sqdn, 7th Cavalry Regiment	28 Sep '90/'91~ Ret'rd to CONUS 14 Apr (Con't next era)
"HHT", 1st Sqdn, 7th Cav Regt	
"A" Trp, 1st Sqdn, 7th Cav Regt	
"B" Trp, 1st Sqdn, 7th Cav Regt	
"C" Trp, 1st Sqdn, 7th Cav Regt	
"D" Trp, 1st Sqdn, 7th Cav Regt	
"E" Trp, 1st Sqdn, 7th Cav Regt	
"F" Trp, 1st Sqdn, 7th Cav Regt	

PERSIAN GULF WAR TIME LINE	16 Sep 1990 '91
4th Battalion, 5th Air Defense	14 Oct '90/'91~ Ret'rd to CONUS 06 Apr (Con't next era)
"HHB", 4th Bn, 5th Air Defense	
"A" Btry, 4th Bn, 5th Air Dfnse	
"B" Btry, 4th Bn, 5th Air Dfnse	
"C" Btry, 4th Bn, 5th Air Dfnse	
"D" Btry, 4th Bn, 5th Air Dfnse	
8th Engineer Battalion	24 Sep '90/'91~ Ret'rd to CONUS 15 Mar (Con't next era)
"HHC", 8th Engineer Bn	
"A" Co, 8th Engineer Bn	
"B" Co, 8th Engineer Bn	
"C" Co, 8th Engineer Bn	
Service Plt, 8th Engineer Bn	
13th Signal Battalion	23 Sep '90/'91~ Ret'rd to CONUS 15 Mar (Con't next era)
"HHC", 13th Signal Bn	
"A" Co, 13th Signal Bn	
"B" Co, 13th Signal Bn	
"C" Co, 13th Signal Bn	
"D" Co, 13th Signal Bn	
15th Finance Battalion	
15th Personnel Service Battalion	
"A" Company, 17th Engineer Battalion	
44th Chemical Co (NBC Defense)	
68th Chemical Company	14 Oct '90/'91~ Ret'rd to CONUS 21 Apr (Con't next era)
123rd Engineer Detachment (Terrain)	
312th Mltry Intelligence Battalion	04 Oct '90/'91~ Ret'rd to CONUS 06 Apr (Con't next era)
"HHC", 312th Mltry Intel Bn	
"A" Co, 312th Mltry Intel Bn	
"B" Co, 312th Mltry Intel Bn	
"C" Co, 312th Mltry Intel Bn	
"D" Co, 312th Mltry Intel Bn	
1st Plt, 502nd Miltry Police Company	

PERSIAN GULF WAR TIME LINE 16 Sep 1990 '91

545th Mltry Police Company 15 Sep '90/'91~ Ret'rd to CONUS 20 Apr (Con't next era)

577 Engineer Detachment (Terrain)

Trial Defense Service

US Air Force, Teams 1 & 2

ATTACHED UNITS

Cavalry
2nd Sqdn, 1st Cavalry Regt, 2nd Armored Div ' '90/'91;

Ordnance
157th Ordnance Det (TOW/Dragon)

Artillery
"C" Btry, 26th Field Artillery '90/'91;
42nd Field Artillery Brigade '90/'91;
75th Field Artillery Brigade '90/'91;
212th Field Artillery Brigade '90/'91;

ASSIGNED UNITS
4th Finance Battalion '90/'91;
50th Ordnance Co '90/'91;
73rd Maintenance Co '90/'91;
571st Medical Det (Air Ambulance) '90/'91;

NON-DIVISIONAL UNITS
2nd Sqdn, 1st Cavalry Regt, 2nd Armored Div '90/'91;
4th and 13th Public Affairs Teams '90/'91;
Recon Plt, 22nd Chem Co, 3rd Armored Div '90/'91;
59th Military Police Co '90/'91;
145th Public Affairs Det (OKARNG) '90/'91;
159th Mobile Army Surgical Hospital '90/'91;
300th Adjutant General Postal Co '90/'91;
413th Civil Affairs Co (USARTX) '90/'91;
476th Transportation Co (Prov) '91;
GSR Section, 522nd Mltry Intel '90/'91;
ATC Section, 2nd Bn, 58th Aviation '90/'91;
ARCENT Liaison Team '90/'91;

PERSIAN GULF WAR TIME LINE	16 Sep 1990 '91
ASSIGNED TO ARCENT	
2nd Sqdn, 1st Cavalry, 2nd Armored	'90/'91;
68th Transportation Battalion	'90/'91;
360th Transportation Company	'90/'91;
382nd Transportation Detachment	'90/'91;
586th Transportation Detachment	'90/'91;
Recon Plt, 22nd Chemical Co, 3rd Armored	'90/'91;

FORCE RESTRUCTURING
08 Mar 1991 - 01 October 1999

FORCE RESTRUCTURING TIME LINE	08 Mar 1991 '92 '93 '94 '95 '96 '97 '98 '99
"HHC", 1st Cavalry Division	08 Mar '91 — — — — — — — — — —'99 (Con't next era)
1st BRIGADE, 1st Cavalry Division	
"HHC", 1st Brigade - "IronHorse"	08 Mar '91 — — — — — — — — — —'99 (Con't next era)
2nd Battalion, 5th Cavalry Regiment	08 Mar '91 — — — — — — — — — —'99 (Con't next era)
"HHC", 2nd Bn, 5th Cavalry Regt	
"A" Co, 2nd Bn, 5th Cavalry Regt	
"B" Co, 2nd Bn, 5th Cavalry Regt	
"C" Co, 2nd Bn, 5th Cavalry Regt	
"D" Co, 2nd Bn, 5th Cavalry Regt	
2nd Battalion, 8th Cavalry Regiment	08 Mar '91 — — — — — — — — — —'99 (Con't next era)
"HHC", 2nd Bn, 8th Cavalry Regt	
"A" Co, 2nd Bn, 8th Cavalry Regt	
"B" Co, 2nd Bn, 8th Cavalry Regt	
"C" Co, 2nd Bn, 8th Cavalry Regt	
"D" Co, 2nd Bn, 8th Cavalry Regt	
3rd Battalion, 32nd Armor Regiment	08 Mar '91/'92; Rdsgn'd 16 Dec as 1st Bn, 12th Cav Regt, below
"HHC", 3rd Bn, 32nd Armor Regt	
"A" Co, 3rd Bn, 32nd Armor Regt	
"B" Co, 3rd Bn, 32nd Armor Regt	
"C" Co, 3rd Bn, 32nd Armor Regt	
"D" Co, 3rd Bn, 32nd Armor Regt	

FORCE RESTRUCTURING TIME LINE	08 Mar 1991 ’92 ’93 ’94 ’95 ’96 ’97 ’98 ’99
1st Battalion, 12th Cavalry Regiment	16 Dec ’92 — — — — — — — — —’99 (Con’t next era)
“HHC”, 1st Bn, 12th Cavalry Regt	
“A” Co, 1st Bn, 12th Cavalry Regt	
“B” Co, 1st Bn, 12th Cavalry Regt	
“C” Co, 1st Bn, 12th Cavalry Regt	
“D” Co, 1st Bn, 12th Cavalry Regt	
2nd BRIGADE, 1st Cavalry Division	
“HHC”, 2nd Brigade - “BlackJack”	08 Mar ’91 — — — — — — — — — —’99 (Con’t next era)
1st Battalion, 5th Cavalry Regiment	08 Mar ’91 — — — — — — — — — —’99 (Con’t next era)
“HHC”, 1st Bn, 5th Cavalry Regt	
“A” Co, 1st Bn, 5th Cavalry Regt	
“B” Co, 1st Bn, 5th Cavalry Regt	
“C” Co, 1st Bn, 5th Cavalry Regt	
“D” Co, 1st Bn, 5th Cavalry Regt	
1st Battalion, 8th Cavalry Regiment	08 Mar ’91 — — — — — — — — — —’99 (Con’t next era)
“HHC”, 1st Bn, 8th Cavalry Regt	
“A” Co, 1st Bn, 8th Cavalry Regt	
“B” Co, 1st Bn, 8th Cavalry Regt	
“C” Co, 1st Bn, 8th Cavalry Regt	
“D” Co, 1st Bn, 8th Cavalry Regt	
1st Battalion, 32nd Armor Regiment	08 Mar ’91/’92; Rdsgn’d 16 Dec as 2nd Bn, 12th Cav Regt, below
“HHC”, 1st Bn, 32nd Armor Regt	
“A” Co, 1st Bn, 32nd Armor Regt	
“B” Co, 1st Bn, 32nd Armor Regt	
“C” Co, 1st Bn, 32nd Armor Regt	
“D” Co, 1st Bn, 32nd Armor Regt	
2nd Battalion, 12th Cavalry Regiment	16 Dec ’92— — — — — — — — —’99 (Con’t next era)
“HHC”, 2nd Bn, 12th Cavalry Regt	
“A” Co, 2nd Bn, 12th Cavalry Regt	
“B” Co, 2nd Bn, 12th Cavalry Regt	
“C” Co, 2nd Bn, 12th Cavalry Regt	
“D” Co, 2nd Bn, 12th Cavalry Regt	
3rd BRIGADE, 1st Cavalry Division	
“HHC”, 3rd Brigade - “GreyWolf”	21 May ’91 — — — — — — — — — —’99 (Con’t next era)

FORCE RESTRUCTURING TIME LINE	08 Mar 1991 ’92 ’93 ’94 ’95 ’96 ’97 ’98 ’99
3rd Battalion, 41st Inf Regiment	08 Mar ’91/’92; Dsgn’d 29 Nov as 1st Bn, 9th Cav Regt, below
“HHC”, 3rd Bn, 41st Inf Regt	
“A” Co, 3rd Bn,41st Inf Regt	
“B” Co, 3rd Bn,41st Inf Regt	
“C” Co, 3rd Bn,41st Inf Regt	
“D” Co, 3rd Bn,41st Inf Regt	
1st Battalion, 67th Armor Regiment	08 Mar ’91/’92; Dsgn’d 29 Nov as 3rd Bn, 8th Cav Regt, below
“HHC”, 1st Bn, 67th Armor Regt	
“A” Co, 1st Bn, 67th Armor Regt	
“B” Co, 1st Bn, 67th Armor Regt	
“C” Co, 1st Bn, 67th Armor Regt	
“D” Co, 1st Bn, 67th Armor Regt	
3rd Battalion, 67th Armor Regiment	08 Mar ’91/’92; Relv’d 29 Nov from 1st Cav Div
“HHC”, 3rd Bn, 67th Armor Regt	
“A” Co, 3rd Bn, 67th Armor Regt	
“B” Co, 3rd Bn, 67th Armor Regt	
“C” Co, 3rd Bn, 67th Armor Regt	
“D” Co, 3rd Bn, 67th Armor Regt	
4th Battalion, 6th Infantry Regiment	’93; Redsgn’d 11 Aug as 2nd Bn, 7th Cavalry Regt, below
“HHC”, 4th Bn, 6th Infantry Regt	
“A” Co, 4th Bn, 6th Infantry Regt	
“B” Co, 4th Bn, 6th Infantry Regt	
“C” Co, 4th Bn, 6th Infantry Regt	
“D” Co, 4th Bn, 6th Infantry Regt	
2nd Battalion, 7th Cavalry Regiment	11 Aug ’93 — — — — — — — —’99~ (Con’t next era)
“HHC”, 2nd Bn, 7th Cavalry Regt	
“A” Co, 2nd Bn, 7th Cavalry Regt	
“B” Co, 2nd Bn, 7th Cavalry Regt	
“C” Co, 2nd Bn, 7th Cavalry Regt	
“D” Co, 2nd Bn, 7th Cavalry Regt	
3rd Battalion, 8th Cavalry Regiment	29 Nov ’92 — — — — — —’99~ (Con’t next era)
“HHC”, 3rd Bn, 8th Cavalry Regt	
“A” Co, 3rd Bn, 8th Cavalry Regt	
“B” Co, 3rd Bn, 8th Cavalry Regt	
“C” Co, 3rd Bn, 8th Cavalry Regt	
“D” Co, 3rd Bn, 8th Cavalry Regt	

FORCE RESTRUCTURING TIME LINE	08 Mar 1991 '92 '93 '94 '95 '96 '97 '98 '99
1st Battalion, 9th Cavalry Regiment	29 Nov '92 — — — — — — — — —'99~ (Con't next era)
"HHC", 1st Bn, 9th Cavalry Regt	
"A" Co, 1st Bn, 9th Cavalry Regt	
"B" Co, 1st Bn, 9th Cavalry Regt	
"C" Co, 1st Bn, 9th Cavalry Regt	
"D" Co, 1st Bn, 9th Cavalry Regt	
COMBAT AVIATION BRIGADE, 1st Cavalry Division	
"HHC", Combat Aviation - "Warriors"	
1st Sqdn, 7th Cavalry Regiment	14 Apr '91 — — — — — — — — — —'99~ (Con't next era)
"HHT", 1st Sqdn, 7th Cavalry Regt	
"A" Trp, 1st Sqdn, 7th Cavalry Regt	
"B" Trp, 1st Sqdn, 7th Cavalry Regt	
"C" Trp, 1st Sqdn, 7th Cavalry Regt	
"D" Trp, 1st Sqdn, 7th Cavalry Regt	
"E" Trp, 1st Sqdn, 7th Cavalry Regt	
"F" Trp, 1st Sqdn, 7th Cavalry Regt	
1st Battalion, 227thth Aviation Regt	Apr '91 — — — — — — — — — —'99~ (Con't next era)
"HHC", 1st Bn, 227th Avn Regt	
"A" Co, 1st Bn, 227th Avn Regt	
"B" Co, 1st Bn, 227th Avn Regt	
"C" Co, 1st Bn, 227th Avn Regt	
"D" Co, 1st Bn, 227th Avn Regt	Apr '91/'92~ Dsgn'd 06 Nov as 4th Bn, 227th Avn, below
4th Battalion, 227th Aviation (Prov)	06 Nov '92/'93~, below
4th Battalion, 227th Aviation	16 Dec '93 — — — —'97~ Dsgn'd 25 Jun 2nd Bn, 227th Avn, below
2nd Battalion, 227th Aviation Regiment	25 Jun '97 —'99~ (Con't next era)
"HHC", 2nd Bn, 227th Avn Regt	
"A" Co, 2nd Bn, 227th Avn Regt	
"B" Co, 2nd Bn, 227th Avn Regt	
"C" Co, 2nd Bn, 227th Avn Regt	
"D" Co, 2nd Bn, 227th Avn Regt	
"F" Co, 1st Bn, 58th Aviation (ATGC)	
ARTILLERY, 1st Cavalry Division	
"HHB", Division Arty - "Red Team"	21 Apr '91 — — — — — — — — — —'99~ (Con't next era)
"A" Battery, 21st Field Artillery	21 Apr '91 — — — — — — — —'97; Reorgn'd 16 Sep, below
"A" Battery, 333rd Field Artillery	21 Apr '91/'92; Redsgn'd 16 Dec as "B" Btry, 26th FA, below
"B" Battery, 26th Field Artillery	16 Dec '92 — — — — —'97~ Deactv'd 16 Sep; Actv'd as part of 21st FA, below

FORCE RESTRUCTURING TIME LINE	08 Mar 1991 '92 '93 '94 '95 '96 '97 '98 '99
1st Battalion, 21st Field Artillery	16 Sep '97 ——'99~ (Con't next era)
"HHS", Battery, 1st Bn, 21st Field Arty	
"A" Battery, 1st Bn, 21st Field Arty	
"B" Battery, 1st Bn, 21st Field Arty	
"C" Battery, 1st Bn, 21st Field Arty	
"D" Battery, 1st Bn, 21st Field Arty (TXARNG)	
1st Battalion, 82nd Field Artillery	06 Apr '91 — — — — — — — — —'99~ (Con't next era)
"HHB", 1st Bn, 82nd Field Arty	
"A" Btry, 1st Bn, 82nd Field Arty	
"B" Btry, 1st Bn, 82nd Field Arty	
"C" Btry, 1st Bn, 82nd Field Arty	
Service Btry, 1st Bn, 82nd Field Arty	
2nd Battalion, 82nd Field Artillery	16 Dec '92 — — — — — — — —'99~ (Con't next era)
"HHB", 2nd Bn, 82nd Field Arty	
"A" Btry, 2nd Bn, 82nd Field Arty	
"B" Btry, 2nd Bn, 82nd Field Arty	
"C" Btry, 2nd Bn, 82nd Field Arty	
Service Btry, 2nd Bn, 82nd Field Arty	
3rd Battalion, 82nd Field Artillery	06 Apr '91 — — — — — — — — —'99~ (Con't next era)
"HHB", 3rd Bn, 82nd Field Arty	
"A" Btry, 3rd Bn, 82nd Field Arty	
"B" Btry, 3rd Bn, 82nd Field Arty	
"C" Btry, 3rd Bn, 82nd Field Arty	
Service Btry, 3rd Bn, 82nd Field Arty	
68th Chemical Company	21 Apr '91 — — — — — — — — —'99~ (Con't next era)
ENGINEER BRIGADE, 1st Cavalry Division	
"HHD", Engineer Brigade - "Snappers"	16 Oct '92 — — — — — — — —'99~ (Con't next era)
8th Engineer Battalion	15 Mar '91 — — — — — — — — —'99~ (Con't next era)
"HHC", 8th Engineer Batallion	
"A" Co, 8th Engineer Bn	
"B" Co, 8th Engineer Bn	
"C" Co, 8th Engineer Bn	
Service Plt, 8th Engineer Bn	
20th Engineer Battalion	16 Oct '92 — — — — — — — —'99~ (Con't next era)
"HHC", 20th Engineer Bn	
"A" Co, 20th Engineer Bn	
"B" Co, 20th Engineer Bn	
"C" Co, 20th Engineer Bn	
Service Plt, 20th Engineer Bn	

FORCE RESTRUCTURING TIME LINE	08 Mar 1991 '92 '93 '94 '95 '96 '97 '98 '99
91st Engineer Battalion	16 Oct '92 — — — — — — — — —'99~ (Con't next era)
"HHC", 91st Engineer Bn	
"A" Co, 91st Engineer Bn	
"B" Co, 91st Engineer Bn	
"C" Co, 91st Engineer Bn	
Service Plt, 91st Engineer Bn	
DIVISION SUPPORT COMMAND, 1st Cavalry Division	
"HHC", Support Cmd - Wagon Masters"	06 Apr '91 — — — — — — — — — —'99~ (Con't next era)
15th Support Battalion (FWD)	21 Apr '91 — — — — — — — — — —'99~ (Con't next era)
"HHC", 15th Support Bn (FWD)	
"A" Co, 15th Support Bn (FWD)	
"B" Co, 15th Support Bn (FWD)	
"C" Co, 15th Support Bn (FWD)	
27th Support Battalion (MAIN)	18 Apr '91 — — — — — — — — — —'99~ (Con't next era)
"HHC", 27th Support Bn (MAIN)	
"A" Co, 27th Support Bn (MAIN)	
"B" Co, 27th Support Bn (MAIN)	
"C" Co, 27th Support Bn (MAIN)	
"D" Co, 27th Support Bn (MAIN)	
"E" Co, 27th Support Bn (MAIN)	
115th Support Battalion (FWD)	21 Apr '91 — — — — — — — — — —'99~ (Con't next era)
"HHC", 115th Support Bn (FWD)	
"A" Co, 115th Support Bn (FWD)	
"B" Co, 115th Support Bn (FWD)	
"C" Co, 115th Support Bn (FWD)	
215th Support Battalion (FWD)	16 Dec '91 — — — — — — — — — —'99~ (Con't next era)
"HHC", 215th Support Bn (FWD)	
"A" Co, 215th Support Bn (FWD)	
"B" Co, 215th Support Bn (FWD)	
"C" Co, 215th Support Bn (FWD)	
615th Support Battalion (AVN) (Prov)	08 Jul '96~ Actv'd 16 Sep, below
615th Support Battalion (AVN)	16 Sep '96— —'99~ (Con't next era)
"HHC", Support Bn (AVN)	
"A" Co, 615th Support Bn (AVN)	
"B" Co, 615th Support Bn (AVN)	
Division Material Management Center (DMMC)	

FORCE RESTRUCTURING TIME LINE	08 Mar 1991 ’92 ’93 ’94 ’95 ’96 ’97 ’98 ’99
SEPARATE BNs/COs, 1st Cavalry Division - “Maverick”	
“HHC”, 1st Cavalry Division	08 Mar ’91 — — — — — — — — —’99 (Con’t next era)
Band, 1st Cavalry Division	08 Mar ’91 — — — — — — — — —’99 (Con’t next era)
1st Cavalry Horse Detachment	08 Mar ’91 — — — — — — — — —’99 (Con’t next era)
4th Battalion, 5th Air Defense	06 Apr ’91 — — — — — — — — —’99 (Con’t next era)
“HHB”, 4th Bn, 5th Air Defense	
“A” Btry, 4th Bn, 5th Air Defense	
“B” Btry, 4th Bn, 5th Air Defense	
“C” Btry, 4th Bn, 5th Air Defense	
“D” Btry, 4th Bn, 5th Air Defense	
13th Signal Battalion	15 Mar ’91 — — — — — — — — —’99 (Con’t next era)
“HHC”, 13th Signal Bn	
“A” Co, 13th Signal Bn	
“B” Co, 13th Signal Bn	
“C” Co, 13th Signal Bn	
“D” Co, 13th Signal Bn	
312th Mltry Intelligence Battalion	06 Apr ’91 — — — — — — — — —’99 (Con’t next era)
“HHC”, 312th Mltry Intelligence Bn	
“A” Co, 312th Mltry Intelligence Bn	
“B” Co, 312th Mltry Intelligence Bn	
“C” Co, 312th Mltry Intelligence Bn	
“D” Co, 312th Mltry Intelligence Bn	
545th Mltry Police Company	20 Mar ’91 — — — — — — — — —’99 (Con’t next era)
DIRECT SUPPORT	
9th Air Support Operations	08 Mar ’91 — — — — — — — — —’99 (Con’t next era)
15th Finance Battalion	08 Mar ’91 — — — — — — — — —’99 (Con’t next era)
15th Personnel Service Battalion	08 Mar ’91 — — — — — — — — —’99 (Con’t next era)

DESERT PEACEKEEPERS
1992 - 2000

DESERT PEACEKEEPERS TIME LINE 20 Oct 1992 '93 '94 '95 '96 '97 '98 '99 '00

OPERATION INTRINSIC ACTION 92-3 1992
Task Force 3-41 Infantry 20 Oct '92/departed 15 Dec
"HHC", 3rd Battalion, 41st Infantry Regiment
"A" Company, 3rd Battalion, 41st Infantry Regiment
"B" Company, 1st Battalion, 67th Armor Regiment
"C" Company, 3rd Battalion, 67th Armor Regiment
"B" Battery, 3rd Battalion, 82nd Field Artillery
"B" Battery, 4th Battalion, 5th Air Defense Artillery
"B" Company, 8th Engineer Battalion
15th Personnel Support Company
215th Forward Support Battalion

OPERATION IRIS GOLD 93-1 1993
Task Force 1-9 Cavalry 15 Jan '93/departed 15 Mar
"HHC", 1st Battalion, 9th Cavalry Regiment
"A" Company, 1st Battalion, 9th Cavalry Regiment
"C" Company, 1st Battalion, 9th Cavalry Regiment
"B" Company, 3rd Battalion, 8th Cavalry Regiment
"C" Company, 3rd Battalion, 8th Cavalry Regiment
"C" Company, 3rd Battalion, 67th Armor Regiment
"B" Battery, 3rd Battery, 82nd Field Artillery
"B" Battery, 4the Battalion, 5th Air Defense Artillery
"B" Company, 8th Engineer Battalion
15th Personnel Support Company
215th Forward Support Battalion

OPERATION INTRINSIC ACTION 95-3 1995
Task Force 1-5 22 Aug '95/departed Nov
1st Battalion, 5th Cavalry Regiment
2nd Battalion, 12th Cavalry Regiment
1st Battalion, 8th Cavalry Regiment
3rd Battalion, 82nd Field Artillery
15th Forward Support Battalion
27th Main Support Battalion
3rd Battalion, 29th Field Artillery
2nd Battalion, 77th Armor
91st Engineer Battalion
"A" Battery, 21st Field Artillery
"C" Battery, 4th Battalion, 5th Air Defense Artillery
545th Military Police Company

DESERT PEACEKEEPERS TIME LINE 20 Oct 1992 '93 '94 '95 '96 '97 '98 '99 '00

OPERATION INTRINSIC ACTION 96-3 1996

Task Force 1-9, — 26 Aug '96/departed

"HHC", 1st Battalion, 9th Cavalry Regiment
"D" Company, 1st Battalion, 9th Cavalry Regiment
"B" Company, 1st Battalion, 12th Cavalry Regiment
"D" Company, 1st Battalion, 12th Cavalry Regiment
"A" Battery, 2nd Battalion, 82nd Field Artillery
"A" Company, 8th Engineer Battalion
215th Forward Support Battalion
15th Finance Battalion

3rd Brigade Combat Team — 19 Sep '96/

3rd Battalion, 8th Cavalry
1st Battalion, 5th Cavalry
82nd Field Artillery
"A" Battery, 21st Multiple Rocket System
68th Chemical Company
13th Signal Battalion

OPERATION INTRINSIC ACTION 97-01 1997

Task Force 1-7 Cavalry — Feb '97/departed May

"HHT", 1st Squadron, 7th Cavalry
"A" Troop, 1st Squadron, 7th Cavalry
"B" Troop, 1st Squadron, 7th Cavalry
"C" Troop, 1st Squadron, 7th Cavalry
"B" Battery, 1st Battalion, 82nd Field Artillery
"C" Company, 20th Engineer Battalion
615th Aviation Support Battalion
"A" Battery, 4th Battalion, 5th Air Defense Artillery
545th Military Police Company
312th Military Intelligence Battalion

OPERATION INTRINSIC ACTION 97-02 1997

Task Force 2-12 (Charger) — Jun '97/departed Oct

"HHC", 2nd Battalion, 12th Cavalry Regiment
"A" Company, 2nd Battalion, 12th Cavalry Regiment
"D" Company, 2nd Battalion, 12th Cavalry Regiment
"B" Company, 2nd Battalion, 5th Cavalry Regiment
"D" Company, 2nd Battalion, 5th Cavalry Regiment
"A" Battery, 3rd Battalion, 82nd Field Artillery
"A" Company, 91st Engineer Battalion
15th Forward Support Battalion
4th Platoon, 545th Military Police Company
2nd Platoon, "B" Battery, 4th Battalion, 5th Air Defense Artillery
312th Military Intelligence Detachment

DESERT PEACEKEEPERS TIME LINE 20 Oct 1992 '93 '94 '95 '96 '97 '98 '99 '00

OPERATION INTRINSIC ACTION 98-02 1998
Task Force 2-7 May '98/departed Aug
"HHC", 2nd Battalion, 7th Cavalry Regiment
"A" Company, 2nd Battalion, 7th Cavalry Regiment
"D" Company, 2nd Battalion, 7th Cavalry Regiment
"A" Company, 3rd Battalion, 8th Cavalry Regiment
"D" Company, 3rd Battalion, 8th Cavalry Regiment
"B" Company, 8th Engineer Battalion
"C" Battery, 2nd Battalion, 82nd Field Artillery
215th Forward Support Battalion

OPERATION DESERT SPRING 00-01 2000
Task Force 1-9
"HHC", 1st Battalion, 9th Cavalry Regiment Apr '00/departed Jul
"A" Company, 1st Battalion, 9th Cavalry
"B" Company, 1st Battalion, 9th Cavalry
"B" Company, 2nd Battalion, 8th Cavalry
"C" Company, 2nd Battalion, 8th Cavalry
"B" Battery, 2nd Battalion, 82nd Field Artillery
"A" Company, 20th Engineer Battalion
2nd Platoon, "B" Battery, 4th Battalion, 5th Air Defense Artillery
3rd Platoon, 68th Chemical Company (DECON)
5th Platoon. 545th Military Police Company
4th Platoon, "D" Company, 312th Military Intelligence Battalion (GSR)

OPERATION INTRINSIC ACTION 01-01 2001
Task Force 2-5 Dec '01/'02 departed Apr
"HHC" 2nd Battalion, 5th Cavalry
"A" Company, 2nd Battalion, 5th Cavalry
"B" Company, 2nd Battalion, 5th Cavalry
"D" Company, 2nd Battalion, 5th Cavalry
"D" Company, 2nd Battalion, 8th Cavalry
"A" Company, 20th Engineer Battalion
2nd Platoon, "A" Battery, 4th Battalion, 5th Air Defense Artillery
Decon Platoon, 68th Chemical Company
"A" Battery. 1st Battalion, 82nd Field Artillery
115th Forward Support Battalion (-)
"A" Company, 312th Military Intelligence Battalion (-)
6th Platoon, 545th Military Police Company
Enlisted Terminal Attack Controllers (ETAC), 1st Cavalry Division

BOSNIA PEACEKEEPERS
NATO Stabilization Forces (SFOR), "Task Force Eagle"
20 August 1998 - 01 October 1999

1st BRIGADE COMBAT TEAM (20 August 1998 to 15 Mar 1999)

EAGLE BASE

"HHC", 1st Cav Div
"HHC", 1st Bde Cmbt Tm
"HHB", 1-82nd FA Bn
"B" Btry, 1-82 FA
"C" Btry, 1-82nd FA
"C" Btry, 1-21st FA
"HHC", 13th Signal Bn
"C" Co, 13th Signal Bn
"D" Co, 13th Signal Bn
15th Finance Det
15th Pers Serv Det
"HHC", 312th MI Bn
"B" Co, 312th MI Bn
545th MP Company
410th MP Company

CAMP McGOVERN

"HHC", 1-8th Cavalry
"A" Co, 2-5th Cavalry
"B" Co, 1-8th Cavalry
"C" Co, 1-8th Cavalry
"D" Co, 1-8th Cavalry
"B" Co, 20th Engineer Bn

CAMP COMANCHE

"HHC", 4th Bde Cmbt Tm
"HSC", 2-227th Avn Regt
"A" Co, 2-227th Avn Regt
"B" Co, 2-227th Avn Regt
"C" Co, 2-227th Avn Regt
"D" Co, 2-227th Avn Regt
"HHC", 615th Avn Spt Bn
"HHC", 115th Fwd Spt Bn
"A" Co, 2-8th Cavalry
"C" Co, 2-5th Cavalry
"HHC", 3-229th Avn Regt **
"A" Co, 3-229th Avn Regt **
"B" Co, 3-229th Avn Regt **
"C" Co, 3-229th Avn Regt **
"D" Co, 3-229th Avn Regt **
"I/K" Co, 4-159th Avn Regt **
301st Atk Sqn, RNAF **

CAMP DOBOL

"HHC", 2-5th Cavalry (-)
"D" Co, 2-5th Cavalry
"A" Co, 1-8th Cavalry
"B" Co, 2-8th Cavalry
"A" Co, 20th Engineer Bn
"A" Co, 4-5th ADA (-)

CAMP BEDROCK

"HHC", 2-8th Cavalry
"C" Co, 2-8th Cavalry
"D" Co, 2-8th Cavalry
"C" Co, 20th Engineer Bn
"A" Co, 52nd Engineer Bn

CAMP DEMI

"HHC", 2-5th Cavalry
"B" Co, 2-5th Cavalry
"A" Btry, 1-82nd FA

** Units attached 20 Aug 1998 and not organic to 1st Cavalry Division.
Relieved from attachment 30 Mar 1999

2nd BRIGADE COMBAT TEAM (15 March 1999 to 01 October 1999)

EAGLE BASE

"HHC", 1st Cav Div *
"HHC", 2nd Bde Combat Tm
"HHB", 3-82nd FA
"A" Btry, 3-82nd FA
"B" Btry, 3-82nd FA
"C" Btry, 1-21st FA *
"HHC", Engineer Bde
"HHC", 91st Engineer Bn (-)
"C" Co, 91st Engineer Bn
545th MP Co (-) *
"HHC", 312th MI Bn *
"B" Co, 312th MI Bn *
"HHC", 13th Signal Bn *
"B" Co, 13th Signal Bn
15th Finance Det *
15th Pers Serv Det *
401st MP Co (-) **

CAMP McGOVERN

"HHC", 1-12th Cavalry
"A" Co, 1-12th Cavalry
"C" Co, 1-12th Cavalry
"D" Co, 1-12th Cavalry
"C" Co, 1-5th Cavalry
1st Plt, "C" Btry, 3-82nd FA
"A" Co, 91st Engineer Bn
2nd Plt, 401st MP Co **

CAMP COMANCHE

"HSD", 4th Bde
"HHC", 1-4th Avn Regt **
"A" Co, 1-4th Avn Regt **
"C" Co, 1-4th Avn Regt **
"D" Co, 1-4th Avn Regt **
"D" Trp, 1-7th Cavalry
"E" Trp, 1-7th Cavalry
"F" Trp, 1-7th Cavalry
"HHC", 1-159th Avn Regt **
"A" Co, 1-159th Avn Regt **
"B" Co, 1-159th Avn Regt **
"D" Co, 1-159th Avn Regt **
"HHC", 15th Fwd Spt Bn
"A" Co, 15th Fwd Spt Bn
"B" Co, 15th Fwd Spt Bn
"C" Co, 15th Fwd Spt Bn
"A" Co, 62nd Engineer Bn **
"HHC", 404th Avn Spt Bn **
"C" Btry, 4-5th ADA Bn

CAMP DOBOL

"HHC", 1-5th Cavalry
"B" Co, 1-5th Cavalry
"D" Co, 1-5th Cavalry
"B" Co, 1-12th Cavalry
"B" Co, 2-12th Cavalry
"C" Co, 2-12th Cavalry
"B" Co, 91st Engineer Bn
1st Plt, 401st MP Co **

CAMP BEDROCK

** CLOSED **

CAMP DEMI

"HHC", 2-12th Cavalry
"A" Co, 2-12th Cavalry
"D" Co, 2-12th Cavalry
"A" Co, 1-5th Cavalry
2nd Plt, "C" Btry, 3-82nd FA
1st Plt, "C" Co, 91st Engineer Bn

* Holdover units from fall 1998 deployment
** Units attached and not organic to 1st Cavalry Division

BOSNIA PEACEKEEPERS TIME LINE 20 Aug 1998 '99

The following have been identified as candidate units that were either attached to the MND-N or 1st Cavalry Division during the Bosnian Peacekeeping Operation.

Avn Units	
"HHC", 1st Bn, 4th Avn Regt, 4th ID	01 Apr '99
"A" Co, 1st Bn, 4th Avn Regt	01 Apr '99
"C" Co, 1st Bn, 4th Avn Regt	01 Apr '99
"D" Co, 1st Bn, 4th Avn Regt	01 Apr '99
"HHC", 1st Bn, 159th Avn Regt	01 Apr '99
"A" Co, 1st Bn, 159th Avn Regt	01 Apr '99
"B" Co, 1st Bn, 159th Avn Regt	01 Apr '99
"D" Co, 1st Bn, 159th Avn Regt	01 Apr '99
"HHC", 3rd Bn, 229th Avn Regt	20 Aug '98/'99; Relv'd from Atchm't 30 Mar
"A" Co, 3rd Bn, 229th Avn Regt	20 Aug '98/'99; Relv'd from Atchm't 30 Mar
"B" Co, 3rd Bn, 229th Avn Regt	20 Aug '98/'99; Relv'd from Atchm't 30 Mar
"C" Co, 3rd Bn, 229th Avn Regt	20 Aug '98/'99; Relv'd from Atchm't 30 Mar
"D" Co, 3rd Bn, 229th Avn Regt	20 Aug '98/'99; Relv'd from Atchm't 30 Mar
"I/K" Co, 4th Bn, 159 Avn Regt	20 Aug '98/'99; Relv'd from Atchm't 30 Mar
Royal Netherlands Air Force 301st Attack Sqdn,	20 Aug '98/'99; Relv'd from Atchm't 30 Mar
"HHC", 404th Avn Spt Battalion	01 Apr '99
Engineer Units	
NMCB-40 Seabee Construction Bn	27 Sep '98/'99; Relv'd from Atchm't 02 May
"A" Co, 62nd Engr Bn, 13th Corps Spt	
249th Engineer Bn (Prime Power)	07 Oct '98
463rd Engineer Det (Fire Fighters)	'98/'99; Relv'd from Atchm't 02 Apr
765th Ordnance Co (EOD)	15 Mar '99

ATTACHED UNITS

K-9 Det
2nd Battalion, 12th Infantry Regt
13th Psychological Operations Bn (MNARNG)
22nd Mobile Public Affairs Det
24th Medical Det (NEARNG)
36th Medical Evacuation Battalion
52nd Engineer Battalion, Ft. Carson, CO

61st Area Medical Support Battalion
- 41st Combat Support Hospital
- 43rd Medical Det (Veterinarian Services)
- 85th Medical Det (CSC)
- 126th Medical Co (Air Ambulance) (CAARNG)
- 147th MEDLOG Det (-)
- 224th Medical Det (Preventative Medicine)
- 502nd Medical Det (Dental)
- 565th Medical Co (GA)
- Swedish Medical Co

3rd Battalion, 82nd Infantry Regt
1st Battalion, 87th Infantry Regt, 10th Mtn Div
133rd Mobile Public Affairs Det
151st Postal Co
183rd Maintenance Support Team, III Corps, Ft. Carson, CO
215th Finance Battalion
233rd Preventative Medicine Det
301st Information Support Team (MIST)
314th Mobile Public Affairs Det
319th Mobile Public Affairs Det
329th Postal Co
351st Postal Co
354th Civil Affairs Brigade
358th Civil Affairs Brigade
376th Engineer Det (FF) (MOARNG)
401st Mltry Police Battalion, Ft. Hood, TX
401st Expeditionary Weather Sqdn
403rd Civil Affairs Battalion
404th Avn Support Battalion
410th Mltry Police Battalion
413th Civil Affairs Battalion (TX)
414th Civil Affairs Battalion
416th Civil Affairs Battalion (PA)
416th Air Traffic Services
431st Civil Affairs Battalion (AR)
454th Civil Affairs Brigade
478th Civil Affairs Battalion (FL)
482nd Engineer Det
486th Civil Affairs Battalion (OK)
546th Personnel Services Battalion

BOSNIA PEACEKEEPERS TIME LINE 20 Aug 1998 '99

546th Adjutant General (Postal) Co
175th Adjutant General (Postal) Co
336th Adjutant General (Postal) Co
806th Adjutant General (Postal) Co

615th Avn Support Battalion, XVIII Corps
744th Engineer Det (EOD)
755th Postal Co
American Red Cross Contingent

International Trps
Nordic-Polish Brigade
Finnish Mechanized Inf Battalion
Norwegian Mechanized Inf Batallion
Polish Mechanized Inf Battalion
Swedish Mechanized Inf Battalion
Swedish Medical Co
Military Police Co
Nordic-Polish Support Group
Danish Battalion
Baltic Contingent
Lithunania Plt
Latvian Plt
Estonian Plt

Russian Separate Airborne Brigade
Turkish Brigade - Zenica

TODAY'S CAVALRY
01 October 1999 +

TODAY'S CAVALRY TIME LINE 01 Oct 1999 '00 '01 '02 '03 '04 '05 '06 '07 '08 '09 '10 '11 '12 '13 '14 '15 '16 '17 '18 '19 '20

"HHC", 1st Cavalry Division 01 Oct '99/'00~

1st BRIGADE, 1st Cavalry Division
"HHC", 1st Brigade - "IronHorse" 01 Oct '99/'00~
"C" Trp, 10th Cavalry Regt 18 Sep '01/'02~ Actv'd at Ft Hood, TX as the Bde Recon Unit

TODAY'S CAVALRY TIME LINE	01 Oct 1999 '00 '01 '02 '03 '04 '05 '06 '07 '08 '09 '10 '11 '12 '13 '14 '15 '16 '17 '18 '19 '20
2nd Battalion, 5th Cavalry Regiment	01 Oct '99/'00~
"HHC", 2nd Bn, 5th Cavalry Regt	
"A" Co, 2nd Bn, 5th Cavalry Regt	
"B" Co, 2nd Bn, 5th Cavalry Regt	
"C" Co, 2nd Bn, 5th Cavalry Regt	
"D" Co, 2nd Bn, 5th Cavalry Regt	
1st Battalion, 8th Cavalry Regiment	01 Oct '99/'00~
"HHC", 1st Bn, 8th Cavalry Regt	
"A" Co, 1st Bn, 8th Cavalry Regt	
"B" Co, 1st Bn, 8th Cavalry Regt	
"C" Co, 1st Bn, 8th Cavalry Regt	
"D" Co, 1st Bn, 8th Cavalry Regt	
2nd Battalion, 8th Cavalry Regiment	01 Oct '99/'00~
"HHC", 2nd Bn, 8th Cavalry Regt	
"A" Co, 2nd Bn, 8th Cavalry Regt	
"B" Co, 2nd Bn, 8th Cavalry Regt	
"C" Co, 2nd Bn, 8th Cavalry Regt	
"D" Co, 2nd Bn, 8th Cavalry Regt	
2nd BRIGADE, 1st Cavalry Division	
"HHC", 2nd Brigade "BlackJack"	01 Oct '99/'00~
1st Battalion, 5th Cavalry Regiment	01 Oct '99/'00~
"HHC", 1st Bn, 5th Cavalry Regt	
"A" Co, 1st Bn, 5th Cavalry Regt	
"B" Co, 1st Bn, 5th Cavalry Regt	
"C" Co, 1st Bn, 5th Cavalry Regt	
"D" Co, 1st Bn, 5th Cavalry Regt	
1st Battalion, 12th Cavalry Regiment	01 Oct '99/'00 ~
"HHC", 1st Bn, 12th Cavalry Regt	
"A" Co, 1st Bn, 12th Cavalry Regt	
"B" Co, 1st Bn, 12th Cavalry Regt	
"C" Co, 1st Bn, 12th Cavalry Regt	
"D" Co, 1st Bn, 12th Cavalry Regt	
2nd Battalion, 12th Cavalry Regiment	01 Oct '99/'00~
"HHC", 2nd Bn, 12th Cavalry Regt	
"A" Co, 2nd Bn, 12th Cavalry Regt	
"B" Co, 2nd Bn, 12th Cavalry Regt	
"C" Co, 2nd Bn, 12th Cavalry Regt	
"D" Co, 2nd Bn, 12th Cavalry Regt	

TODAY'S CAVALRY TIME LINE	01 Oct 1999 '00 '01 '02 '03 '04 '05 '06 '07 '08 '09 '10 '11 '12 '13 '14 '15 '16 '17 '18 '19 '20
3rd BRIGADE, 1st Cavalry Division	
"HHC", 3rd Brigade - "GreyWolf"	01 Oct '99/'00~
2nd Battalion, 7th Cavalry Regiment	01 Oct '99/'00~
"HHC", 2nd Bn, 7th Cavalry Regt	
"A" Co, 2nd Bn, 7th Cavalry Regt	
"B" Co, 2nd Bn, 7th Cavalry Regt	
"C" Co, 2nd Bn, 7th Cavalry Regt	
"D" Co, 2nd Bn, 7th Cavalry Regt	
3rd Battalion, 8th Cavalry Regiment	01 Oct '99/'00~
"HHC", 3rd Bn, 8th Cavalry Regt	
"A" Co, 3rd Bn, 8th Cavalry Regt	
"B" Co, 3rd Bn, 8th Cavalry Regt	
"C" Co, 3rd Bn, 8th Cavalry Regt	
"D" Co, 3rd Bn, 8th Cavalry Regt	
1st Battalion, 9th Cavalry Regiment	01 Oct '99/'00~
"HHC", 1st Bn, 9th Cavalry Regt	
"A" Co, 1st Bn, 9th Cavalry Regt	
"B" Co, 1st Bn, 9th Cavalry Regt	
"C" Co, 1st Bn, 9th Cavalry Regt	
"D" Co, 1st Bn, 9th Cavalry Regt	
COMBAT AVIATION BRIGADE, 1st Cavalry Division	
"HHC", Combat Aviation - "Warriors"	01 Oct '99/'00~
1st Sqdn, 7th Cavalry Regiment	01 Oct '99/'00~
"HHT", 1st Sqdn, 7th Cavalry Regt	
"A" Trp, 1st Sqdn, 7th Cavalry Regt	
"B" Trp, 1st Sqdn, 7th Cavalry Regt	
"C" Trp, 1st Sqdn, 7th Cavalry Regt	
"D" Trp, 1st Sqdn, 7th Cavalry Regt	
"E" Trp, 1st Sqdn, 7th Cavalry Regt	
"F" Trp, 1st Sqdn, 7th Cavalry Regt	
1st Battalion, 227th Aviation Regt	01 Oct '99/'00~
"HHC", 1st Bn, 227th Avn Regt	
"A" Co, 1st Bn, 227th Avn Regt	
"B" Co, 1st Bn, 227th Avn Regt	
"C" Co, 1st Bn, 227th Avn Regt	
"D" Co, 1st Bn, 227th Avn Regt	

TODAY'S CAVALRY TIME LINE	01 Oct 1999 '00 '01 '02 '03 '04 '05 '06 '07 '08 '09 '10 '11 '12 '13 '14 '15 '16 '17 '18 '19 '20
2nd Battalion, 227th Aviation Regt	01 Oct '99/'00~
"HHC", 2nd Bn, 227th Avn Regt	
"A" Co, 2nd Bn, 227th Avn Regt	
"B" Co, 2nd Bn, 227th Avn Regt	
"C" Co, 2nd Bn, 227th Avn Regt	
"D" Co, 2nd Bn, 227th Avn Regt	
"F" Co, 1st Bn, 58th Avn (ATGC)	01 Oct '99/'00~
ARTILLERY, 1st Cavalry Division	
"HHB", Division Arty - "Red Team"	01 Oct '99/'00~
1st Battalion, 21st Field Artillery	01 Oct '99/'00~
"HHS", Battery, 1st Bn, 21st Field Arty	
"A" Battery, 1st Bn, 21st Field Arty	
"B" Battery, 1st Bn, 21st Field Arty	
"C" Battery, 1st Bn, 21st Field Arty	
"D" Battery, 1st Bn, 21st Field Arty (TXARNG)	
1st Battalion, 82nd Field Artillery	01 Oct '99/'00~
"HHB", 1st Bn, 82nd Field Arty	
"A" Btry, 1st Bn, 82nd Field Arty	
"B" Btry, 1st Bn, 82nd Field Arty	
"C" Btry, 1st Bn, 82nd Field Arty	
Service Btry, 1st Bn, 82nd Field Arty	
2nd Battalion, 82nd Field Artillery	01 Oct '99/'00~
"HHB", 2nd Bn, 82nd Field Arty	
"A" Btry, 2nd Bn, 82nd Field Arty	
"B" Btry, 2nd Bn, 82nd Field Arty	
"C" Btry, 2nd Bn, 82nd Field Arty	
Service Btry, 2nd Bn, 82nd Field Arty	
3rd Battalion, 82nd Field Artillery	01 Oct '99/'00~
"HHB", 3rd Bn, 82nd Field Arty	
"A" Btry, 3rd Bn, 82nd Field Arty	
"B" Btry, 3rd Bn, 82nd Field Arty	
"C" Btry, 3rd Bn, 82nd Field Arty	
Service Btry, 3rd Bn, 82nd Field Arty	
68th Chemical Company	01 Oct '99/'00~
ENGINEER BRIGADE, 1st Cavalry Division	
"HHD", Engineer Brigade - "Snappers"	01 Oct '99/'00~

TODAY'S CAVALRY TIME LINE	01 Oct 1999 '00 '01 '02 '03 '04 '05 '06 '07 '08 '09 '10 '11 '12 '13 '14 '15 '16 '17 '18 '19 '20
8th Engineer Battalion	01 Oct '99/'00~
"HHC", 8th Engineer Bn	
"A" Co, 8th Engineer Bn	
"B" Co, 8th Engineer Bn	
"C" Co, 8th Engineer Bn	
Service Plt, 8th Engineer Bn	
20th Engineer Battalion	01 Oct '99/'00~
"HHC", 20th Engineer Bn	
"A" Co, 20th Engineer Bn	
"B" Co, 20th Engineer Bn	
"C" Co, 20th Engineer Bn	
Service Plt, 20th Engineer Bn	
91st Engineer Battalion	01 Oct '99/'00~
"HHC", 91st Engineer Bn	
"A" Co, 91st Engineer Bn	
"B" Co, 91st Engineer Bn	
"C" Co, 91st Engineer Bn	
Service Plt, 91st Engineer Bn	
DIVISION SUPPORT COMMAND, 1st Cavalry Division	
"HHC", Support Comd - Wagon Masters"	01 Oct '99/'00~
15th Support Battalion (FWD)	01 Oct '99/'00~
"HHC", 15th Support Bn (FWD)	
"A" Co, 15th Support Bn (FWD)	
"B" Co, 15th Support Bn (FWD)	
"C" Co, 15th Support Bn (FWD)	
27th Support Battalion (MAIN)	01 Oct '99/'00~
"HHC", 27th Support Bn (MAIN)	
"A" Co, 27th Support Bn (MAIN)	
"B" Co, 27th Support Bn (MAIN)	
"C" Co, 27th Support Bn (MAIN)	
"D" Co, 27th Support Bn (MAIN)	
"E" Co, 27th Support Bn (MAIN)	
115th Support Battalion (FWD)	01 Oct '99/'00~
"HHC", 115th Support Bn (FWD)	
"A" Co, 115th Support Bn (FWD)	
"B" Co, 115th Support Bn (FWD)	
"C" Co, 115th Support Bn (FWD)	

TODAY'S CAVALRY TIME LINE	01 Oct 1999 '00 '01 '02 '03 '04 '05 '06 '07 '08 '09 '10 '11 '12 '13 '14 '15 '16 '17 '18 '19 '20
215th Support Battalion (FWD)	01 Oct '99/'00~
"HHC", 215th Support Bn (FWD)	
"A" Co, 215th Support Bn (FWD)	
"B" Co, 215th Support Bn (FWD)	
"C" Co, 215th Support Bn (FWD)	
615th Support Battalion (AVN)	01 Oct '99/'00~
"HHC", 615th Support Bn (AVN)	
"A" Co, 615th Support Bn (AVN)	
"B" Co, 615th Support Bn (AVN)	
Division Material Management Center (DMMC)	
SEPARATE BNs/COs, 1st Cavalry Division - "Maverick"	
"HHC", 1st Cavalry Division	01 Oct '99/'00~
Band, 1st Cavalry Division	01 Oct '99/'00~
1st Cavalry Horse Detachment	01 Oct '99/'00~
4th Battalion, 5th Air Defense	01 Oct '99/'00~
"HHB", 4th Bn, 5th Air Defense	
"A" Btry, 4th Bn, 5th Air Defense	
"B" Btry, 4th Bn, 5th Air Defense	
"C" Btry, 4th Bn, 5th Air Defense	
"D" Btry, 4th Bn, 5th Air Defense	
13th Signal Battalion	01 Oct '99/'00~
"HHC", 13th Signal Bn	
"A" Co, 13th Signal Bn	
"B" Co, 13th Signal Bn	
"C" Co, 13th Signal Bn	
"D" Co, 13th Signal Bn	
312th Mltry Intelligence Battalion	01 Oct '99/'00~
"HHC", 312th Mltry Intel Bn	
Analysis Control Element (Prov)	15 Oct '99/'00~
"A" Co, 312th Mltry Intel Bn	
"B" Co, 312th Mltry Intel Bn	
"C" Co, 312th Mltry Intel Bn	
"D" Co, 312th Mltry Intel Bn	
545th Mltry Police Company	01 Oct '99/'00~

TODAY'S CAVALRY TIME LINE	01 Oct 1999 '00 '01 '02 '03 '04 '05 '06 '07 '08 '09 '10 '11 '12 '13 '14 '15 '16 '17 '18 '19 '20
DIRECT SUPPORT	
9th Air Support Operations	01 Oct '99/'00~
15th Finance Battalion	01 Oct '99/'00~
15th Personnel Service Battalion	01 Oct '99/'00~
Staff Judge Advocate	
ATTACHED UNITS	
4th Public Affairs Detachment	01 Oct '99/'00~
123rd Engineer Battalion	01 Oct '99/'00~
557th Engineer Battalion	01 Oct '99/'00~

NOTE: The Following Abbreviations Apply:

Absrb'd	Absorbed	Det	Detachment	Reasgn'd	Reassigned
Actv'd	Activated	Dfnse	Defense	Regt	Regiment
Arty	Artillery	Hdqtrs	Headquarters	Relv'd	Relieved
Asgn'd	Assigned	Inactv'd	Inactivated	Reorg'd	Reorganized
Asgm't	Assignment	Intel	Intelligence	Replc'd	Replaced
Atchm't	Attachment	Maint	Maintenance	Ret'rd	Returned
Avn	Aviation	Med	Medium	Spc'l	Special
Bn	Battalion	Mltry	Military	Sqdn	Squadron
Btry	Battery	Plt	Platoon	Supt	Support
Co	Company	Prov	Provisional	Xfer'd	Transferred
Cmd	Command	QtrMstr	Quartermaster	Trans	Transportation
Dsng'd	Designated	Recon	Reconnaissance	Trp	Troop

Appendix I: GarryOwen

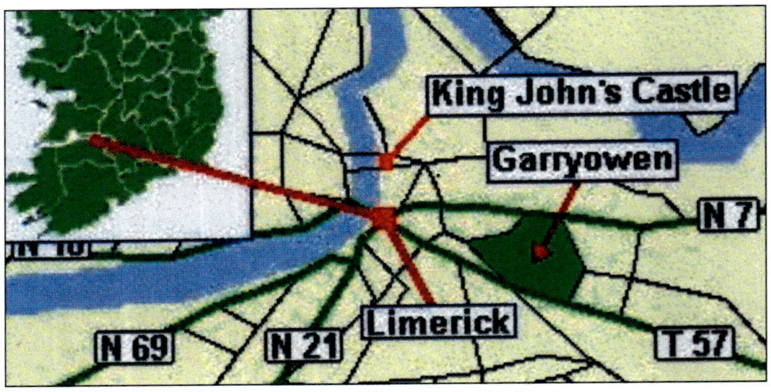

GarryOwen - Limerick, Ireland

"GarryOwen" is an old Irish quick-step that can be traced back to the early 1860s. In 1867 "GarryOwen" was adopted by the 7th Cavalry Regiment as the official Air (tune) of the Regiment, and the historical nickname given to the 7th Cavalry Regiment and troopers. It became the Official tune of the 1st Cavalry Division in 1981. "GarryOwen" has become undoubtedly the most famous of all the regimental marches in the Army.

The geographical area that provided the inspiration and the name of one of the most popular, rollicking folk songs of Ireland is situated on the upward slope of a hill in Limerick County, near the City of Limerick. Local traditions and folk lore have preserved the historical significance of the area and the origin of its name "GarryOwen", a compounded word composed of two Irish words, which means "Garden of Owen".

The terrain features of Garryowen gave the loyal patrons of the garden a broad, commanding view of the richly cultivated surrounding countryside, the old town of Limerick and the valley of the Shannon River which gently washes the battered, fortified towers of King John's castle which was constructed in the late 1180's to control traffic along the river. The cottage of Owen and surrounding plot of ground soon became a favorite holiday resort with the near-by citizens of Limerick because the atmosphere and accommodations were somewhat similar to those offered to the London mechanic by the Battersea tea-gardens.

A review of Irish literature reveals that *"Owen's garden was a general rendezvous for those who sought simple pleasure and amusement. The elderly drank together under the shade of trees and the young played at ball, goal, or other athletic activities on the green; while a few lingered in the near-by hedge-rows with their fair acquaintances. Owen's garden was soon to became as famous for scenes of strife as it was for mirth and humor; and broken arms, legs and heads became a staple article of manufacture in the neighborhood."*

"These new diversions were encouraged by a number of young people having a greater supply of animal spirits than wisdom to control themselves. The young gentlemen being fond of wit, amused themselves by having parties at night to wring the heads off all the geese, and tearing knockers off the doors in the neighborhood. They sometimes suffered their genius to soar as high as the breaking of a street lamp, and even resorting to the physical violence of a watchman. But, this type of joking was found a little too serious to be repeated very frequently, for few achievements of so daring a violence were documented in the records. They were obliged to content themselves with less ambitious distinction of destroying the door knockers and store-locks, annoying the peace of the neighborhood, with long continued assaults on the front doors, terrifying the quiet onlookers with every species of insult and provocation, and indulging their fratricidal propensities against all the geese in GarryOwen."

"The fame of the 'GarryOwen Boys' soon spread far and wide. Their deeds were celebrated by some inglorious minstrel of the day in that melody which has, since, resounded over the world; and even symbolically competed for national popularity with 'St. Patrick's Day'. A string of verses were appended to the tune which soon enjoyed equal notoriety. The name of Garryowen was as well known as that of the city of Limerick, itself, and Owen's garden became almost a synonym for Ireland."

"GarryOwen" is known to have been used by Irish regiments as a drinking song. As the story goes, one of the Irish "melting pot" troopers of the 7th Cavalry, under the influence of "spirits", was singing the song. By chance Custer heard the melody, liked the cadence, and soon began to hum the tune himself. The tune has a lively beat, that

King John's Castle - Limerick

accentuates the cadence of marching horses, and for that reason was adopted as the regimental song soon after Custer arrived at Fort Riley, Kansas to take over command of the 7th Cavalry Regiment. It was the last song played for Custer's men as they left General Terry's column at the Powder River and rode into history.

During First Team ceremonies the song is not sung; however, it is customary for the song to be played at the conclusion of the activities and the guests stand and clap.

Lyrics of the song are as follows:

[Verse 1]
Let Bacchus's sons be not dismayed,
but join with me each jovial blade,
come booze and sing and lend your aid,
to help me with the chorus:

"Chorus"
Instead of spa we'll drink down ale
and pay the reckoning on the nail,
for debt no man shall go to jail
from GarryOwen in glory

[Verse 2]
We are the boys who take delight
in smashing Limerick lamps at night,
and through the street like sportsters fight,
tearing all before us *(Chorus)*

[Verse 3]
We'll break windows, we'll break doors,
the watch knock down by threes and fours,
then let the doctors work their cures,
and tinker up our bruises *(Chorus)*

[Verse 4]
We'll beat the bailiffs out of fun,
we'll make the mayor and sheriffs run,
we are the boys no man dare dun,
if he regards a whole skin *(Chorus)*

[Verse 5]
Our hearts so stout have got us fame,
for soon 'tis known from whence we came,
where're we go they dread the name,
of GarryOwen in glory *(Chorus)*

There was a special set of lyrics written for the 7th Cavalry Regiment in 1905. For those special troopers of the 7th:

[Verse 1]
We are the pride of the Army
and a regiment of great renown,
Our name's on the pages of history.
From sixty-six on down.
If you think we stop or falter
While into the fray we're going
Just watch the steps with our heads erect,
While our band plays GarryOwen. *(Chorus)*

"Chorus"
In the Fighting Seventh's the place for me,
Its the cream of all the Cavalry;
No other regiment ever can claim
Its pride, honor, glory and undying fame.

[Verse 2]
We know fear when stern duty
Calls us far away from home,
Our country's flag shall safely o'er us wave,
No matter where we roam.
"Tis the gallant 7th Cavalry
It matters not where we are going"
Such you'll surely say as we march away;
And our band plays GarryOwen. *(Chorus)*

[Verse 3]
Then hurrah for our brave commanders!
Who led us into the fight.
We'll do or die in our country's cause,
And battle for the right.
And when the war is o'er,
And to our home we're goin
Just watch your step, with our heads erect,
When our band plays GarryOwen. *(Chorus)*

Reference: "From Custer to MacArthur, the 7th US Cavalry"

Author's Note: "GarryOwen" was also the Regimental March of another famous fighting unit, *The Royal Irish Regiment*, that was organized in 1684 from Irish Pikemen and Musketeers by the Earl of Granard to fight for King William. This Regiment has seen service in all parts of the world. For their outstanding valor at the Battle of Namur, they received the title of "The Royal Regiment of Foot of Ireland". In addition, in recognition of its deeds on this occasion, King William conferred the right of displaying the badge of the Harp and Crown, and that of the Lion of Nassau, with the explanatory legend.

The Royal Irish showed noble courage and performed gallant service throughout the Crimean War. On their colors are inscribed "Egypt"; "China"; "Blenheim"; "Ramillies"; "Oudenarde"; "Malpaquet"; "Pegu"; "Savastopol"; "New Zealand"; "Afghanistan, 1879-80"; "Egypt, 1882"; "Tel-el-Kebir"; "Nile, 1884-85"; "South Africa, 1900-02"; "Flanders, 1914" and "Gallipoli, 1915." The Royal Irish Regiment was disbanded in 1922 on the formation of the Irish Free State.

One cannot wonder how many of the former members of the Royal Irish Regiment emigrated to the United States and enlisted as troopers of the US Cavalry Units. Could this have been how Custer heard this song?

Part of the mystery may be solved. The history of the 69th New York Infantry reflects the history and progress of the Irish in America. From unwelcome immigrants escaping famine and persecution, they were assimilated and integrated into the society of America. Its ranks were filled with heroes, priests, poets, politicians, laborers, lawyers, in short a cross section of the Ireland's greatest export - her sons.

The "Fighting Sixty-Ninth" had its origin in early 1851, when the Irish citizens in New York City formed a militia regiment known locally as the Second Regiment of Irish Volunteers. Unanimously, the group selected "GarryOwen" as their official regimental marching song. On 12 October 1851, the Regiment was officially accepted as part of the New York Militia and designated as the Sixty-Ninth Regiment. In 1858, the Regiment would have its first call to duty. Their many subsequent calls to duty included the Civil War, Spanish American War, the Mexican War, World War I and World War II. Today, the 69th is now officially the 1st Battalion, 69th Infantry (Mechanized), and is part of the 42nd Infantry Division.

The 42nd Infantry Division received the name "Rainbow Division" in 1917 during its organization at Camp Mills, Long Island, New York. The Chief of Staff of the Division, at that time, was Colonel Douglas MacArthur. While discussing the organization of the Division and reviewing the National Guard units from 26 states that would make up the Division, MacArthur commented "The 42nd Division stretches like a Rainbow from one end of America to the other." The comment caught the interest of those present and they decided to call it the "Rainbow Division".

1st Cavalry Division Band

Appendix J: Fiddler's Green

[Verse 1]
Halfway down the trail to Hell
In a shady meadow green
Are the Souls of all dead Troopers camped
Near a good old-time canteen,
And this eternal resting place is
know as Fiddlers' Green

[Verse 2]
Marching past straight through to Hell
The Infantry are seen
Accompanied by the Engineers,
Artillery and Marines,
For none but the shades of Cavalrymen
Dismount at Fiddlers' Green

[Verse 3]
Though some go curving down the trail
To seek a warmer scene,
No Trooper ever gets to Hell
Ere he's emptied his canteen.
And so rides back to drink again
With friends at Fiddlers' Green

[Verse 4]
And so when man and horse go down
Beneath a sabre keen,
Or on roaring charge of fierce melee
You stop a bullet clean.
And the hostiles come to get your scalp
Just empty your canteen,
And put your pistol to your head
And go to Fiddlers' Green.

The origin and author of Fiddlers' Green is unknown. It was believed to have originated in the 1800's and was sung by the soldiers of the 6th and 7th Cavalry. Its first known appearance in published form was in a 1923 Cavalry Journal.

Appendix K: Museum

1st Cavalry Division Museum

The mission of the 1st Cavalry Museum is to collect, preserve, interpret, exhibit and educate military personnel and the general public on the history of the 1st Cavalry Division and its subordinate units.

The site for the museum includes a large outside exhibit of the broad variety of heavy weapon systems platforms utilized and encountered by elements of the 1st Cavalry Division in the accomplishment of its assigned missions as well as an indoor facility that houses a gallery of tableaux vivants that depicts various historical events and battles in the history of the Division and a Souvenir Shop.

The 1st Cavalry Museum, Building #2218, is located just past III Corps Headquarters on Headquarters Avenue. When you are in the Fort Hood area, you are invited to tour both the Museum and adjacent Souvenir Shop. The Museum will provide the opportunity for you to share a piece of your past with your family.

Building 2218
P.O. Box 5187
Fort Hood, TX 76545
Telephone: (254) 287-3626 or DSN# 737-3626
Facsimile: (254) 287-6423

HOURS

Monday through Friday 9:00 AM - 4:00 PM
Saturday and Sunday 12:00 AM - 4:00 PM
Christmas, New Year's and Thanksgiving - Closed

RESOURCES OVERVIEW

Exhibits of the Museum include an antecedent period gallery which traces the beginnings of the horse soldier regiments to the formation of the 1st Cavalry Division in 1921. Subsequent galleries include World War II, the Korean Conflict, Vietnam War, Gulf War and the most recent Operation Joint Forge, in Bosnia. Artifacts and text tell the story of the major role played by the First

Team in national and world events. In the Hall of Honor, the Museum pays tribute to members of the 1st Cavalry Division who were awarded the Medal of Honor for valiant action. Finally, the "Soldier of the "80's" gallery provides a look at cavalry soldiers of tomorrow and the future mission roles to be assigned to modern "horse soldiers".

SCOPE

2,000 vintage and 19,000 copy photos, including panoramic prints, lantern slides, and film transparencies.

FINDING AIDS

Inventory, descriptive listing, and card index; access by personal and proper name, geographic location, date, and photographer.

ASSISTANCE

A professional staff supports inquiries. Written and phone inquiries accepted on a limited basis. Electrostatic copies available. Photoduplications must be ordered through the museum. Appointments recommended for use of reserve collections.

PUBLICATION

No publication fee. Collection credit required. A limited number of items require third-party permission for publication.

SUBJECTS

Activities of the 1st Cavalry Division with broad coverage of cavalry soldiers on maneuvers, training, and the social life and customs of military life; also included are a few military combat scenes from the various wars in which the cavalry has seen action; Pershing's Punitive Expedition, WW II, Korean War, Vietnam and the Persian Gulf of Southwest Asia.

PEOPLE

Prominent military figures in the US Army, especially those connected with the 1st Cavalry Division; Major General William C. Chase, General George A. Custer, Brigadier General Hugh F. T. Hoffman, General Douglas MacArthur. Most other members of the division are represented but may not be identified.

GEOGRAPHIC

Areas where the 1st Cavalry Division has seen military action, including Mexico, Australia, the Admiralty Islands, Philippines, Japan, Korea, Vietnam, Southwest Asia and in the United States, including maneuvers in Louisiana, New Mexico, Oklahoma, South Dakota, and Wyoming. Texas locales in West, South, and Central Texas such as Marfa, El Paso (Fort Bliss), Fort Hood, Killeen and San Antonio.

RECORD GROUPS

Photographs are arranged as a single collection, classified in grouped as described above.

Preserve the Legend

Appendix L: Weapon Systems

The development of the weapon system platforms has evolved over the years by the constant search for designs and solutions for equipment that is capable of supporting rapid responses and dominance across a full spectrum of operations. To bridge the gap between the weapon capabilities of today and those that will be introduced as the Objective Force of a digitized "system of all systems" land combat capability, several variants of the weapon system platforms having tested, proven improvements will phased into production.

The following sections describe the characteristics and features of those equipments that will be introduced and assigned to the 1st Cavalry Division.

Abrams - M1A2, Main Battle Tank

MISSION

Provide heavy armor superiority on the battlefield.

DESCRIPTION AND SPECIFICATIONS

Initial production of the M1 began 1978, the first M1A1 in 1985 and the first M1A2, workhorse of the Gulf War, in 1986. The Abrams accommodates a crew of four, commander, driver, gunner and loader.

The Abrams tank modernization strategy supports the Army Vision. The Abrams tank closes with and destroys enemy forces on the integrated battlefield using mobility, firepower, and shock effect. The 120mm main gun on the M1A1 and M1A2 combined with the powerful 1,500 hp turbine engine and special armor, make the Abrams tank particularly suitable for attacking or defending against large concentrations of heavy armor forces in a highly lethal battlefield.

Features of the M1A1 include increased armor protection; suspension improvements; and a nuclear, biological, and chemical (NBC) protection system that increases survivability in a contaminated environment. The M1A1D modification scheduled for 1535 M1A1s, consists of an M1A1 with integrated Applique computer and a far-target-designate capability.

The M1A2 program provides the Abrams tank with the necessary improvements in lethality, survivability, and fightability required to defeat advanced threats. The M1A2 includes an independent thermal viewer for the commander, an improved commander's weapon station, position navigation equipment, a distributed data and power architecture, an embedded diagnostic system, and improved fire control system. The M1A2 System Enhancement Program (SEP) adds second generation thermal sensors and a Thermal Management System (TMS). The SEP includes upgrades to processors/memory that enable the M1A2 to use the common command and control software of the Army, enabling the rapid transfer of digital situational data and overlays. The Abrams modernization strategy also includes a new engine program, the Abrams Integrated Management (AIM) Overhaul Program, and parts obsolescence program, which will reduce the operational and support costs, and logistical footprint associated with the Abrams.

Bradley - M2/M3, Tracked Armored Fighting Vehicle

MISSION

Provide infantry and cavalry fighting vehicles with digital command and control capabilities, significantly increased situational awareness, enhanced lethality, survivability, and improved sustainability and supportability.

DESCRIPTION AND SPECIFICATIONS

The Bradley Fighting Vehicle (BFV) System includes two vehicles, the M2 Infantry Fighting Vehicle (IFV) and the M3 Cavalry Fighting Vehicle (CFV) and is in service with the armies of the USA and Saudi Arabia. Since 1981, 6720 vehicles of both types have been fielded.

The role of the Bradley M2 IFV is to transport infantry to critical locations on the battlefield, to provide fire cover to dismounted troops and to suppress enemy tanks and fighting vehicles. The M2 carries three crew members, commander, gunner and driver plus six fully equipped infantrymen. The M3 CFV performs scout missions and carries three crew members plus two scouts.

The Bradley M2A3/M3A3 (IFV/CFV) incorporates the following Vehicle Improvement Features:

Two Second Generation Forward Looking Infrareds (FLIRs) in the Improved Bradley Acquisition System (IBAS) and Commanders Independent Sight (CIV) provide "Hunter-Killer target hand-off" capability with ballistic fire control system; Embedded diagnostics; Integrated combat command and control digital communications suite hosting Force XXI Battle Command Brigade-and-Below (FBCB2) package with digital maps, messages and friend/foe situational awareness; Position navigation system with GPS and inertial navigation system; Enhanced squad situational awareness with squad leader display integrated into vehicle digital images and IC3.

Current models/variants include: Bradley M2/M3 A0, A2, A2 ODS (Operation Desert Storm), A3 IFV/CFVs, Bradley Fire Support Team (BFIST) Vehicle, Bradley Linebacker, and Manportable Air Defense System (MANPADS) Under Armor.

M113A1, Armored Personnel Carrier (APC)

MISSION

Provide a highly mobile, survivable, and reliable tracked vehicle platform.

DESCRIPTION AND SPECIFICATIONS

The M113 Family of Vehicles (FOV) is designed to maintain pace with Abrams and Bradley-equipped units and to adapt to a wide range of current and future battlefield tasks through the integration of specialized mission modules at minimum operational and support cost.

The M113A1 is a lightly armored full tracked air transportable personnel carrier designed to carry personnel and certain types of cargo. The vehicle is capable of: amphibious operations in streams and lakes; extended cross country travel over rough terrain and high speed operation on improved roads and highways.

The M113 APC was the first modern "battle taxi"; developed to transport infantry forces on the mechanized battlefield. It is fitted with a 2 stroke six cylinder Detroit diesel providing power through a 3 speed automatic gearbox and steering differential. The main armament is a single .50 caliber heavy barrel machine gun, and the secondary armament is a single .30 caliber machine gun. The M113 is built of aircraft quality aluminum which allows it to possess some of the same strengths as steel at a much lighter weight. This distinct weight advantage allows the M113 to utilize a relatively small engine to power the vehicle, as well as carry a large payload cross-country. The vehicle is capable of "swimming" bodies of water.

The M113 Family includes thirteen variants of light armored tracked vehicles used in a variety of combat and combat support roles. More than 80,000 M113 Family of Vehicle (FOV) systems have been produced. The M113 FOVs is composed of a mix of derivative systems consisting of the A1, A2, and A3 configurations.

- M58, Wolf Smoke Generator Carrier
- M106, Self-propelled 107mm Mortar
- M113, Armored Medical Evacuation Vehicle (AMEV)
- M125, Self-propelled 81mm Mortar
- M548, Cargo Carrier
- M577, Command Post Carrier
- M730, Guided Missile Equipment Carrier (Chaparral}
- M901, Improved TOW Vehicle
- M981, Fire Support Team Vehicle (FISTV)
- M1059, Smoke Generator Carrier (Lynx)
- M1064, Self-propelled 120mm Mortar
- M1068, Standard Integrated Command Post System (SICPS) Carrier
- M1108, Universal Carrier

Apache - AH64D, Longbow Attack Helicopter

MISSION

Conduct rear, close, and deep operations and deep precision strikes; provide armed reconnaissance and security when required in day, night, and adverse weather conditions.

DESCRIPTION AND SPECIFICATIONS

Apache Longbow is a development and acquisition program for a millimeter-wave radar air/ground targeting system capable of being used day, night, in adverse weather, and through battlefield obscurants. Longbow integrates a mast-mounted millimeter-wave fire control radar (FCR), a radar frequency interferometer, and a radar frequency fire-and-forget HELLFIRE missile on the Apache. The digitized target acquisition system of the Longbow provides automated detection, location, classification, prioritization, and target handover.

The AH-64D cockpit is redesigned to digitize and multiplex all systems. MANPRINT crew stations have multi-function displays to reduce pilot work-load and increase effectiveness. The modernized Apache heavy attack team will now be able to provide a truly coordinated rapid-fire capability (servicing 16 separate targets within one minute) to the maneuver force commander on a 24-hour basis in day, night, and adverse weather conditions.

Apache Longbow adds significant warfighting capability to the combined arms team through increased survivability, lethality, versatility, and long-term reliability improvements and be an operationally effective, suitable weapon system, far more effective in defeating threat armored vehicles and more survivable against threat air defense weapons than the AH-64A.

Kiowa Warrior - OH58D, Reconnaissance/Attack Helicopter

MISSION

Conduct armed reconnaissance, security, target acquisition and designation, command and control, light attack, and defensive air combat (defensive) missions in support of combat and contingency operations.

DESCRIPTION AND SPECIFICATIONS

The Kiowa Warrior is the a rapidly deployable, lightly armed reconnaissance helicopter. The Kiowa Warrior includes advanced visionics, navigation, communication, weapons, and cockpit integration systems. The Mast-Mounted Sight (MMS) houses a thermal imaging system, lowlight television, and a laser rangefinder/designator. These systems allow target acquisition and engagement at stand-off ranges and in adverse weather conditions. The highly accurate navigation system of the Kiowa Warrior provides precise target location that can be sent digitally to other aircraft or artillery via its advanced communications system. Battlefield imagery can be transmitted to provide "near real time" situational awareness to command and control elements.

The Kiowa Warrior is rapidly deployable by air and can be fully operational within minutes of arrival. The armament systems combine to provide anti-armor, anti-personnel, and anti-aircraft capabilities at standoff ranges.

Although Kiowa Warrior fielding is complete, the Army is currently installing a series of safety and performance modifications to keep the aircraft safe and mission effective until it is retired.

Blackhawk - UH60L, Utility Tactical Transport Helicopter

MISSION

Provide air assault, general support, aeromedical evacuation, command and control, electronic warfare, and special operations support to combat and stability and support operations.

DESCRIPTION AND SPECIFICATIONS

The Blackhawk, along with the Comanche, Apache, and Chinook, is a principal component of the multi-function battalion, the building block of the Aviation Force Structure for the new millennium. The Blackhawk (UH-60) utility tactical transport helicopter has enhanced the overall mobility of the Army, due to dramatic improvements in troop capacity and cargo lift capability, compared to the UH-1 Huey it replaces. On the asymmetric battlefield, it provides the commander the agility to get to the fight quicker and to mass effects throughout the battlespace across the full spectrum of conflict.

An entire 11-person, fully equipped infantry squad can be lifted in a single Blackhawk, transported faster than in predecessor systems, in most weather conditions. The Blackhawk can reposition a 105mm Howitzer, its crew of six, and up to 30 rounds of ammunition in a single lift. The critical components and systems of the aircraft are armored or redundant, and its airframe is designed to progressively crush on impact to protect the crew and passengers.

This year, a program was to recapitalize and upgrade the now over twenty year old UH-60A model Blackhawks to the UH-60M configuration. The UH-60M will include a digitized cockpit to provide situational awareness, a standardized configuration to include the T701C engine and power train, and a zero-time, zero-mile service life extension.

Comanche - RAH66, Reconnaissance/Attack Helicopter

MISSION

Perform the armed reconnaissance mission for attack helicopter and air cavalry units.

DESCRIPTION AND SPECIFICATIONS

The Comanche (RAH-66) is the next-generation helicopter for the Army. It is designed to perform the armed reconnaissance and light-attack reconnaissance mission. The Comanche will significantly expand the capability of the Army to conduct reconnaissance operations in all battlefield environments, day or night, and during adverse weather. The Comanche will protect the force with its advanced electro-optical sensors, aided-target recognition, and sensor/weapons integration.

The digital communications capacity of the Comanche will enhance the capability of the Army to win the battlefield information war, and allow interface with the Joint Surveillance and Target Attack Radar System (JSTARS) and other joint sensors and weapons platforms. The design for rapid rearm, refuel, and repair of the Comanche will increase the tempo of operations. With low observability, target recognition, and digitized communications, the Comanche can conduct deep, precision-strike missions against time-sensitive targets. The Comanche will replace three types of helicopters (AH-1, OH-58, and OH-6) that currently perform the armed reconnaissance mission.

The armament mission equipment package consists of a 20mm turreted gatling gun, air-to-ground rockets and missiles along with an advanced electro-optical target acquisition and designation system, aided target recognition, and helmet-mounted display. One third of aircraft will include the Fire Control Radar (FCR), however, every aircraft will have provisions to incorporate the FCR.

In 2001, the Comanche program entered engineering and manufacturing development (EMD) phase which is the last phase of development, with two prototype aircraft in active flight test status. During EMD, the program plan is to build an additional five test aircraft to complete development testing, followed by eight aircraft that will be devoted to initial pilot training and operational testing.

M1025/114, High Mobility Multipurpose Wheeled Vehicle (HMMWV)

MISSION

Provide a common light tactical vehicle capability.

DESCRIPTION AND SPECIFICATIONS

The High Mobility Multipurpose Wheeled Vehicle (HMMWV) is a light, highly mobile, diesel-powered, four-wheel-drive vehicle that uses a common chassis. Using common components and kits, the HMMWV can be configured to become a troop carrier, armament carrier, S250 shelter carrier, ambulance, TOW missile carrier, and a Scout vehicle. The heavy variant, with a payload of 4,400 pounds, was developed as the prime mover for the light howitzer, towed VULCAN system, and heavier shelter carriers. The HMMWV replaced the 1/4 ton jeep, the M718A1 ambulance, 1/2-ton mule, 1-1/4-ton Gamma Goat, and M792 ambulance.

Since its inception, the HMMWV has undergone numerous design and configuration updates and changes. These changes have included technological, environmental, operational, and safety improvements, such as higher payload capability, radial tires, In 1995, two major improvements were introduced, The A2 configuration which incorporates a four-speed, electronic transmission; 6.5 liter diesel engine; and improvements in transportability and the expanded capacity vehicle (ECV) vehicle which has a payload of 5,100 pounds, including the crew. The ECV chassis is used for the M1114 "up-armored" HMMWV and serves as a platform for mission payloads for systems that exceed 4,400 pounds.

The "up-armored" M1114 HMMWV, developed to provide increased ballistic and blast protection primarily for the Military Police (MP) and contingency forces incorporates the following:

• An M2 or M60 machine gun, or a 40mm grenade launcher mounted on the roof. The locking bifold door opens from a 360-degree turret to offer rearward protection for the gunner at all times.

• A heavy-duty blast protection system protects the crew from the impact of a 12-pound anti-tank mine detonated in the front of the vehicle, or a 4-pound device detonated in the rear.

• High light-transmission transparent armor, or "white glass," is used for all windows. It offers distortion-free performance by day, and enhances the crew's use of infrared vision equipment at night.

• Fully armored doors that offer 7.62mm armor piercing perimeter protection at 100-meter standoff. Each features operational windows equipped with "white glass", and in a fixed position can be used as gun portals.

• Armor in the roof assembly is rolled on the edges to overlap the perimeter armor, offering the crew protection from all angles and providing overhead burst protection at 100 meters from the fragmentation of a 155mm shell.

Wolverine - M1, Heavy Assault Bridge (AHB)

MISSION

Provide the heavy brigade combat team with a military-load-class (MLC) 70-ton gap-crossing capability of up to 24 meters with the mobility, survivability, sustainability, and supportability of the maneuver force.

DESCRIPTION AND SPECIFICATIONS

The Wolverine launcher is mounted on a M1A2 Abrams system enhancement program chassis and is operated by a two-man crew. The bridge is 26 meters long and can span gaps up to 24 meters. It supports a MLC, 70-ton crossing at 16 kilometers per hour. The bridge is launched from under armor in five minutes and retrieved in less than ten minutes.

A critical force modernization item for the engineer regiment, the Wolverine enables heavy force maneuver by allowing units to rapidly transit tank ditches, road craters, and partially damaged bridge sections. This enabler increases the survivability of the maneuvering force by reducing the time spent in vulnerable areas waiting to cross. The current Armored Vehicle Launched Bridge (AVLB) minimally supports Abrams tank unit crossings at reduced-gap length (15 meters) and reduced crossing speeds.

M270, Multiple Launch Rocket System (MLRS)

MISSION

Provide counterbattery fire and suppression of enemy air defenses, light materiel, and personnel targets.

DESCRIPTION AND SPECIFICATIONS

The Multiple Launch Rocket System (MLRS) is an artillery weapon system that supplements cannon artillery fires by delivering large volumes of firepower in a short time against critical, time-sensitive targets. The MLRS is capable of supporting and delivering all freeflight basic and Extended Range (ER-MLRS) rockets and Army Tactical Missile System (ATACMS) Block I missiles.

The basic warhead carries improved conventional submunitions. Growth programs are underway to extend the range and accuracy of the rockets and to upgrade the launcher to fire precision guided rockets and missiles to include Guided MLRS (GMLRS) and ATACMS/Brilliant Anti-Armor Submunition (BAT) Block II weapons.

The M270A1 upgrade starts with rebuilding the carrier and launcher loader module. The rebuilt M270 is then upgraded by adding the Improved Fire Control System (IFCS) and the Improved Launcher Mechanical System (ILMS) modifications. The IFCS will mitigate electronic obsolescence and will provide growth for future weapon systems. The ILMS will provide rapid response to time-sensitive targets by reducing the aiming time by seventy percent and reducing the reload time by forty percent. The M270A1 upgrade launcher supports the Army Transformation by providing overmatch capabilities to the Counterattack Corps through 2020.

MISSION

Provide the primary artillery support for armored and mechanized infantry divisions.

DESCRIPTION AND SPECIFICATIONS

The M109A6 (Paladin) Howitzer is the most technologically-advanced self-propelled cannon system in the U.S. Army. The A6 designation identifies several

Paladin - M109A6, Self Propelled Howitzer - 155mm

changes to the standard model that provide improvements to weapon survivability, responsiveness, RAM, armament, and terminal effects. The fire control system is fully automated, providing accurate position location and azimuth reference, and on-board ballistic solutions of fire missions. The howitzer has a computer-controlled gun drive through servos with manual backup. Paladin uses state-of-the-art components to achieve dramatic improvements in the following:

- Survivability: "shoot and scoot" tactics; improved ballistic and nuclear, biological, and chemical protection.
- Responsive fires: capable of firing within 45 seconds from a complete stop with on-board communications, remote travel lock, and automated cannon slew capability.
- Accurate fires: on-board Position Navigation (POSNAV) and technical fire control.
- Extended range: 30 km with High Eexplosive - Rocket Assisted Projectiles (HE-RAP) and M203 propellant.
- Increased reliability: improved engine, track, and diagnostics
- Upgrades include: Global positioning system-aided self-location, M93 Muzzle Velocity System, and commercial off-the-shelf-based computer processor.

Crusader - M2001, Self Propelled Howitzer - 155mm

MISSION

Provide direct and general support fires to maneuver forces on the future battlefield; become the indirect fire support "system of systems" howitzer and resupply vehicle.

DESCRIPTION AND SPECIFICATIONS

The Crusader Self-Propelled Howitzer (SPH) is the cornerstone system of the Fire Support Modernization Strategy and the strategic hedge necessary for the Legacy Force while the Army transforms itself to the Objective Force. The Crusader is a 155mm self-propelled howitzer system that will increase artillery survivability, lethality, mobility, and operational capability and effectiveness through use and integration of advanced technology. The Crusader program has been restructured to improve its deployability and relevance to the transformation and the Objective Force while retaining all of its key performance parameters (range, rate-of-fire, mobility, and resupply).

Key features of Crusader are lower weight (38,9642 tons), smaller size (two howitzers transportable in the C-17), and a change in resupply vehicle philosophy. The lower weight and smaller size are achieved through the extensive use of titanium and composites, reduction in onboard ammunition, employment of reconfigurable armor, elimination of one road wheel, and incorporation of the Abrams-Crusader Common Engine, the LV 100-5. Now, Crusader howitzers will be resupplied by one-tracked and one-wheeled resupply vehicle for each two howitzers, versus one-tracked resupply vehicle per howitzer. SPH critical technologies and capabilities include the XM297E2 integral mid-wall cooled cannon, modular artillery charge system, automatic inductive settable multi-option fuze, fully automated ammunition handling and loading systems, composite armor, enhanced survivability, improved mobility, and embedded training and diagnostics.

The tracked and wheeled resupply vehicles (RSV-T and RSV-W) enable automatic transfer of ammunition and fuel to the SPH or another RSV to meet required firing rates; enable autonomous operation; and capitalize on cost and operational advantages of component commonality. RSV critical technologies and capabilities include an automated docking boom, ammunition resupply system, fuel transfer system, improved mobility, and embedded training and diagnostics. The Crusader will displace the M109A6 Paladin Self-Propelled Howitzer and the M992A2 Field Artillery Ammunition Supply Vehicle (FAASV) beginning in 2nd Quarter of Fiscal 1908.

MISSION

Detect, identify, and mark areas of nuclear and chemical contamination; sample for nuclear, biological, and chemical (NBC) contamination; and report accurate information to supported commanders in real time.

DESCRIPTION AND SPECIFICATIONS

The Nuclear, Biological and Chemical Reconnaissance System (NBCRS)-Fox Block I Modification (M93A1) contains an enhanced NBC sensor suite consisting of the M21 Remote Sensing Chemical Agent Alarm (RSCAAL), MM1 Mobile Mass Spectrometer, Chemical Agent Monitor/Improved

Fox - M93A1, Nuclear, Biological and Chemical Recon System (NBCRS)

Chemical Agent Monitor (CAM/ICAM), AN/VDR-2 Beta Radiac, and M22 Automatic Chemical Agent Detector/Alarm (ACADA). The NBC sensor suite has been digitally linked with the communications and navigation subsystems by a dual-purpose central processor system known as the Multipurpose Integrated Chemical Agent Detector (MICAD). The MICAD processor fully automates NBC warning and reporting functions and provides the crew commander with full situational awareness of the NBC sensors, navigation, and communications systems. The M93A1 Fox is also equipped with an advanced navigation system Global Positioning System (GPS) and Autonomous Navigation System (ANAV) that enables the system to accurately locate and report agent contamination. The mobility platform is a six-wheeled, all-wheel-drive armored vehicle capable of cross-country operation at speeds up to sixty-five mph. The Fox System is fully amphibious with swimming speeds up to six mph.

As a reconnaissance vehicle, it can locate, identify, and mark chemical/biological agents on the battlefield. The Fox usually accompanies the scouts or motorized reconnaissance forces when performing its NBC mission. It has an over-pressure filtration system that permits the crew to operate the system in a shirt-sleeve environment that is fully protected from the effects of NBC agents and contamination.

The M93A1 system is operated by a three-person crew (legacy systems require a four-person crew). The M93A1 will be one of the few systems fielded with a fully interactive class 4/5 electronic technical manual (IETM). The IETM is a single multimedia CD that contains the twelve-manual library and is structured to incorporate advanced diagnostics that support the system.

The Block II Modification to the M93A1 Fox NBCRS will incorporate enhanced technologies that will enable on-the-move standoff chemical agent detection. The Chemical Biological Mass Spectrometer (CBMS) will improve the detection and identification of liquid chemical agents while providing a first-time biological agent detection capability to the reconnaissance platform. An open architecture approach for sensor suite integration will enable expansion/ upgrade of sensors and on-board computers at minimal cost as well as significant improvements in NBC reconnaissance reporting and situational awareness for commanders.

Single Channel Ground and Airborne Radio System (SINCGARS)

MISSION

Provide commanders with a highly reliable, secure, easily maintained combat net radio (CNR) with voice and data handling capability, in support of command and control operations.

DESCRIPTION AND SPECIFICATIONS

The Single Channel Ground and Airborne Radio System (SINCGARS), with its internet controller, provides the communications link for the digitized force, SINCGARS configurations include manpack, vehicular (both low and high power), and airborne models. Communications security is integrated into currently produced versions of the ground and airborne radios. System improvement program models provide upgrades to enhance operational capability in the tactical internet environment. The advanced system improvement program (ASIP) models (of a reduced size and weight) provide further enhancements to operational capability in the tactical internet environment.

Land Warrior (LW)

MISSION

The Land Warrior System (LW), developed for the individual combat soldier, provides for significant improvements in lethality, survivability, battle command, mobility, sustainment and training-mission rehearsal.

DESCRIPTION AND CAPABILITIES

The LW is a first generation modular, integrated fighting system for the dismounted combat soldiers that incorporates an assortment of systems, components and technologies into a lethal, survivable soldier. LW systems/components includes:

- A lightweight helmet with integral ballistic/laser eye protection, monocular computer display, day/night target sensor unit and an adaptable chemical/biological mask.
- A mission-tailored computer software subsystem addresses that soldier's core battlefield functions, display management, and mission equipment and supply ensuring that Land Warrior will be FBCB2 compatible and interoperable on the digital battlefield.
- A modular, back mounted, load-carrying unit with integrated computer and battery, soldier and squad Computer/Radio System (CRS) radio system, GPS and a video picture capture unit. The squad radio is based on an improved SINCGARS radio and will be fully compatible with SINCGARS Service Improvement Package (SIP).
- A modified M-4 carbine/M-16 rifle with an aiming light, video camera, Thermal Weapon Sight (TWS), special close-combat optics and Laser RangerFinder/Digital Compass (LRF/DCA). The LRF/DCA provides the soldier with range and direction information. When coupled with his individual location from GPS, the soldier has accurate target location when calling for indirect fire and combat identification.
- Protective clothing enhancements of chemical biological garments and modular body armor. The new Land Warrior body armor, designed of modular ballistic plates, protects against small arms fire even at close range and provides improved ballistic protection at a reduced weight.

The integrated squad and soldier radios and the GPS within the CRS provide digital control and command and situational awareness capability previously unavailable to the combat soldier. The GPS provides the soldier's location to the computer, integrates the soldier's position with location reports from other soldiers and displays the information on the digital map on the integrated helmet mounted display.

The radios of the CRS, controlled by the computer, provides both digital and voice capabilities to the dismounted soldier. The soldier radio is provided to all LWs. Squad radios, provided to all team leaders and above, are SINCGARS compatible. The soldier radio is based on a repackaged handheld commercial radio made by Motorola. This gives the soldier the ability to communicate on the digitized battlefield with others in his squad, greatly improving situational awareness and survivability through increased command and control.

MISSION

Provide battle command and situational awareness information from brigade level down to the soldier or weapon platform level.

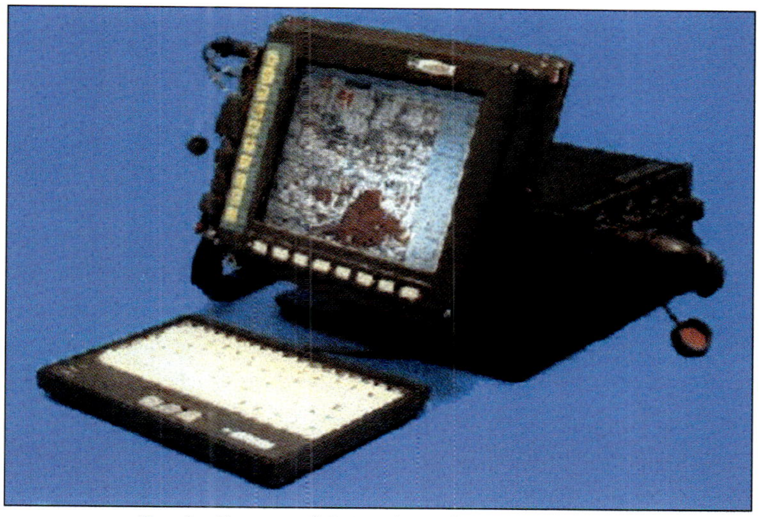

Battle Command Brigade and Below (BCB2)

DESCRIPTION AND CAPABILITIES

The Force XXI Battle Command Brigade-and-Below (BCB2) forms the principal digital command and control system for the Army at brigade levels and below. Applique hardware and software of the BCB2 are integrated into the various platforms at brigade and below, as well as appropriate division and corps slices necessary to support brigade operations. The BCB2 provides situational awareness by integrating information from Global Positioning System (GPS), weapon sensors aboard the host weapon platform, and from external updates from the radio link to the tactical internet.

The BCB2 system is designated ACAT 1D and is in the engineering and manufacturing development (EMD) phase. The system features the interconnection of platforms through a communications infrastructure called the Tactical Internet to transmit situational awareness data. The BCB2 will be fitted on every weapon platform from the individual soldier LW, the Abrams Tank, the Bradley Linebacker air defense system and other battalion level weapon systems.

Avenger - Low Level Air Defense System

MISSION

Provide mobile, short-range, air defense protection against cruise missiles, unmanned aerial vehicles, low-flying fixed-wing aircraft, and helicopters to divisions, armored cavalry regiments, separate brigades, and corps/theater air defense brigades.

DESCRIPTION AND SPECIFICATIONS

The Avenger system is a lightweight, highly mobile, and transportable surface to air missile/gun weapon system mounted on a High Mobility Multipurpose Wheeled Vehicle (HMMWV). It has a two-man crew and can operate in day or night and in clear or adverse weather conditions. The system incorporates a fully rotatable turret, operator's position with displays, fire control electronics, and standard vehicle-mounted launchers which support and launch multiple Stinger missiles.

The Avenger can be operated remotely up to 50 meters from the fire unit and can shoot on the move. A notable upgrade is the slew-to-cue (STC), which is embedded into a new Avenger fire control computer (AFCC). The STC accepts digital radar track data from external sources, then automatically slews the turret to place an aerial target into the sighting field of view of the gunner. The AFCC, an approved Warfighting Rapid Acquisition Program (WRAP), corrects significant system obsolescence issues and allows for additional system growth at minimal cost.

Bradley Linebacker - M6, Short Range Air Defense Vehicle

MISSION

Provide dedicated forward-area air defense for heavy-maneuver forces; engage and defeat a variety of threat platforms including rotary-wing aircraft, unmanned aerial vehicles, cruise missiles, and fixed-wing aircraft.

DESCRIPTION AND SPECIFICATIONS

The M6 Bradley Linebacker is a Bradley M2A2 Operation Desert (OD) Storm vehicle, modified by replacing the TOW missile launcher with a four-missile STINGER launcher (Standard Vehicle Mounted Launcher [SVML]). With this modification, the crew can conduct a ground-to-air engagement while remaining under armor protection. The linebacker also incorporates the Forward Area Air Defense Command and Control (FAAD C2) software on a handheld terminal unit. By integrating Global Positioning System (GPS) and FAAD C2, the Linebacker provides an automated slew-to-cue function.

Sentinel - AN/MPQ-64, Search and Track Radar System

MISSION

Provide critical air surveillance of the forward areas; automatically detect, track, classify, identify, and report targets (cruise missiles, unmanned aerial vehicles, rotary- and fixed-wing aircraft) to air defense weapons systems.

DESCRIPTION AND SPECIFICATIONS

The Sentinel is used with the Forward Area Air Defense Command and Control (FAAD C2) system. The Sentinel accomplishes its primary mission by providing key target data to Short Range Air Defense (SHORAD) weapon systems and battlefield commanders via the FAAD C2 data link or directly from the Sentinel, using Enhanced Position Location Reporting System (EPLRS) or Single Channel Ground and Airborne Radio System (SINCGARS) data radios.

The Sentinel system consists of the High Mobility Multipurpose Wheeled Vehicle (HMMWV) group and the Antenna Transceiver Group (ATG) mounted on a one-ton, wide-track trailer, its identification friend-or-foe (IFF), and FAAD C2 interfaces. The sensor is an advanced three-dimensional battlefield X-band air defense phased-array radar with an acquisition range of 40 km. The Sentinel is capable of operating both day and night, in adverse weather conditions, and in battlefield environments of dust, smoke, aerosols, and enemy countermeasures. It provides 360-degree azimuth coverage for acquisition and tracking.

The Sentinel automatically detects, tracks, classifies, identifies, and reports targets including cruise missiles, unmanned aerial vehicles, rotary-wing aircraft, and fixed-wing aircraft. It can identify targets moving at speeds from hovering to fast-moving and located at positions from nap of the earth to the maximum engagement altitude of SHORAD weapons. Accurate and fast reacting, Sentinel acquires targets sufficiently forward of the forward line of troops to improve air defense weapon reaction time and allow engagement at optimum ranges. The Sentinel-integrated IFF reduces the potential for fratricide of friendly aircraft. The Sentinel is transported by an M1097A1 HMMWV using standard transportation methods.

Heavy Equipment Transporter System (HETS)

MISSION

Transport, deploy, recover and evacuate combat-loaded M1-Series Abrams Tanks or other vehicles of similar weight in support of legacy equipped and digitized combat units.

DESCRIPTION AND SPECIFICATIONS

The Heavy Equipment Transporter System (HETS) consists of the M1070 truck tractor and M1000 semi-trailer. HETS is capable of transporting vehicles in the 70-ton weight range, primarily M1-Series Abrams Tanks. It operates on highways (worldwide with permits), secondary roads, and cross-country. HETS has a number of features that significantly improve the mobility and overall performance of the system in a tactical environment. The M1070 tractor has front and rear-axle steering, a central tire-inflation system, and cab space for six personnel to accommodate the two HETS operators and four tank crewmen. The M1000 semi-trailer has automatically steerable axles and a load-leveling hydraulic suspension. The M1070 and M1000 are procured under separate acquisition programs.

Palletized Load System (PLS)

MISSION

Perform line and local haul, unit resupply, and related missions in both operational and tactical environments (Corps-Brigade areas of operation) in support of legacy equipped, digitized, and transformation force combat units.

DESCRIPTION AND SPECIFICATIONS

The Palletized Load System (PLS) consists of a prime mover (10 x 10 ft configuration), equipped with an integral load-handling system providing selfload/unload capability, and capable of transporting a 16.5-ton payload. PLS carries equipment/ammunition/supply loads on demountable "flatrack" cargo beds, 8 x 8 x 20 ft International Standards Organization (ISO) containers, or shelters, and is able to tow a 16.5-ton payload trailer that is also capable of carrying flatracks. Two additional pieces of equipment enhance PLS flexibility. The Containerized Roll-in/out Platform (CROP) is an A-frame type flatrack which fits inside a 20-ft ISO container. A Container Handling Unit (CHU) enables PLS to pick up and transport ISO containers without using a flatrack. Flatracks and CROP are interchangeable between PLS and the HEMTT-LHS. The PLS prime mover features a central tire inflation system that significantly improves off-road mobility. Current NATO agreements require PLS to maintain interoperability with comparable British, German, and French systems through the use of a common flatrack. PLS is a major enabler in the Army's drive to achieve a distribution-based logistics system.

Forward Repair System (FRS)

MISSION

Provide a forward, mobile maintenance/repair vehicle to return heavy force systems to an operational condition.

DESCRIPTION AND SPECIFICATIONS

The Forward Repair System (FRS) is a "must-have" enabler for Force XXI and is a new start in FY00. The FRS is a high mobility, forward maintenance system that reduces man hours for maintenance personnel, enabling conversion to the Force XXI design. The FRS will minimize the current practice of using recovery vehicles, the five-ton wrecker and the M88 for maintenance lift (e.g., removing engines), and replace the M113A2/A3 armored personnel carrier for transporting maintenance equipment. The current practice causes a severe shortage of recovery assets. Force XXI, with its expanded battlespace, only increases the need for dedicated recovery assets. The FRS consists of a Palletized Load System (PLS) truck for mobility. It is overhead-lift capable, and it has a mission module that contains the tools, diagnostic equipment, and repair parts needed to return heavy force systems to battle or into a self-recovery mode. The emphasis in developing the system is on procuring non-developmental components.

Tactical Unmanned Aerial Vehicle (TUAV)

MISSION

Provide reconnaissance, surveillance, and target acquisition (RSTA) information at an initial range of 50 km, day or night, in limited adverse weather conditions with a future, objective range extending to 200 km.

DESCRIPTION AND SPECIFICATIONS

The Tactical Unmanned Aerial Vehicle (TUAV) is intended for use in military environments where real-time information feedback is needed, but manned aircraft are unavailable, or excessive risk or other conditions render use of manned aircraft imprudent. A TUAV system consists of two ground control stations (GCSs), one portable ground control station, one portable ground data terminal, four remote video terminals (RVTs), a minimum of three air vehicles (AVs), modular mission payloads, and launch and recovery equipment.

The GCS collects, processes, analyzes, and distributes digitized battlefield information by interfacing with present and planned service command, control, communications, computers and intelligence (C4I) systems. Flight and mission commands are sent to the AVs from the GCS. RSTA imagery and AV position data are downlinked directly to the GCS or RVTs located in tactical operations centers of the brigade, its subordinate battalions, and/or the direct support artillery or supporting aviation units. The complete TUAV system is transportable by two C-130 aircraft. Mission capability will be enhanced as advanced mission payloads become available, maximizing battlefield digitization to increase the effectiveness of other weapon systems.

Appendix M: The Association

On 17 July 1944, a group of First Team soldiers in the Admiralty Islands agreed to form an Association of 1st Cavalry Division soldiers and veterans to preserve old friendships and conduct periodic reunions. The first reunion of the 1st Cavalry Division Association was held in 1948 at El Paso, TX - Ft. Bliss, where the 1st Cavalry Division was organized in 1921.

Today, the 1st Cavalry Division Association is the association of veterans, forming a fraternal organization with some 24,000 of it's 55,000 alumni on the active rolls with a period of duty in the 1st Cavalry Division ranging from the 1930's to today's active duty trooper. The organization is nonsectarian, nonprofit and nonpolitical.

Who is eligible for membership? Anyone who has been assigned or attached to the 1st Cavalry Division any time, anywhere is eligible for life membership. Other Military Personnel who served along side the 1st Cavalry Division and friends of the cavalry who have not served with the division can become Associate Members.

Belonging to the 1st Cavalry Division Association preserves and strengthens friendships shared over the years through chapter activities, annual reunions and the publication of the monthly newspaper, "The Saber". There is the opportunity to meet other people and make new friendships through association activities. From horse cavalry through ground and airmobile infantry to heavy armor, we share a great division heritage and tradition.

Former members of the 1st Cavalry Division may enroll and obtain tangible evidence of membership in this great fraternity, i.e. a lapel pin, membership card and certificate, decals, a one-year subscription to the "Saber" and finally to be involved in special projects such as the Foundation of the 1st Cavalry Division Association that provides educational assistance to the children of First Team soldiers killed or disabled while serving in the Division, to descendents of survivors of the Ia Drang battles, and to active duty soldier Association members assigned to the Division, their spouses and children.

1st Cavalry Division Association

Appendix N: Bibliography

Reference Books and Documentation

Thomas B. Allen, F. Clinton Berry and Norman Polar; "CNN - War in the Gulf", Turner Publications, Inc - Atlanta, GA 1991

Appleman, Roy E.; "South to the Naktong, North to the Yalu", Center of Military History , US Government Printing Office - Washington, DC 1961

Andrade, Dale; "Luzon; December 1944 - July 1945", Center of Military History, US Government Printing Office - Washington, DC 1998

Bluhm, Col Raymond K. Jr., Historical Services Branch Chief; "1st Cavalry Division Organizational History", Center of Military History, US Government Printing Office - Washington, DC 22 September 1992

Bradley, Major (Retired) John H.; "From the Dark Side of the Moon - Raiding by Ground Forces in the Southwest Pacific during the Second World War", Rice University Press - Houston, TX 1970.

Cole, LTC (Retired) Hilbert H.; "9th (Air) Cavalry Brigade (Provisional) Combat Operations After Action Report", Minerva Press - London, UK 1995

Committee on Veteran's Affairs, US Senate; "Medal of Honor Recipients 1863-1973, US Government Printing Office - Washington DC 1973

Conn, Stetson, editor; "Army Lineage Series", US Government Printing Office - Washington, DC 1974

Daily, Edward L.; "From Custer to MacArthur", Turner Publications - Paducah, KY 1995

Daily, Edward L.; "History of the 70th Tank Battalion", Turner Publications - Paducah, KY 2000

Department of Military History; "Lineage and Honors", Department of the Army

Dunn, Si; "The First Cavalry Division - A Historical Overview 1921-1983", Taylor Publishing Co. - Dallas, TX 1984

Dzwonchyk, Wayne E., editor; "Army Lineage Series, Aviation", US Government Printing Office - Washington, DC 1986

Eyre, Michael A.; Private Collection - "Surrender of Iraqi Soldiers", Desert Storm, On site - Iraq Desert 1991

Field, James J. Jr,; "History of United States Naval Operations: Korea", US Government Printing Office - Washington, DC 1962

Gurney, Gene Col USAF (Ret); "VIETNAM - The War in the Air", Crown Publishers Incorporated - New York, NY 1985

Henk, Edward; Private Collection - "Recollections of the Camp Evans Rocket Attack", On Site - Vietnam 1968

Hermes, Walter G; "Truce Tent and Fighting Front", Center of Military History, US Government Printing Office - Washington, DC 1966

Historical Division; "The Admiralities - Operations of the 1st Cavalry Division; 29 February - 18 May 1944", War Department - Washington, DC 1945

Jones; Ralph; Private Collection - "Recollections of the Fire Support Base Illingsworth Attack", On Site - Vietnam 1970"

Kail, William E., Lt; Unit Historical Report - "Operation Bullseye V", Office of Information, 1st Cavalry Division, Republic of Vietnam 1967

Maclean, John A.; Private Collection - "United Nations Command Military Armistice Commission", Administration of the Army - Japan

Maclean, John A.; Private Collection - "1st Cavalry Division - The First Team", Information Section, Headquarters 1st Cavalry Division - Korea 1960

McKenny, Janice E., editor; "Army Lineage Series, Field Artillery - Regular Army and Army Reserve", US Government Printing Office - Washington, DC 1984

Metz, Leon C.; "Desert Army - Fort Bliss on the Texas Boarder", Mangan Books - El Paso, TX 1995

Morrison Samuel E.; "Leyte, June 1944 - January 1945", Little, Brown and Company - Boston, MA 1958

Morrison Samuel E.; "The Liberation of the Philippines - Luzon, Mindanbio and Visayas", Little, Brown and Company, Boston 1959

Morrocco, John; "Rain of Fire", Boston Printing Company - Boston, MA 1985

Nolan, Keith William; "Into Cambodia", Presidio Press - Novato, CA 1990

Philbrick, Jack F.;Private Collection - "Recollections of the 8th Cavalry Regiment", Camp Crawford, JA 1954

Phillips, J. E. and Gregory, R. M.; "America's First Team in the Gulf", Taylor Publishing Co. - Dallas, TX 1992

Raines, Edgar F.; "Eyes of Artillery - The Origins of Modern Unites States Army Aviation in World War II", Center of Military History, US Government Printing Office - Washington, DC 2000

Rees, David ed; "The Korean War - History and Tactics", Orbis Publishing Co. - London, UK 1984

Schubart, Frank N. and Kraus, Theresa L.; "The Whirlwind War", Center of Military History, US Government Printing Office - Washington, DC 1995

Scott, Gorden C.; Private Collection - "Recollections of the 1st Cavalry Division on the DMZ - 1961/63"

Shane, Jerry W.; "Private Data Base Collection - the 1st Cavalry Division, attached units 1950 - 1953", - Scottsdale, AZ 19 May, 1998

Shelby, Stanton L.; "Anatomy of a Division - The 1st Cavalry in Vietnam", Presidio Press - Novato, CA 1987.

Shelby, Stanton L.; "Vietnam Order of Battle", Galahand Books, New York, NY 1987

Silver, Col. (Ret) Benjamin S.; "Ride At A Gallop", Davis Brothers Publishing Company - Waco, TX 1990

Smith, Gene; "The Dark Summer", Macmillian Publishing Co. - New York, NY 1987

Stubbs, Mary Lee and Connor, Stanley Russell, editors; "Army Lineage Series", Armor/Cavalry - Regular Army and Army Reserve, US Government Printing Office - Washington, DC 1984

Tennant, Larrie; Private Collection - "Recollections of the DMZ Police Company", On site - Korea, 1958/59"

Tolson, John J., Lt Gen; "Vietnam Studies, Air Mobility 1961 - 1971", Center of Military History , US Government Printing Office - Washington, DC 1973

The 1st Air Cavalry Division; "Vietnam 1965 - 1969", Published by Dia Printing Company, Ltd. - Tokyo, Japan 1970

Trotola, Alfred; Private Collection - "Recollections of the The Sisters Outpost of the 8th Cavalry Regiment, Korea - 21 November 1951"

Voyles, James A.; Private Collection - "Firebase Moe, III Corps", Vietnam War, On site - III Corps Area 1970

Walker, Harold D.; Private Collection - "Memories of the 1st Cavalry Division on the DMZ - 1957/58"

Wilson, John B,; "Maneuver and Firepower - The Evolution of Divisions and Separate Brigades", Center of Military History , US Government Printing Office - Washington, DC 1998

Wright, Major B. C.; "The 1st Cavalry Division in World War II, 1st Cavalry Division, Toppan Printing Company Ltd. - Tokyo, Japan 1947

First Team Public Affairs Office

Staff Document; "World War II Souvenir Battle Diary - 1st Cavalry Division", Tokyo, Japan 1945. Reprinted by Greenlee Printing - Calumet, MI

Staff Document; "1st Cavalry Division"; Prepared by Information Section, Headquarters, 1st Cavalry Division - Korea 1958 (Revised 1960)

Staff Document; "First Cavalry Division - Interim Report of Operations, July 1965 to December 1966" Published by 1st Cavalry Division Association - Albuquerque, NM Undated

Staff Document; "Historical Presentation", - Fort Hood, TX Undated

Staff Document; "Change of Command - 1st Brigade, 1st Cavalry Division", - Fort Hood, TX 09 June 1995

Staff Document; "Change of Command - 2nd Brigade, 1st Cavalry Division", - Fort Hood, TX 06 February 1986

Staff Document; "Change of Command - 3rd Brigade, 1st Cavalry Division", - Fort Hood, TX 07 February 1997

Staff Document; "Change of Command - Combat Aviation Brigade, 1st Cavalry Division", - Fort Hood, TX 16 June 1988

Staff Document; "Change of Command - Aviation Brigade, 1st Cavalry Division", - Fort Hood, TX 24 June 1992

Staff Document; "Change of Command - Aviation Brigade, 1st Cavalry Division", - Fort Hood, TX 15 July 1994

Staff Document; "Change of Command - Aviation Brigade, 1st Cavalry Division", - Fort Hood, TX 15 July 1996

Staff Document; "History of the 4th Aviation Brigade Headquarters", - Fort Hood, TX Undated

Staff Document; "1st Cavalry Division Artillery", - Fort Hood, TX 1997

Staff Document; "Change of Command - 1st Cavalry Division Artillery", - Fort Hood, TX 23 June 1997

Staff Document; "Change of Command - Engineer Brigade, 1st Cavalry Division", - Fort Hood, TX 03 July 1996

Staff Document; "Change of Command - 1st Cavalry Division Support Command", - Fort Hood, TX 12 July 1996

Staff Document; "History of the 5th Cavalry Regiment", - Fort Hood, TX Undated

Staff Document; "Change of Command - 1st Battalion, 5th Cavalry Regiment", - Fort Hood, TX 22 May 1996

Staff Document; "Change of Command - 2nd Battalion, 5th Cavalry Regiment", - Fort Hood, TX 05 June 1996

Staff Document; "History of the 8th Cavalry Regiment", - Fort Hood, TX Undated

Staff Document; "History of the 7th Cavalry Regiment", - Fort Hood, TX Undated

Staff Document; "Change of Command - 1st Squadron, 7th Cavalry Regiment", - Fort Hood, TX 06 July 1994

Staff Document; "Change of Command - 2nd Battalion, 7th Cavalry Regiment", - Fort Hood, TX 02 October 1997

Staff Document; "Change of Command - 1st Battalion, 8th Cavalry Regiment", - Fort Hood, TX undated

Staff Document; "Change of Command - 2nd Battalion, 8th Cavalry Regiment", - Fort Hood, TX 10 April 1996

Staff Document; "Change of Command - 3rd Battalion, 8th Cavalry Regiment", - Fort Hood, TX 17 April 1996

Staff Document; "History of the 9th Cavalry Regiment", - Fort Hood, TX Undated

Staff Document; "Change of Command - 1st Battalion, 9th Cavalry Regiment", - Fort Hood, TX 02 May 1997

Staff Document; "History of the 12th Cavalry Regiment", - Fort Hood, TX Undated

Staff Document; "Change of Command - 1st Battalion, 12th Cavalry Regiment", - Fort Hood, TX 04 June 1996

Staff Document; "Change of Command - 2nd Battalion, 12th Cavalry Regiment", - Fort Hood, TX 02 April 1997

Staff Document; "Change of Command - 1st Battalion, 227th Aviation Regiment, Aviation Brigade, 1st Cavalry Division", - Fort Hood, TX 04 February 1997

Staff Document; "Change of Command - 4th Battalion, 227th Aviation Regiment, Aviation Brigade, 1st Cavalry Division", - Fort Hood, TX 06 November 1996

Staff Document; "Activation Ceremony - 1st Battalion, 21st Field Artillery", - Fort Hood, TX 16 September 1997

Staff Document; "Change of Command - 1st Battalion, 82nd Field Artillery", - Fort Hood, TX 07 June 1995

Staff Document; "Change of Command - 2nd Battalion, 82nd Field Artillery", - Fort Hood, TX 02 May 1995

Staff Document; "Change of Command - 3rd Battalion, 82nd Field Artillery", - Fort Hood, TX 05 June 1997

Staff Document; "Change of Command - 8th Engineer Battalion", - Fort Hood, TX 16 June 1997

Staff Document; "Change of Command - 20th Engineer Battalion", - Fort Hood, TX 19 June 1994

Staff Document; "Change of Command - 91st Engineer Battalion", - Fort Hood, TX 09 July 1996

Staff Document; "Change of Command - 15th Forward Support Battalion", - Fort Hood, TX 11 June 1996

Staff Document; "Change of Command - 27th Main Support Battalion", - Fort Hood, TX 14 June 1995

Staff Document; "Change of Command - 115th Forward Support Battalion", - Fort Hood, TX 23 June 1995

Staff Document; "Change of Command - 215th Forward Support Battalion", - Fort Hood, TX 21 May 1997

Staff Document; "Activation Ceremony - 615th Aviation Support Battalion", - Fort Hood, TX 17 September 1996

Staff Document; "Change of Command - 9th Air Support Operations, 3rd Air Support Operations Group", - Fort Hood, TX 31 May 1996

Staff Document; "Change of Command - 15th Finance Battalion", - Fort Hood, TX 09 October 1996

Staff Document; "Change of Command - 15th Personnel Service Battalion", - Fort Hood, TX 05 August 1997

InterNet WEB Sites

ARCENT; "United States Forces Command HQ, Third US Army"

Brooks, Jacob A. SPC, "A" Co, 2nd Bn, 12th Cavalry; "Ode to the Desert"

Desert Voice; "The Weekly Newspaper of Camp Doha, Kuwait"

First Cavalry Division Association; "Photograph Album Collection"

Griffin, Gerald; Irish Prose - "The Collegians, Chapter 1"

Hourigan, Joseph; "The History of the Fighting 69th Infantry"

Kline, Joseph D.; Artist, "Missile! Missile! Missile!, - Copyright 1993"

Mississippi State University; "Wars of the United States"

Mississippi State University; "The Vietnam War"

National Archives and Records Administration; "Still Pictures Branch"

US Army; "Talon - The On-Line Magazine of Task Force Eagle"

US Army Center of Military History; "Lineage and Honors"

US Army Center of Military History; "Medal of Honor Citations"

US Army Pamphlet 672-3; "Decorations, Awards and Honors, 29 January 1988"

USCENTCOM; "United States Central Command Mission"

Appendix O: Contributors

CONTRIBUTORS and ADVISORY EDITORS

Coleman J. D., LTC. (Ret) "B" Company, 2nd Battalion, 8th Cavalry Regiment and 1st Cavalry Division PIO, Vietnam.

Hodes, John T., COL (Ret), Platoon Leader "D" Company, 1st Battalion and "F" Company, 2nd Battalion, 8th Cavalry Regiment, Korea.

Laporte, Leon J., GEN, USA, G3 then Chief of Staff 1st Cavalry Division Desert Shield/Desert Storm.

Milano, James M., COL, USA, G3 1st Cavalry Division, Operation Joint Forge, Bosnia.

Shoemaker, Robert M., GEN (Ret) Commander. 1st Cavalry Division TRICAP period Ft. Hood, Texas.

Walton, Donald H., COL (Ret) Commander. 302nd Reconnaissance Company. 1st Cavalry Division WWII period.

TECHNICAL CONTRIBUTORS

Arthur J. Junot, BG, USA (Ret), Executive Director of the 1st Cavalry Division Association.

General Junot enlisted in the Army in 1947 and later attended Officer Candidate School at Ft. Riley, KS where he was commissioned a Second Lieutenant of Cavalry in 1950. He served with the 23rd Infantry Regiment during the Korean War and during his two tours of duty in Vietnam he served with the Military Assistance Command Vietnam, the 16th Aviation Group of the Americal Division and commanded a battalion of the 34th General Support Group. During the 1970s he commanded the 1st Cavalry Division Support Command at Fort Hood and the US Army Troop Support Command in St. Louis from which he retired in 1978 after 31 years of service having served in Cavalry, Infantry, Armor, Artillery and Transportation branches.

Robert F. Litle, Jr., COL, USA (Ret),

Colonel Robert Litle entered service through the US Army Air Corps Aviation Cadet program in WW II. By the end of the war, he was a B-29 pilot and he returned to college at Purdue University. Upon graduation with a degree in Air Transportation, he transferred from the US Air Force Reserve to the Army National Guard and returned to active service in 1949 as a Liaison Pilot in the Infantry.

During his military service he commanded at all levels-from infantry platoon through aviation group, including command of the 2nd Battalion, 7th Cavalry in Vietnam.

Following his retirement in 1973 he was elected to the Army Aviation Hall of Fame.

Jeffery E. Phillips, LTC, USAR

Commissioned in 1979, Lieutenant Colonel Phillips has seen duty in armor units in the United States and Europe and commanded an M-1 Abrams tank company in the 1st "Tiger" Brigade of the 2nd Armored Division.

Phillips served as the 1st Cavalry Division Public Affairs Officer throughout the Southwest Asia Campaign. His team of Army photojournalists extensively recorded, captured vivid images of the combat operations of the 1st Cavalry Division in the Gulf and later wrote "America's First Team in the Gulf," the illustrated history of the division in operations Desert Shield and Desert Storm.

Before joining the 1st Cavalry Division, Phillips served with United Nations peacekeeping forces in the Middle East, sharing the 1988 Nobel Peace Prize awarded to the Peacekeepers. As an Army Reservist he now serves annually as an Army spokesman, Army Public Affairs, the Pentagon. He lives in Washington, DC and is currently Director of Public Affairs for the Department of Veteran Affairs.

Steven C. Draper, Director, 1st Cavalry Division Museum

Mr. Draper is a graduate of Michigan State University earning a Bachelor Degree in history and interpretive services.

Mr. Draper began his museum career in 1975, as Curatorial Assistant and Museum Interpreter with the Michigan Historical Museum. Following a successful consulting career, in 1983, Mr. Draper became the Director/Curator of the Lowry Heritage Museum, Lowry Technical Training Center, Lowry AFB, Colorado, where he was responsible for overall operations of the museum and the museum's aircraft restoration program.

Since 1992, Mr. Draper has been the Director of the 1st Cavalry Division Museum, 1st Cavalry Division, Fort Hood, Texas. and also serves as advisor to the Division's Commanding General on matters of the Division's history and preservation of it's material culture. He is responsible for the collecting, preservation, and exhibition of the museum's collection of over 3,000 objects and 100 military vehicles.

In 1999, Mr. Draper was deployed to Bosnia with 1st Cavalry Division to collect artifacts, papers, photographs and other materials to document to the Division tour there.

Mr. Draper has been awarded the Army Achievement Medal for Civilian Service and Army's Commander's Award for Civilian Service.

Mr. Draper is member of the American Association of Museums and its Curator's standing committee, Mountain Plains Association of Museums, Texas Association of Museums and the Central Texas Association of Museums. He has served on a number of museum professional committees both state, regional and national including chairman of the Curator's Committee for the Air Force Museum Program.

TECHNICAL EDITOR

Lyman Chan Duryea, COL, USA (Ret)

Colonel Duryea is a West Point graduate and a former Commander of "C" Company of the 2nd Battalion, 7th Cavalry Regiment in Vietnam. Colonel Duryea, with the reputation of a fine troop commander, has since specialized in French language and history, teaching at West Point and serving overseas in diplomatic positions. He holds a Masters degree in Military Arts and Science from the Army command and General Staff College, a Masters degree in French from Middleberg College and a Doctorate in American Military History from Temple.

AUTHOR

William H. Boudreau, President - Cavalry Outpost Publications.

William H. Boudreau served with "F" Troop, 8th Cavalry Regiment during the Occupation of Japan. Following his military service, Mr. Boudreau, earned a Mechanical Engineer Degree, completed Graduate Studies in Control Systems and followed a technical career of systems development which closely paralleled the weapon innovation and advancements of those utilized by the 1st Cavalry Division.

In 1962, Mr. Boudreau, lead an engineering team which developed a helicopter radar terrain following system which was successfully tested at Ft. Huachuca, AZ and Ft. Rucker, AL and accepted by the Army as a basic flight instrument requirement. He shadowed the journey of the 1st Cavalry Division to Vietnam, monitoring the evolution of tactical aviation technology through the mid and late '60's.

His next encounter with the 1st Cavalry Division came in the mid '70's with the early research, development and testing of helicopter night vision and flight director control systems. In the '80's, the response to the threat of the "Winds of War" bought about the development of the M-21 Remote Chemical Sensor Monitor and M-1 Main Battle Tank Environmental Protection System.

In the '90's, Mr. Boudreau followed the 1st Cavalry Division, in spirit, to the air and ground war of the Persian Gulf with the fielding of ruggedized laser guidance systems for the M-109 Howitzer and the M-1 Main Battle Tank and advanced Heads Down CRT instrumentation for OH-58 and AH-64 Helicopters.

Now retired, Mr. Boudreau conducts military history research, which he publishes on the Internet. He is currently the Historian for the 1st Cavalry Division Association.

Appendix P: Glossary

ABBREVIATIONS and ACRONYMS

A2C2 - Army Airborne Command and Control System
AA - Assembly Area
AAN - Army After Next
ABCS - Army Battle Command System
ACADA - Automatic Chemical Agent Detector Alarm
ACCB - Air Cavalry Combat Brigade
AFATDS - Advanced Field Artillery Tactical Data System
AH - Assault Helicopter
AHB - Assault Helicopter Battalion
AIM - Abrams Integrated Management
AKA - Also Known As
ANAV - Autonomous Navigation System
AO - Area of Operation
AOR - Area of Responsibility
APD - High Speed Transport
ARA - Aerial Rocket Artillery
ARCENT - Army Forces Central Command
APC - Armored Personnel Carrier
ARVN - Army of the Republic Vietnam
ASAS - All Sources Analysis Systems
ASHB - Assault Support Helicopter Battalion
ASH - Assault Support Helicopter
ASIP - Advanced System Improvement Program
ATM - Asynchronous Transfer Mode
AV - Air Vehicle
AVLB - Armored Vehicle Launched Bridge
ATG - Antenna Tranceiver Group
AWE - Army Warfighting Experiments
AWR-5 - Army War Reserves, Set-5
BAT - Brilliant Anti-Armor Submunition
BCB2 - Battle Command Brigade and Below
BDU - Battle Dress Uniform
BFA - Battlefield Functional Area
BFIST - Bradley Fire Support Team
BFV - Bradley Fighting Vehicle
C4I - Command, Control Communications, Computers and Intelligence
C4ISR - Command, Control, Commmunications, Computers, Intelligence, Surveillance & Recon
ACADA - Automatic Chemical Agent Detector/Alarm
CAM - Chemical Agent Monitor
CBMS - Chemical Biological Mass Spectrometer
CCC - Civilian Conservation Corps
CCF - Chinese Communist Forces
CD - Compact Disk
CENTCOM - United States Central Command
CFV - M3 Cavalry Fighting Vehicle
CGS - Common Ground Station
CIDG - Civilian Irregular Defence Group
CIV - Commanders Independent Sight
COHORT - Cohesion, Operational Readiness Training
CONUS - Continental United States
CP - Command Post
CPX - Command Post Exercise
CROP - Containerized Roll-in/out Platform
CRS - Computer/Radio System
CRT - Cathode Ray Tube
CSM - Command Sergeant Major
CSS - Combat Service Support
CSSCS - Combat Service Support Control System
CTP - Common Tactical Picture
DA - Department of the Army
DBDU - Desert Battle Dress Uniform
DISCOM - Division Support Command
DMZ - Demilitarized Zone
DRRF - Deployment Readiness Reaction Force
DRS - Division Restructuring Study
DTSS - Digital Topographic Support System
EAF - Entity Armed Forces
EOD - Explosive Ordnance Disposal
EMD - Engineering Manufacturing Development
FAADC2 - Forward Area Air Defense Command Control
FBCB2 - Forward Battle Command Brigade-and-Below
ECV - Expanded Capacity Vehicle
EPA - Environmental Protection Agency
FARRP - Forward Area Refueling Rearming Points
FCS - Future Combat Systems
FDC - Fire Detection Center
FEAF - Far Eastern Air Force
FLIR - Forward Looking Infrared
FMC - Field Mission Capability
FOV - Family of Vehicles
FRS - Forward Repair System
FRY - Federal Republic of Yugoslavia
FSB - Fire Support Base
FSB - Forward Support Battalion
FSSE - Forward Service Support Elements
GBS - Global Broadcast System
GCCS - Global Command and Control System
GCS - Ground Control Station

GFAP - General Framework Agreement for Peace
GO - General Order
GPS - Global Positioning System
HCD - Horse Cavalry Detachment
HEMTT - Heavy Expanded Multi-purpose Tactical Truck
HE-RAP - High Eexplosive-Rocket Assisted Projectiles
HET - Heavy Equipment Transport
HMMWV - High Mobility Multi-purpose Wheeled Vehicle
IFCS - Improved Fire Control System
IFF - Identification Friend-or-Foe
IFV - Infantry Fighting Vehicle
IBAS - Improved Bradley Acquisition System
ICAM - Improved Chemical Agent Monitor
ID - Infantry Division
IETM - Electronic Technical Manual
ILMS - Improved Launcher Mechanical System
IMETS - Integrated Metrological System
IO - Information Operations
ISO - International Standards Organization
I&R - Intelligence and Reconnaissance
JCDB - Joint Common Data Base
JTRS - Joint Tactical Radio System
KCOMZ - Korean Communications Zone
KKMC - King Khalid Military City
KMAG - United States Army Advisory Group, Korea
LCM - Landing Craft, Medium
LCPR - Landing Craft, Personnel Ramped
LCV - Landing Craft, Vehicle
LCVP - Landing Craft, Vehicle, Personnel
LRF/DCA - Laser RangerFinder/Digital Compass
LRRP - Long Range Reconnaissance Patrol
LSD - Landing Ship, Dock
LST - Landing Ship, Tank
LTF - Logistics Task Force
LW - Land Warrior
LZ - Landing Zone
MACV - Military Assistance Command, Vietnam
MANPADS - Manportable Air Defence System
MASSTER - Modern Army Selected Systems Test, Evaluation and Review
MCS - Maneuver Control System
MET - Mission Essential Tasks
MICAD - Multipurpose Integrated Chemical Agent Detector
MILES - Multi Integrated Laser Engagement System
MLC - Military Load Class
MLR - Main Line of Resistance
MLRS - Multiple Launched Rocket Systems
MND-N - MultiNational Division - North
MMS - Mast Mounted Sight
MOPP - Mission Oriented Protective Posture
MOS - Military Occupational Specialties
MOUT - Military Operations on Urbanized Terrain
MRE - Meals, Ready to Eat
MSB - Main Support Battalion
MSE - Mobile Subscriber Equipment
MST - Maintenance Support Team
MTMC - Military Traffic Management Command
MTS - Movement Tracking System
NATO - North Atlantic Treaty Organization
NBC - Nuclear, Biological and Chemical
NCO - Non Commissioned Officer
NK - North Korean
NLF - National Liberation Front
NMCB - Navy Mobile Construction Battalion
NTC - National Training Center
NVA - North Vietnamese Army
O&S - Operations and Support
OD - Olive Drab
ODS - Operation Desert Storm
OP - Observation Post
OPCON - Operational Control
OPFOR - Opposition Forces
OPLR - Outpost Line of Resistance
OPTEMPO - Operating Tempo
OSCE - Organization for Security and Cooperation in Europe
PID - Provisional Infantry Division
POL - Petroleum, Oil & Lubricants
POMCUS - Prepositioned Material Configured into Unit Sets
POW - Prisoner of War
PT - Patrol, Torpedo
QRF - Quick Reaction Force
REFORGER - REturn of FORces to GERmany
RCT - Regimental Combat Team
RCP - Regimental Command Post
ROCID - Reorganized Objective Combat Infantry Division
ROK - Republic of Korea
ROM - Refuel on the Move
ROWPU - Reverse Osmosis Water Purification Unit
RSOI - Reception, Staging, Onward movement and Integration
RSUT - Tracked Resupply Vehicle
RSTA - Reconnaissance, Surveillance, and Target Acquisition
RSUW - Wheeled resupply vehicle
SA - Surface to Air
SAC - Strategic Air Command
SEP - System Enhancement Program
SFOR - NATO Stabilization FORce
RSCALL - Remote Sensing Chemical Agent Alarm
RVT - Remote Video Terminals
SCU - Special Ceremonial Unit
STC _ Slew to Cue
SHORAD - Short Range Air Defense
SINGARS - Single Channel Ground to Air Communication System
SIP - Service Improvement Package
SMARTT - Secure Mobile Anti-Jam Reliable Tactical Terminal
SPH - Self-Propelled Howitzer
STX - Situational Tactical Exercise

SVML - Standard Vehicle Mounted Launcher
TA - Transfer of Authority
TAA - Tactical Assembly Area
TAACOM - Theater Area Army Command
TACFIRE - Tactical Fire
TF - Task Force
TFE - Task Force Eagle
TO&E - Table of Organization and Equipment
TOW - Tube-launched Optically-tracked Wire-guided missile
TRICAP - Triple-Capability
TUAV - Tactical Unmanned Aerial Vehicle
TWS - Thermal Weapon Sight
US - United States
USARCENT - United States Army Forces Central Command
USCENTCOM - United States Central Command
USNS - United States Navy Ship
USS - United States Ship
USSR - Union of Soviet Socialist Republics
UN - United Nations
UNC - United Nations Command
UNCACK - United Nations Civil Assistance Command, Korea.
VC - Viet Cong
VRS - Voijska Republic Srpska
WIN - Warfighter Information Network
WRAP - Warfighting Rapid Acquisition Program
XO - Executive Officer
ZOS - Zone of Separation

Appendix 2: Index

NUMERIC KEYWORDS

ALPHABETICAL KEYWORDS AND PHRASES

- A -

- B -

- C -

- D -

- E -

- F -

- I -

- J -

- K -

- L -

- M -

- N -

- O -

- P -

- Q -

- R -

- S -

- T -

- U -

- V -

- W -

- X -

- Y -

Personal Memoires

Autographs

5th Cavalry Regiment

7th Cavalry Regiment

8th Cavalry Regiment

9th Cavalry Regiment

12th Cavalry Regiment

227th Aviation Regiment

"F" Company
1st Battalion
58th Aviation

21st Field Artillery

82nd Field Artillery

68th Chemical Company

8th Engineer Battalion

20th Engineer Battalion

91st Engineer Battalion

15th Support Battalion